Shinkansen

The image of the shinkansen – or 'bullet train' – passing Mount Fuji is one of the most renowned images of modern Japan. Yet, despite its international reputation for speed and punctuality, little is understood about what makes it work so well and what its impact is. This book provides a comprehensive account of the history of the shinkansen, from its planning during the Pacific War, to its launch in 1964 and subsequent development. It goes on to analyse the reasons behind the shinkansen's success, and demonstrates how it went from being simply a high-speed rail network to attaining the status of iconic national symbol. It considers the shinkansen's relationship with national and regional politics and economic development, its financial viability, the environmental challenges it must cope with, and the ways in which it reflects and influences important aspects of Japanese society. It concludes by considering whether the shinkansen can be successful in other countries developing high-speed railways. Overall, this book provides a thorough examination of the phenomenon of the shinkansen, and its relationship with Japanese society.

Christopher P. Hood is Director of the Cardiff Japanese Studies Centre, Cardiff University and Associate Fellow at Chatham House. He is the author of *Japanese Education Reform: Nakasone's Legacy* (Routledge 2001), co-editor with G. Bownas and D. Powers of *Doing Business with the Japanese* (2003), and regularly handles media enquiries relating to Japan.

Routledge Contemporary Japan Series

1 **A Japanese Company in Crisis**
   Ideology, strategy, and narrative
   Fiona Graham

2 **Japan's Development Aid**
   Old continuities and new perspectives
   Edited by David Arase

3 **Japanese Apologies for World War II**
   A rhetorical study
   Jane W. Yamazaki

4 **Linguistic Stereotyping and Minority Groups in Japan**
   Nanette Gottlieb

5 **Shinkansen**
   From bullet train to symbol of modern Japan
   Christopher P. Hood

Shinkansen

From bullet train to symbol of modern Japan

Christopher P. Hood

LONDON AND NEW YORK

First published 2006
by Routledge
2 Park Square, Milton Park, Abingdon, Oxon, OX14 4RN

Simultaneously published in the USA and Canada
by Routledge
270 Madison Ave, New York NY 10016

Routledge is an imprint of the Taylor & Francis Group

Transferred to Digital Printing 2007

© 2006 Christopher P. Hood

Typeset in Times New Roman by
Newgen Imaging Systems (P) Ltd, Chennai, India

All rights reserved. No part of this book may be reprinted or
reproduced or utilised in any form or by any electronic,
mechanical, or other means, now known or hereafter
invented, including photocopying and recording, or in any
information storage or retrieval system, without permission in
writing from the publishers.

British Library Cataloguing in Publication Data
A catalogue record for this book is available
from the British Library

Library of Congress Cataloging in Publication Data
Hood, Christopher P. (Christopher Philip), 1971–
    Shinkansen : from bullet train to symbol of
modern Japan / Christopher P. Hood.
      p. cm. – (Routledge contemporary Japan series)
    Includes bibliographical references and index.
    1. High speed trains – Japan. I. Title. II. Series.

TF1450.H66 2006
385'.22'0952–dc22   2005019926

ISBN10: 0–415–32052–6 (hbk)
ISBN10: 0–415–44409–8 (pbk)

ISBN13: 978–0–415–32052–8 (hbk)
ISBN13: 978–0–415–44409–5 (pbk)

**To Miella and
Mr & Mrs Gotoh**

Contents

	List of figures	ix
	List of tables	xi
	Notes on style	xiii
	Preface and acknowledgements	xv
	Map of Japan and shinkansen lines	xix
1	Introduction	1
2	From bullet train to low flying plane	18
3	Ambassador of Japan	44
4	Whose line is it anyway?	71
5	The bottom line	91
6	The need for training	130
7	Mirror of Japan	162
8	Conclusion	196
	Appendix 1: chronology of significant dates in the history of the shinkansen	211
	Appendix 2: shinkansen lines and stations	220
	Appendix 3: shinkansen types	231
	Glossary and abbreviations	236
	Notes	240
	Bibliography	250
	Index	259

Figures

1.1	300-series shinkansen passing tea fields in Shizuoka Prefecture	10
1.2	100-series shinkansen passing Hamanako	12
1.3	Train hierarchy in Japan	17
2.1	Views in and around Tōkyō station	28
2.2	Journey times between Tōkyō and Ōsaka (1926–1964)	29
2.3	Shin-Aomori station	33
2.4	Views around Kyūshū	36
2.5	(a) Distance of lines, (b) travelling times (1963), (c) travelling times (2004), (d) travelling times (all shinkansen lines completed)	38
2.6	E3-series shinkansen pictures	40
2.7	0-series shinkansen near Higashi-Hiroshima	41
3.1	0-series-renewal shinkansen	46
3.2	E1-series shinkansen pictures	49
3.3	700-series shinkansen pictures	55
3.4	300-series shinkansen passing Mount Fuji	57
3.5	Using the shinkansen's image	63
4.1	Political stations?	74
4.2	Signs calling for shinkansen construction work	84
5.1	Construction of shinkansen lines	92
5.2	A shinkansen passing the monorail link to Haneda Airport	97
5.3	Yamagata shinkansen	98
5.4	Upgrading a line	99
5.5	Shinkansen construction budget since fiscal 1989	103
5.6	Valuation of the shinkansen lines	106
5.7	Central towers in Nagoya	110
5.8	E4-series shinkansen pictures	111
5.9	Linear shinkansen pictures	120
5.10	500-series shinkansen pictures	121
5.11	Experimental shinkansen	123
6.1	Dr Yellow pictures	132
6.2	ATC and digital-ATC	134

6.3	E2-series shinkansen pictures	141
6.4	Professionalism of shinkansen staff	142
6.5	Union membership in JR companies (January 2004)	150
6.6	Cultures of the shinkansen-operating JR companies	153
6.7	100-series-renewal shinkansen	157
6.8	200-series shinkansen	158
6.9	200-series-renewal shinkansen	159
6.10	The Ambitious Japan! Campaign and opening of Shinagawa station	160
7.1	Average precipitation in selected cities	163
7.2	Responding to the challenge of snow	167
7.3	The shinkansen's environmental record	172
7.4	Noise pollution	175
7.5	Reducing the shinkansen's weight	176
7.6	Visual pollution	179
7.7	800-series shinkansen pictures	180
7.8	JR Tōkai's Nagoya Shinkansen dispatch office	186
7.9	*Hikari Rail Star* pictures	191
8.1	Various shinkansen pictures	210
A1.1	Passengers carried by the shinkansen	219
A2.1	Shinkansen lines and population of cities and towns served	230

All photographs in the book are by the author. Colour versions can be seen at www.hood-online.co.uk/shinkansen/

Tables

1.1	Formation of JR companies on 1 April 1987	16
2.1	The shinkansen network	32
2.2	Kyūshū Shinkansen construction	35
2.3	Top air routes in the world	37
5.1	Cost of shinkansen line construction	94
5.2	Comparison of major construction works	96
5.3	Comparison of railway line types	98
5.4	Upgrading a line	101
5.5	MLIT budget for public works (Fiscal 2004)	103
5.6	Importance of the shinkansen for JR companies	109
5.7	Population of cities and towns along the Jōetsu Shinkansen	113
5.8	The fastest passenger trains in the world (2003)	122
5.9	Japan's fastest experimental trains	124
6.1	Shinkansen maintenance	137
7.1	Snowfall on the shinkansen	165
7.2	Shinkansen's noise pollution in priority areas	174
7.3	Activities on shinkansen	187

Notes on style

Japanese names are given in their proper order, with the surname first and personal name second. With names of Westerners of Japanese descent, Western order is preserved.

Macrons are used to denote long vowel sounds, which are twice the length of a short vowel in pronunciation. The only exceptions to this are for names where some use the option of Gotoh rather than Gotō, for example. Although many authors prefer not to use macrons on common place names, I use them here as this is the convention used by the railway companies themselves on the signs at those stations (e.g. Tōkyō rather than Tokyo). Macrons have been added as appropriate to words in quotations where they were omitted in the original text. An apostrophe is used between combinations 'n' and 'yo', for example, to distinguish between the sounds 'n'yo' and 'nyo', with the exception of 'Sanyō', which strictly should be written as 'San'yō'. Japanese nouns do not have plural forms; so 'shinkansen' could be singular or plural depending on context.

Where translations have not been attributed to other sources, they are my own.

Throughout the text 'shinkansen' is written in lower case to indicate when the train itself or the system as a whole is being referred to. Capitalization of the first letter is only used with the name of a particular line. Trains types are referred to using the system '0-Series', for example, rather than 'Series 0' or 'Class 0'.

British English is used throughout the book. The only exception to this is the usage of 'billion' and 'trillion' where the universal system is used, that is, 1 billion = 1,000,000,000 and 1 trillion = 1,000,000,000,000.

In 2001, the Ministry of Transport merged with some other ministries to become the Ministry of Land, Infrastructure and Transport (MLIT). For simplicity, the acronym MLIT is used throughout this book.

Throughout the text the Japan Railway (JR) companies that were formed following the reform of Japan National Railways (JNR) are referred to by their abbreviated, and more familiar, names. For example, JR Tōkai instead of Central Japan Railway Company and JR East instead of East Japan Railway Company. A full list of these and other commonly used terms are provided in the Glossary and abbreviations. Only in the Bibliography are their standard English names used.

Metric measurements are used in the text. The following table provides the basis for common conversions:

  1 kilometre (km) = 0.62 miles
  1 metre (m) = 3.28 feet or 1.09 yards
  1 centimetre (cm) = 0.39 inches
  1 metric ton (t) = 0.98 imperial ton or 1.10 US ton
  1 kilogram (kg) = 2.21 pounds
  1 kilowatt (kW) = 1.34 horsepower

Where conversions are given, an exchange rate of £1 = ¥195 and $1 = ¥110 has been used.

Preface and acknowledgements

Conducting the research for this book has been quite a journey. Given that one of the main themes of the book is the symbolism of the shinkansen, it is apt that I am writing this Preface on board a shinkansen exactly 40 years to the day since the shinkansen began passenger operations. At the commemorative ceremony before I boarded this train, it was easy to feel the emotion of many of those around me, ranging from pride amongst the JR Tōkai staff to excitement amongst the assembled media, railway enthusiasts, and the five special guests who were born on the day that the Tōkaidō Shinkansen started. That Japan's media was well represented and so many railway enthusiasts turned up for the ceremony that took place around 6:00 a.m. is clearly a sign of the importance of the event that occurred 40 years ago. As a reminder of that event a large photograph of the opening ceremony stood behind the assembled dignitaries. Further promotional material, including a detailed display covering the history and plans for the future of the shinkansen, also filled Tōkyō Station.

My original plan to conduct research about the shinkansen dates back more than 12 years. However, the formal fieldwork for this book started about five years ago with my first visit JR Tōkai. Without the co-operation of the various companies and organizations I have visited, this research would not have been possible. Consequently I would like to take this opportunity to thank all of those that were so helpful in allowing me to visit their facilities, observe their operations and interview their staff. In particular I would like to thank Central Japan Railway Company (JR Tōkai), East Japan Railway Company (JR East), West Japan Railway Company (JR West), Kyūshū Railway Company (JR Kyūshū), the Railway Technical Research Institute (RTRI), the East Japan Railway Company Foundation, Japan Railway Construction, Transport and Technology Agency (JRTT), Hokkaidō Railway Company (JR Hokkaidō), Japan Railway Technical Service (JARTS) and the Ministry of Land, Infrastructure and Transport (MLIT). I wish that I could list the over 200 people from these and other organizations that I have interviewed for this research, but space does not permit it. However, I would like to thank all those that I did interview, and would like to mention a few individuals from the companies and organizations mentioned earlier who have helped to plan visits or to answer my many questions on a number of occasions during the past few years – Kondō Kunihiro, Morimura

xvi *Preface and acknowledgements*

Tsutomu, Morishita Tatsushi (JR Tōkai); Yokoo Takeshi (JR East); Kosuga Ken'ichi (JR West); Yoshida Sayako and Suyama Yōko (JR Kyūshū); Shimizu Kenji (JRTT); Oiyama Tatsuya (JARTS); Yamada Kōji (JR Hokkaidō). I would also like to thank Sekiguchi Masao (JR East) for allowing me to drive a train at the JR East General Education Centre.

There are also various other companies and organizations that I have visited over the past few years or who have provided information whom I would like to thank: All Nippon Airways, Aomori Prefectural Government, First Great Western, Fullers Brewers, Hitachi, Hokkaidō Prefectural Government, Japan Airlines, Kagoshima Prefectural Government, Komoro City Government, Nippon Sharyō, Sanrio, Shinjō City Government, the Taiwan High Speed Railway Corporation (THSRC), and Yamagata Prefectural Government. I would also like to thank NHK, who helped me to see the fortieth anniversary ceremony from such a good location.

There are numerous individuals, some of whom I met at companies mentioned earlier, I would like to thank for providing information and support of various kinds: Aoki Kunio (East Japan Railway Culture Foundation), Geoffrey Bownas, Phil Deans (SOAS, London University), Dave Fossett, Steve Hubbard, Isomura Yōji (formerly at JR East), Eddy Jones, Kawashima Ryōzō, Kay Kondō (Chiba University of Commerce), Mizutani Fumitoshi (Kōbe University), Miyawaki Hideo (Kyōdo News), Mutō Hisashi (formerly at Kokurō), Nakano Tsuneo (Kōbe University), Nakaoka Nozomu, Nakasone Yasuhiro, Naitō Hiroshi, Nishio Gentarow, Emmanuel Ogbonna (Cardiff University), Andrew Potter (who allowed me to sit in on his lectures at Cardiff University to learn about railway economics and how British railways are supposedly run), Christopher and Phillida Purvis, Shima Takashi, Shōji Ken'ichi (Kōbe University), Sogō Shinsaku, Rod Smith (Imperial College, London), Sudō Hiroshi (University of Kitakyūshū), Suga Tatsuhiko (formerly at East Japan Railway Culture Foundation), Tabata Hirokuni (Tōkyō University), Takahashi Kin (Japanese Trade Union Federation), Tanaka Shigeru, Yamana Norio (Ministry of Finance), Yasubuchi Seiji (UBS) and Wada Shigeru (International Transport Workers' Union).

I would also like to thank all those that have helped me to develop this research in other ways – either by providing a forum for me to present my research to date, by asking questions or by merely passing on their enthusiasm. In particular I would like to thank the active members of the jtrains discussion group (especially Dave Fossett, Martin Guest, Naitō Hiroshi, Doug Coster, Mitch Sako, Norman Simpson and Dick Harris), the Japanese Railway Society (especially Anthony Robins, Oliver Mayer, Richard Tremaine and Bill Pearce), Chatham House's Japan Discussion Group members, and the Japan Society. Also thanks to the huge number of anonymous people who have contributed to this work through casual conversations and being there for me to observe. I would also like to thank the publishers, Routledge, and particularly Peter Sowden, for their support of this research as a whole and for allowing me to include so many photographs, which form an integral part of the study.

I would also like to thank the staff at the Cardiff Japanese Studies Centre – David Williams, Rosemary Smith, Yuri Kyōko, Kudara Masakazu, and Jan Richards – and

Preface and acknowledgements xvii

also Roger Mansfield of the Cardiff Business School for their support over the past few years. A particular word of thanks also to David Williams and Dave Fossett for reading through drafts of this book at short notice and offering their suggestions. Also a big thank you to Keith Haines and Robert Rigby at Japan Airlines, who over the years have helped to ensure that my fear of flying became a little bit more bearable and allowed me to travel in comfort on many an occasion between London and Japan.

I would also like to thank the various companies and organizations that have helped me pursue this research through the provision of funding; Cardiff Business School, the Japan Foundation Endowment Committee, the Great Britain Sasakawa Foundation, Japan Airlines and the Daiwa Anglo-Japanese Foundation.

I would like to thank all my family and friends who helped to keep me going. Special thanks go to my parents and my uncle Robin, who have inspired me on many occasions to pursue my studies. Thanks also to my mother-in-law, Teresa, and sister-in-law, Kit, for their support. Many thanks to all those who either provided me with their views on the shinkansen, Japanese society and/or a roof over my head during my many trips to Japan – especially the Gotoh family, Tanaka family, Tanabe Family, Mariko Donnelly, Funatsu Suguru, Antony Hathaway, Iwashita Yōko, Eddy Jones, Kinoshita Midori, Kishi Kanako, Matsunaga Eri, Steve and Mitsuyo Ryles, Sekine Yukio, Sekiya Junko, Tsurumoto Wakako, the Wada family, and Yamada Masaaki and family. Finally a special word of thanks to my wife, Man Yee, and my daughter, Miella (who is too young to remember her first trips on shinkansen and has provided many sleepless nights and challenges to my research schedule) – you both fill up my senses.

Christopher P. Hood
1 October 2004
on-board *Hikari* 313

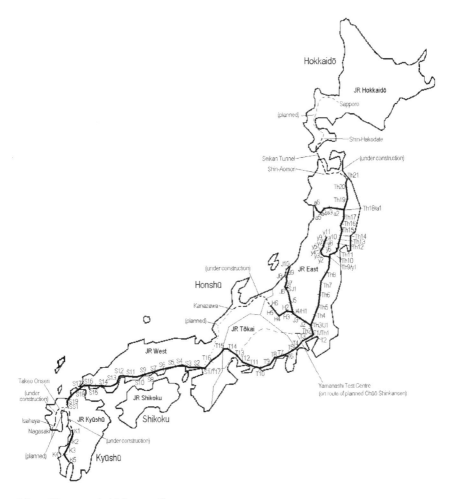

Map of Japan and shinkansen lines.

Notes
This map is not exactly to scale. Place names and proposed stations (names may be altered in the future) refer to those mentioned in the main text which are on planned shinkansen routes. See Appendix 2 for list of station names on lines presently in operation.

1 Introduction

> One that lights a corner of society is a treasure of the nation.[1]

Although the words above refer to Sogō Shinji rather than his greatest legacy, they are words that clearly reflect the significance of the shinkansen project that he helped to initiate. Since the shinkansen began commercial services in 1964, the main lines have carried over 7.6 billion passengers.[2] The total distance travelled by Tōkaidō shinkansen alone is about 1.5 billion kilometres, equivalent of a round trip between Earth and Jupiter or 37,500 laps of the Earth. During this time there have been no passenger fatalities on the main shinkansen lines due to collision, derailment or infrastructural failure. Average delays are less than one minute per train, with nearly all trains arriving on time. The shinkansen remains one of the fastest train services in the world. These facts are impressive and will be analysed in later chapters. However, these statistics, or at least an impression of them, tend to be the limit of the knowledge of the shinkansen outside of Japan and perhaps even within Japan. Although its image, particularly passing in front of Mt Fuji, is internationally renowned, how and why it works *so* well is not common knowledge. Even its name is not well known beyond the shores of Japan. For the majority it is 'the bullet train'.

At one level the shinkansen is merely a means of transportation. However, this study will reveal that the shinkansen is much more significant than this. The shinkansen has become a symbol of Japan. The way it was established, the network has developed, how it is operated, and even the way it looks reflect many different aspects of Japanese society. While its importance as a symbol for some Japanese may have lessened, in this process it has faded into normalcy so that it has become an even better tool by which to study Japanese society. Furthermore, this study also demonstrates that more than merely reflecting aspects of Japanese society, the shinkansen has also been used as a harbinger of change in Japan.

It was no accident that the shinkansen services started when they did. The Japanese economy was expanding at an incredible rate. Prime Minister Ikeda had only two years previously announced the 'Double Your Income Plan', whereby all Japanese were to double their real income by the end of the decade. Not only was this target reached, it was achieved in just seven years (Horsley and Buckley

1990:62). However, more than just the development of the economy, and the start of the shinkansen services, 1964 is also remembered by the Japanese for the hosting of the Olympics. This event signalled Japan's return from the ashes of defeat in the Pacific War to the international community. This was graphically demonstrated through the choice of person to light the Olympic Flame at the Opening Ceremony. Rather than a famous national icon, as has become the norm in recent years, the flame was lit by Sakai Yoshinori, a boy born in Hiroshima on 6 August 1945 (Horsley and Buckley 1990:72). To provide a further link between the shinkansen and the Olympics, some events were staged in Ōsaka, over 500 km from Tōkyō, so that competitors, and perhaps more significantly, the world's media, could experience the marvel of Japan's new technology. Clearly then, there is a potential for the shinkansen to be studied for more than its engineering or technological prowess.

Before discussing the particular focus of this study, I would like to clarify some assumptions that I have made in writing it. First, I have taken into account the interests of three main readership groups. One group are those who already have some knowledge of the shinkansen, but have less knowledge of Japan. Another group are those who have knowledge of Japan, but limited knowledge of the shinkansen. The final group are Japanese whom have an interest in a non-Japanese perspective of the subject, particularly given that I have managed to gain access to some sources that they may not normally be able to. Naturally there may be some overlap between the final group and the other two. Second, I have also worked on the basis that few who read this book have an in-depth knowledge of railway engineering. As much as possible I have tried to present information and ideas that take these assumptions in to account, whilst trying to avoid either over-simplification or too much detail when discussing areas that may be unfamiliar to one of the readership groups.

This introductory chapter sets out how the work is presented, the methodology for research, the problems that I have had to address while conducting the research and presenting the results and why I chose to conduct the research. Finally it presents some background information about the current shinkansen-operating companies and how they were formed. Chapter 2 covers the history and development of the shinkansen, from the initial plans during the Pacific War to the plans for a linear shinkansen in the future. This chapter provides a useful foundation for the later chapters and is intended to be largely descriptive rather than analytical. Chapter 3 considers the issue of the symbolism of the shinkansen, a theme that runs through most of the book. The chapter introduces issues relating to the general study of symbolism, covering issues such as how a symbolic object or concept can lead to the development of national and international images and stereotypes. It further looks at how the shinkansen became a symbol of Japan, the knowledge and use of the image of the shinkansen in Japan and abroad, and issues relating to the terminology of the word 'shinkansen'. Chapter 4 covers perhaps the most infamous aspect of the shinkansen. For almost without fail, when Japanologists speak of Japanese politics, they raise the issue of 'pork-barrel' politics, with the most well-used example of this being the Jōetsu Shinkansen that links Tōkyō to former Prime Minister Tanaka Kakuei's home constituency of Niigata. Chapter 5 looks at the financial viability of the shinkansen and its impact upon the national and regional economies of Japan. Chapter 6 deals with the issues relating to the employees and how the shinkansen works so well. Chapter 7 deals with various significant issues that have not been developed in other parts of the book, but with the focus reversed from the previous chapters, taking significant areas of Japanese society as a starting point and analysing the shinkansen's relationship to them. Chapter 8 draws together all the key points that have been raised throughout the book addresses the issue of whether the shinkansen can be exported or not.

Methodology

This study is a study of the shinkansen and of Japan. This presents a problem. During the course of the research I have gathered far more information than is possible to contain in just one book. This in itself is not a bad thing as it has enabled me to develop a fuller understanding of the shinkansen and its relationship with Japanese society. Indeed only by gaining such a level of understanding can one reach a point where one's instinct, what may be called *haragei* in Japanese, about what information is valid or is invalid be consistently correct. However, as a consequence, in presenting the results of the study compromise is inevitable, especially when considering the assumptions that need to be made about different readership groups. Such compromise is not unusual. However, in Japanese Studies and 'Area Studies' as a whole, it has been perceived as a weakness. Part of the reason for this appears to be that in order to simplify arguments and points, authors have tended to make general statements – particularly in relation to culture. As Morris-Suzuki (1998:3) points out, the result is that 'key terms such as "culture," "ethnicity," and "identity" are often tossed around with such abandon that they themselves have become obstacles rather than aids to better understanding'. Similarly Knipprath (2003:1) criticizes many works for having 'arbitrary definition[s] of culture, over-attribution to national or cultural effects...and the simplistic assumption of homogeneity within a society'. Goldthorpe also argues that

> All too frequently, large issues have been addressed – and large conclusions reached – by reference to a miscellaneous collection of official statistics and reports, case studies, journalism and, where all else fails, personal impressions. Apart from the inevitably patchy nature of evidence of this kind, it imposes little scholarly or scientific discipline. The danger is ever-present that the selection from it that is made reflects as much the needs of an individual author's case as the intrinsic value of particular items.
> (Goldthorpe 1993:xv)

However, one cannot escape the fact that as conclusions have to be reached, generalizations are inevitable and that personal impressions can be highly significant if the research done and the understanding of the focus of the study, whether it be companies, people or a whole society, is of an appropriate level. However, to do this surely means that as well as the readers needing to have confidence that the author has this ability, the readers need to have some understanding of the author themselves. This is particularly the case now that there is an apparent desire for researchers to write in the first person, which has been unfamiliar territory for most British academics.

I have commented elsewhere (Hood 2001:2) that even when academics do not write in the first person, this does not mean that their opinions and views are not presented in the text. This may be done relatively openly or more covertly through not including information that detracts from or is not consistent with their main argument or personal point of view. In attempting to deal with this latter point, I believe that academics should give introduction of themselves and perhaps even some indication of what their position is on key issues. For without an understanding of the author, even if the reader does not agree with the author's bias, it is not be possible for the reader to fully appreciate the thrust of the work. Although all academics strive to be objective and seek 'the truth', it is a position that is unlikely to ever be attained, especially while there is limited openness about the person presenting the information. What one believes to be 'objective' or 'the truth' is likely to be based upon previous experiences, whether direct or learnt from printed materials, for example, as well as on the current research being conducted. There will always be some degree of subjectivity and bias – from those providing the information, who are likely to have vested interests in presenting 'facts' in a particular way, as well as from those presenting the research results. It is impossible to remove the human element from research. We should not even be trying to do so. But we have to acknowledge the subjective element of research and present the results of our work in a way that allows those who read it to appreciate the way in which the research was done and how the results are presented.

With much of my previous research on education reform in Japan, I took a path that not many others have apparently dared take. That was to look at Nakasone Yasuhiro, in what *I* personally considered to be an objective manner. In other words my starting point was to try to understand Nakasone himself and whether his actions in education reform were consistent with trying to have his ideas implemented. Perhaps due to the political bias of many in academia and a degree mistrust that many people, both in and out of academia, have towards Nakasone, as my results revealed an appreciation of his approach taken in the education reforms and his position on issues such as nationalism, some have labelled me as 'a Nakasone fan' and suggested that I have been overly positive. It is regrettable that not presenting the results that the majority would like to read can lead to such a view. I believe that it is necessary that academics are prepared to challenge widely held ideas and preconceptions. Undoubtedly there will be many times when the result of this will be an affirmation of the status quo, but this does not mean that process is without merit. In this study, for example, rather than working on the assumption that Tanaka Kakuei *was* responsible for the building of the Jōetsu Shinkansen, I looked at a range of facts as I did not wish to jump on bandwagons and rely on past assumptions, which may have been based on poorly thought out ideas, research or logic. Although my conclusion may be a disappointment to those who have used this as an example of 'pork-barrel' politics, I hope that the way that this conclusion has been reached, and even the reasons for making the investigation, will be treated appropriately and not be dismissed out of hand or be seen as an example of an apparent overly positive view of Japan and some its conservative elements.

This research is a broad study, covering areas as diverse as politics, economics, sociology, and anthropology. This presents a great challenge to me, as I have had to become acquainted with theories and approaches within each of these fields. I would argue it is only through such an approach that we can truly begin to understand a country. I do not believe that it is practical or desirable to treat each area in isolation. This is probably the single greatest strength of 'area studies'. Although I have become familiar with many concepts within these research fields, it is becoming increasingly difficult to pin-point which works have influenced me most or even which ones I have the greatest regard for. Perhaps this is much as a musician, particularly 'classical musicians', although likely to have been influenced by certain composers and styles in their early years, begin to develop their own styles and methods to the point that they are no longer aware, and it may even be hard for others to identify, what those influences were. While it is common, if not normal, for academics to cover old ground by testing their research against established ideas, I would suggest that those doing original research should not necessarily be bound by these conventions, and that what is of greater importance is to focus on the research itself and present the results as they see appropriate. These results can be tested against established theories and ideas in further works, whether by the same author or by others, potentially in other fields.

This last suggestion is important since a study such as this presents information that is not easily accessible to other specialists. Thereby, it can enrich the debate and understanding of a subject by providing information to those without the means to do such research. For example, in understanding Japanese I have been able to access and use information that many would not be able to comprehend or which would be costly, potentially in terms of money and time, to translate. Furthermore, true comprehension requires a knowledge of the country and its people as well as the language. Translations rarely, if ever, can fully do justice to the 'reality' as it was presented in its original form. It is the skill of the Japanese Studies specialist to present information, taking into account the 'cultural' specifics of Japan, in a meaningful way that can be used by a non-specialist audience or those in different research fields. As well as access to materials, I have managed to gain access to people that even many Japanese would not find possible to do. In this way, I hope that others, including specialists in fields other than Japanese Studies, will be able to use the information that I present to further develop their own research or will be encouraged to study the shinkansen and some aspect of Japanese society for themselves.

Referring back to Knipprath, one of the key criticisms surrounds the use of the word 'culture' and how it is defined. In her analysis of studies of Japanese educational performance, Knipprath (2003:12) found only one study that defined culture, which was as follows;

> Culture provides a general design for life, a covert rationale that guides human behaviour. Members of a culture are usually not conscious of its force, although it gives a direction to their actions and shapes their behavioural and attitudinal patterns. Culture can be compared to the grammar that members of a group follow consciously to communicate with each other. It exerts a powerful and ubiquitous influence on the ways in which individual members act to solve their problems.
> (Shimahara 1986:19)

A similar definition can be found in the work of anthropologist Clyde Kluckhohn, who defined culture as 'The set of habitual and traditional ways of thinking, feeling and reacting that are characteristic of the ways a particular society meets its problems at a particular point in time' (Kluckhohn 1949:17). It is interesting that both of these definitions refer to solution of problems. I would suggest that culture is also often concerned with the preservation of the status quo. Conversely, the introduction of a new culture, or at least the attempt of doing so, can be used as a powerful means of changing the way a society or organization behaves. Shimahara (1986:25) goes on to say 'Culture affects various aspects of human behaviour in a way that is covert, directional, and universal...But culture does not account for all human behaviour. Economic, historical, and ecological conditions constitute an equally powerful explanatory factor of human behaviour.' Naturally this list is not exclusive. I would suggest that education and religion, for example, should be added. Yet, undoubtedly, each of these influence culture and are to some degree also influenced by culture too.

Like 'truth', 'culture' appears to be something that exists, but it is also difficult to be certain that what we think it to be, is in fact the case. It is even hard to be certain that any two individuals have exactly the same definition of what 'culture' is when speaking of a 'shared culture', even though they are likely to be happy using 'culture' as a basis for explaining why behaviour happens in the way that it does. In this study, the term culture will be used on a number of occasions, as I believe that it is not something we can ignore the existence of. Japanese 'culture', as tends to be the case when speaking of any national culture, is an area where stereotypes and clichés are prevalent. Words are used as though based on a common understanding of what they mean, yet often even Japanese themselves would find it hard to identify or explain these terms. But, as Buruma (2001:ix) notes, 'It is hard to avoid the clichés about Japan, because both Japanese and foreigners seem to feel most comfortable with them.' Issues relating to 'culture' will be looked at in greater depth in Chapter 6, but I will work with Shimahara's definition above while keeping in mind its limitations I have already mentioned.

This study is the result of several years of formal study of the shinkansen, on top of many more years of informal study, and of study of Japan. It would be impossible to present all of the information that I have collected. As mentioned earlier, compromise and generalizations are necessary. This may be particularly the case in this study as there have been no similar studies of the shinkansen and its relationship to Japanese society. That I am taking into account the interests, prior knowledge and demands of three readership groups presents a further challenge. For instance, there is a danger that certain examples will appear incredible to some that read this study, while to others it may be simply a normal state of affairs. One of the greatest challenges I have faced is trying to ensure that this study remains focused and balanced, and not overly positivistic or that it appears to be one long promotional material for the shinkansen-operating companies.

A further problem that I have faced is trying to ensure that there is an appropriate balance of information about all of the shinkansen lines, and that there is not an over-concentration on the Tōkaidō Shinkansen, which would perhaps be easy to do. For example, although Semmens' (1997, 2000) studies are about all of the lines, there is a clear imbalance contained within them in favour of the Tōkaidō Shinkansen. One factor that has made finding the balance especially difficult is that it was JR Tōkai that I first made contact with, and as a consequence have spent the greatest time studying in terms of number of visits and period since my first visit to my most recent visit. This is partly to be expected since JR Tōkai has a representative office in London, and so of all the shinkansen-operating companies is the one that British people can most easily gain access to. That the Tōkaidō Shinkansen was the first line, is the busiest line, the most well used by the majority of foreign visitors to Japan, and it runs in front of the iconic Mt Fuji further increases the probability of imbalance in favour of the Tōkaidō Shinkansen.

I have written before (Hood 2001:2, 5) about how it is often chance encounters that can play a significant part in the development of one's understanding, and that often contacts are likely to come from a limited circle of those that are likely to have similar positions to those previously met. This can be a particular problem in researching Japan, where 'cold-calling' is generally not well received. As a consequence, one relies upon those that have already been cooperative to introduce you to those that would not have responded to an initial direct approach. Although I have encountered this problem during the course of this research, it has been less of an issue than in my previous research. The reason for this can largely be put down to the existence of departments at each of the shinkansen-operating companies which are responsible for responding to enquiries from researchers, for example. While local governments seemed to respond well to direct approaches, the national ministries required a more traditional introduction. Unlike my experience at the Ministry of Education[3] during my previous research, I found that the inter-Ministry links, where the common link was railways, was better than intra-Ministry links in Kasumigaseki.[4]

During the course of this research I made several trips to Japan, interviewed relevant people outside of Japan, and attended seminars, conferences and other talks in Britain. Making a number of trips was beneficial for three reasons. First, it allowed me to build up a relationship with the companies and individuals. Japan is a country where developing such trust is particularly important, and being accepted within a circle, becoming part of the *uchi* (see Hendry 2003), of people with whom the company, organization or individual is prepared to divulge more than merely the tip of the iceberg in terms of information was crucial. Second, it allowed me to take time in reviewing information gathered during each trip and consider what areas I would focus on in the next trip. Third, it allowed for gaining a perception of what changes were occurring within some organizations during the period of my research.

The way I conducted my research sometimes varied from trip to trip, and sometimes varied depending on the organization, company or individual visited. Most of my interviews were recorded, allowing them to proceed at a natural pace and for review at a later date. Many of the company visits were also recorded on video recorder or with photographs. Due to the sensitive nature of some of what I was able to observe, there were times when I had to turn off my recording devices. On some trips, particularly the initial visits, I allowed the company or organization to decide what information they wanted to discuss. This was just as revealing as when I set the agenda. When there were times when there was particular information, data, or opinions that I was seeking, I often prepared questions in advance, which were then sent to allow the company, organization or individual time to prepare. Even in these instances, almost all interviews contained extra questions or discussions that either developed out of the answers to previous questions, or were a result of other visits and observations while in Japan. Time spent on trains with personnel from companies when travelling to different facilities often led to moments when they spoke off-the-record which enabled me to develop a better picture of the individuals within the company as well as of the company itself.

I was given incredible access to all of the shinkansen-operating companies, as well as other companies in the shinkansen business. I found all of the companies to be incredibly open and helpful, although, as will be discussed elsewhere, the nature of the companies did vary. I have no doubt that there are things that I still do not know about the companies, but I rarely got the impression that information was deliberately being withheld. The vast majority of the questions I asked were answered in some form, although I was sometimes requested not to publish the answers. One could spend time questioning why the companies would be prepared to be so open, but I do not think that it is beneficial to consider this issue in too much detail here. Suffice to say, that I got the impression that the companies know that in the shinkansen they have something to be proud of, and they are keen to show it off to as wide an audience as possible. With many companies looking to further develop links with companies, organizations and governments abroad, I expect that some are hoping that the publication of my research results will be a form of advertisement.

This point naturally raises the question of whether my results are biased and overly positive. Indeed, this is an issue that has been raised following a number of presentations that I have done. Without doubt, during the information-gathering stage, I have been careful to not be unduly critical, for fear that it may 'burn bridges' and make it harder for me to gain further access to the companies. However, I believe that what is presented in this book is as balanced as is practically possible. It may well be that having read the book, some will still suggest that the book is generally positive in its stance. One reason for this is that there *is* plenty to rave about concerning the shinkansen. There are areas of shinkansen operations that concern me, and, as I was encouraged to do so by the companies directly, I have on occasion discussed this openly with the companies. At no stage did I ever gain the impression that the shinkansen-operating companies believe that the shinkansen is perfect.

On top of this, I have heard many critical comments made by company employees – let alone by independent researchers. These comments by employees are incorporated within the research, whether they are directly cited or not. Due to the nature of how organizations and companies work, as well as the sensitivity of share prices, for example, to information about the way in which companies operate, I have, where I considered it appropriate, left some information in the book anonymous so that the individual or the company that provided the information cannot be identified. Other times the individual or company made the request themselves, with the research operating under the Chatham House Rule.[5] It should also be noted that most of the critical comments contained in the study are systemic rather than being attached to any particular company or organization.

As well as the company visits, visits to governmental organizations, and interviews with other individuals concerned with the shinkansen, I also took the opportunity to make many other observations. One advantage with my previous research on education reform in Japan was that I had spent time 'as a fly on the wall' in the Japanese education system. Indeed, this was the main reason why I chose to conduct that research as my criticism of many works on Japanese education is that the observers at schools are not a normal occurrence, and so one cannot be sure that what is being observed is a true reflection of the normal activities. The popular suggestion that Japanese classrooms are quiet and organized is a prime example of an observation which is very different to experiences of many who teach in Japanese schools on a day-to-day basis. In conducting this research on the shinkansen, it would have been impossible for me to become 'a fly on the wall' in the same way. However, due to the public nature of many parts of shinkansen operations I was still able to observe the reality of shinkansen operations compared to the under-the-spotlight theory. This included, for example, watching the way in which staff operate at stations and on the shinkansen throughout the shinkansen network. Partly due to the support from the publishers in including photographs in the this book, time was also spent making observations line-side. This included some very unforgettable moments, such as taking the famous shot of shinkansen passing Mt Fuji, shinkansen passing through the tea fields in Shizuoka at a very well-known spot amongst Japanese railway enthusiasts (see Figure 1.1), and spending about an hour next to a tunnel on the Sanyō Shinkansen observing the effects of shinkansen entering and exiting the tunnel.

Figure 1.1 A 300-series shinkansen passing tea fields in Shizuoka Prefecture – one of the most popular sites for photographs by railway enthusiasts along the Tōkaidō Shinkansen.

It is worth pausing at this point to discuss one other resource that I used during the course of this research – namely train and railway enthusiasts. I have found such people to be incredibly helpful. In fact, I perceive there to be quite a difference between a railway enthusiast and a train enthusiast, more commonly referred to as a trainspotter. The trainspotter tends to be primarily interested in direct observations. These observations tend to be supported by activities such as the recording of information, for example, noting the registration numbers of particular trains as they pass. Railway enthusiasts, on the other hand, tend to concentrate more on the collection of information about trains and their operations. Although most, if not all, also take part in observation activities at some stage, this is by no means the primary activity and these observations usually take the form of taking photographs rather than the recording of registration numbers. Members of both groups may also keep model trains, although the limited space in Japanese homes has restricted this past-time and led to novel designs and equipment to help overcome this hurdle. Similarly, in Japan, although apparently becoming popular worldwide now, computer software – such as train-driving simulators – have also been developed to respond to the spatial challenges. The nature of these enthusiasts and trainspotters varies a great deal. Some are or have been involved in the railway industry, some are simply casual observers, while for others it is a very significant activity or hobby in their life. There is no lack of materials for these groups of people, such as books and magazines, while the internet provides a forum for these individuals to present their own information and ideas. The content of these internet pages can range from the trivial up to incredibly in-depth and well-researched pieces. The degree to which these sources can be treated as reliable, however, also varies greatly.

I have used the above terminology, 'railway enthusiast' and 'train enthusiast' (or 'trainspotter'), in referring to these groups – which are neither mutually exclusive nor the only groups interested in railway and train matters – quite deliberately. For I have found that while in Britain, it is the trainspotter which would appear to be in greater abundance, in Japan it is the railway enthusiast. Indeed, in Japanese the term usually used is '*tetsudō fan*' (literally 'railway fan'), despite the fact that it is clearly the train that is of interest rather than the railway itself. As Hendry (2003:6–7) discusses the importance of taking photographs (usually with the person in the photograph in front of the famous location) or collecting stamps from tourist spots as evidence of 'proof' of having visited famous sites, something which Buruma (2001:209) also comments on, so the Japanese railway enthusiast appears to need to have photographic evidence, albeit without themselves in the picture, that they have done their activity. Around Japan there must be an abundance of photographs and videos taken by such people, many of which never get viewed more than once I suspect. The taking of the picture appears to be of greater significance than its display or repeated appreciation. That taking pictures is so important is revealed by the fact that many of the magazines for railway enthusiasts in Japan give details about good locations for taking photographs of shinkansen and other trains. So popular are these locations that drivers come to expect to see people at them with cameras ready (JR East interview 2004), and I have even observed one company, JR Tōkai, sending security personnel to a popular spot to speak to photographers to remind them not to get too close to the line and so cause any potential safety hazard.

During the course of my research I have spent time discussing ideas with train and railway enthusiasts. Before many of my research trips I encouraged people on one mailing list to let me know what information they would like me to discover about the shinkansen. The reasons for this are obvious. First, enthusiasts are often aware of what issues are likely to be worthy of investigation. Second, by addressing some of the queries raised by the enthusiasts, although not possible to present the answers to all of the detailed questions that were suggested, I hope that they will find this book a further useful resource.

Naturally the world of the railway enthusiast is a world which I have been drawn into during the course of this research. At times I have enjoyed the challenge of trying to obtain photographs good enough for this publication. Achieving this, particularly when close to the line as the train passes at great speed, must be close to the thrill that hunters get when shooting their targets. Luckily, photography does not involve any blood-letting. However, while using the Japanese magazines for suggestions about some of the locations, partly from the desire to also

12 *Introduction*

get such well-known shots, but also to see how many other people were at that spot, I have also tried to discover different locations by myself. Perhaps this is a part of my non-Japanese individualism rebelling against the norms of Japanese railway enthusiasts. It may also be a reflection of differing opinions about what backdrops are interesting – or for this study – representative of Japan.

Reasons for this study

When I formally began this research in 2000, there were many who questioned my reasoning. Certainly at first glance this study does appear to be an apparent departure from my previous research on Nakasone and education reform in Japan. However, the subjects are not as diverse as one may at first believe. This is a point that will become clear during the course of the book. Here, let me present just two examples of what will be developed further; the recent developments in the shinkansen owe much to the reform of JNR, which was initiated and owed much to Nakasone; and, it is as much due to the quality of the education system and the companies' training that the shinkansen operates so well as to the technical features of the trains and system.

Yet this was not the starting point for this research. One of my earliest encounters with Japan, during a geography class when I was a child, was of a picture of Mt Fuji and a passing shinkansen. It was, therefore, with great anticipation that

Figure 1.2 A 100-series shinkansen passing Hamanako.

I climbed the stairs at Tōkyō Station on my first trip to Japan in 1989 to travel to Ōsaka. Perhaps if it was not for what happened next, my interest in the shinkansen may never have developed into this study. My image of the shinkansen had not altered since childhood and although I was aware that I would be travelling on one of the newer designs of trains, this was no preparation for the sight that confronted me. Rather than the rounded, famous, 'jumbo-jet'-like front that is most associated with the shinkansen, it was a sleek, pointed, *Concorde*-like design (see Figure 1.2). The journey itself, like so many shinkansen journeys, was not particularly memorable – I was barely able to see Mt Fuji – but the sight of that train at Tōkyō Station and the speed and comfort left its mark on me.

Since then I have been to Japan eighteen times, and the journeys on the trains, particularly the shinkansen, have been highlights. As a consequence, when I was an undergraduate student I considered studying the shinkansen or Japanese railways, but at that time found it hard to find sufficient materials. Then, when I decided to embark on a PhD, again I considered studying this subject. However, having seen the education system from the inside on the Japan Exchange and Teaching (JET) Programme, I decided to research that area. On completing that study, I decided that it was time for a fresh start and took the opportunity to finally research this area that has interested me so much over the years. For, although I have not studied it in detail much during the previous ten years or so,[6] I had continued to follow developments of the shinkansen with interest as well as travel extensively throughout Japan. By this time I had also managed to make some contacts at JR Tōkai. During my first, as well as subsequent, research trips I also found that there was now an increasing body of books in Japanese on the shinkansen and Japanese railways that would help me in pursuing my research.

The only question that remained was what aspect of the shinkansen to study. Given that I am not an engineer and that the only other significant English-language study of the shinkansen (Semmens 1997, 2000) had dealt with the technical areas of the shinkansen, I knew that I did not want to concentrate on this area, although this would undoubtedly form some part of the study. The issue that began to interest me was whether it was just me that considered that the image of the shinkansen passing in front of Mt Fuji was a particularly well-known and well-used image of Japan, or whether I personally, for some reason, had a tendency to spot it and remember it more. As an extension of this issue I became concerned with whether the shinkansen itself is a symbol of modern Japan, reflecting aspects of Japanese society. The title of this study clearly sets out what my conclusion is, but the chapters of this book will provide the evidence that led me to this conclusion.

I have often wondered why the issue of symbolism became of such interest to me. Part of it, I suspect, came from my cynicism about modern life, where it often appears that symbolism can count for more than concrete action. However, one cannot ever escape the fact that sometimes it is things closer to home that can often have the greatest influence on us. Humans have a tendency to attempt to rationalise an action, while the real reason can often be as a result of something far less rational. This is a further reason why researching the 'truth' can be so problematic. Although it was my experience of studying with Japanese students at an international school, Concord College, in England that I rationalise as the reason why I first decided to study Japanese formally, I cannot escape the fact that it may have also have been due to other factors go back to my childhood. When considering symbolism, names are particularly important – as will be discussed in Chapter 3. With my own family name having its own famous and symbolic connections, it may be that this was a driving force for this focus of my research. For many the name Hood is most closely related with Robin Hood, whether he really existed or not,[7] and the popularity of his story and what he symbolises has spread around the world. In fact, this is something that I have benefited from myself directly as it has helped me to ensure the correct spelling of my name on a number of occasions.[8] However, in relation to this study, it is another famous Hood that is perhaps of greater significance – *HMS Hood*.[9] The 'Mighty Hood', as the ship was popularly known, was the largest battle cruiser of the Royal Navy. Yet, having been launched at the end of the First World War, it spent much of its active life serving as a symbol of the power and dominance of the British Empire, making several long trips around the world to British colonies (Chesnau 2002:65). Unfortunately it is the event of 24 May 1941 for which the *Hood* is now best remembered, when she was hit by a single strike from the *Bismarck*. With that strike, 'Hood, the pride of the Royal Navy, symbol of Empire and the marvel of her age, had vanished: a legend had disappeared from of the face of the waters' (Chesnau 2002:160). The significance of this devastating loss on the British public, when only three of the 1,418 crew survived, has been likened to an assassination of a British Prime Minister, the flattening of Buckingham Palace, or in recent times, the 2001 attacks on the World Trade Center (HMS Hood Association 2002). Indeed, the shock and significance of the loss was so great that it lead to one the most well known commands in British military history, 'Sink the *Bismarck*'. The effect that *HMS Hood* had during her service and even in her destruction clearly demonstrates the significant role that a man-made object can have on people. I suspect that many would even argue that *HMS Hood* also reflected aspects of British society and the British Empire at that time. It is this side of symbolism, and the potential that the shinkansen plays a similar role, that became of particular interest to me, and as a consequence became the focus of this research.

From JNR to JR

Chapter 2 will consider the history of the shinkansen. However, as an important change happened to the way in which the shinkansen is operated in 1987, rather than break the flow of the next chapter, I would like to briefly introduce some background information on the shinkansen-operating companies here. On 1 April 1987, Japanese National Railways (JNR) was 'broken-up and privatized'. JNR is now a distant memory for many, while others, including myself, had no experience of it. As Kasai (2003:160) says 'As the Japanese saying goes, "Ten years is an age" and for most people JNR is an artefact: its break-up and privatization is just part of ancient history.'

Following its reform, JNR became six regional passenger companies and one nationwide freight company (see Table 1.1). The six passenger companies are the Hokkaidō Railway Company, East Japan Railway Company, Central Japan Railway Company, West Japan Railway Company, Shikoku Railway Company and Kyūshū Railway Company. One freight company, Japan Freight Railway Company, was also created. The companies are commonly referred to as the JR (Japan Railway) companies and together form the JR Group. The companies are commonly referred to as JR Hokkaidō, JR East, JR Central, JR West, JR Shikoku, JR Kyūshū and JR Freight. Further, JR Central is often known as JR Tōkai, the term used throughout this book, due to the prevalence of its business being in the Tōkai region that runs along the east coast of Honshū between Tōkyō and Nagoya. At present the companies that have shinkansen are JR East (operating the Tōhoku Shinkansen, Jōetsu Shinkansen, Hokuriku Shinkansen, Yamagata Shinkansen and Akita Shinkansen), JR Tōkai (operating the Tōkaidō Shinkansen), JR West (operating the Sanyō Shinkansen) and JR Kyūshū (operating the Kyūshū Shinkansen).

Further details about the problems with the way privatization was carried out and how the companies operate will be dealt with at appropriate points throughout the book. However, it is worth noting here that unlike the situation in Britain, for example, each company essentially owns its infrastructure, including the stations, tracks and trains, and is responsible for their maintenance, etc. Through trains, such as the services from Tōkyō to Hakata, see a change of crew from one company's crew to the other at the boundary between JR West and JR Tōkai at Shin-Ōsaka station. The main problem is where companies need to share facilities, such as Tōkyō station in the case of JR East and JR Tōkai, particularly when there is need for any redevelopment work, for example.

It should be further remembered that there are 205 other private railway companies operating in Japan (MLIT interview September 2004).[10] Although less than 150 of these are what would be thought of as rail companies in the conventional sense of the term, it does reveal that the nature of the Japanese railway market is very different to that in many other countries. These private companies derive much of their income from non-railway operating ventures such as housing, retail stores, and other tourist activities. In extreme cases, two or more lines may operate along similar routes. However, these companies are regional, often centred around one or two main cities, and generally do not offer long distance inter-city services that are directly comparable with the shinkansen.

It is also worth noting at this point that the shinkansen sits at the top of a train hierarchy (see Figure 1.3), and although this study argues that the shinkansen reflects many aspects of Japanese society, it would not be totally fair or accurate to suggest that shinkansen is representative of all train services in Japan. Having been to all 47 prefectures in Japan, and most by train, I can say from personal experience that many trains are old, dirty and even do not run to schedule. As Lonsdale (2001:122) says 'Railways are the arteries of Japan, so where the Shinkansen terminates, local trains *chug* off into more remote corners' (emphasis added).

Table 1.1 Formation of JR companies on 1 April 1987

	Passenger railway companies						Sub Total	JR Freight	Total JR
	JR Hokkaidō	JR East	JR Tōkai	JR West	JR Shikoku	JR Kyūshū			
Route km	3,176	7,657	2,003	5,323	880	2,406	21,445	10,010[a]	—
Passenger-km/tonne-km (billion)[b]	3.9	104.5	41.1	45.8	1.7	7.7	204.7	20.0	199,185
Employees	12,719	82,469	21,410	51,538	4,455	14,589	187,180	12,005	199,185
Assets (¥ billion)	976.2	3,884.5	553.0	1,316.3	323.9	738.1	7,792.0	163.8	7,955.8
Long-term debt (¥ billion)	0	3,298.7	319.1	1,015.8	0	0	4,633.8	94.3	4,728.2[c]
Capital (¥ billion)	9.0	200.0	112.0	100.0	3.5	16.0	440.5	19.0	459.5

Source: Aoki et al. (2000:185).

Notes
a Mostly uses passenger companies' tracks.
b Data for Fiscal 1987 (April 1987 to March 1988).
c A further approximately ¥1,200 billion debt to be paid to JRCC to cover their debts.

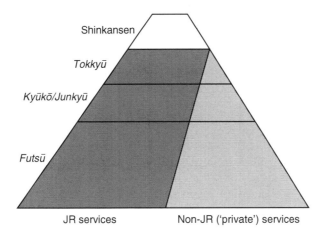

Figure 1.3 Train hierarchy in Japan.

Note
See Glossary for details about different service types.

The shinkansen in many respects represents what JR companies aspire to throughout their operations, although the firms are keen to emphasize publicly that they do not differentiate between shinkansen and other services and that those that work on the shinkansen, for example, should not be regarded as the elite with their company. However, the nature of many commuter or rural services can be so different to the shinkansen that such a high standard is not always achievable.

2 From bullet train to low flying plane

Ladies and gentlemen, welcome to the shinkansen.[1]

Having established the means by which this study was conducted and some background about the companies that operate the shinkansen, it is now possible to look at the history of the shinkansen and to look at the direction in which new developments are proceeding. To be able to see these developments in context, this chapter provides a history of railways in Japan. Each shinkansen line will then be introduced in the chronological order in which construction began. This includes a brief introduction of proposals for other lines that may be constructed or which have been planned. Finally, this chapter provides some brief information on the trains. This chapter is designed to provide a foundation so that the remaining chapters, which address the focus of this study, can be fully understood.

The development of railways in Japan

Although Japan is a, if not *the*, leading railway nation in the world, this was not always the case. The line widely recognized as being the first in Japan was from Shimbashi[2] in Tōkyō to Yokohama (the location of the current Sakuragichō station).[3] The survey work for the 29 km line began in 1870, with the supervisor of the project being a British engineer called Edmund Morel. The line was officially opened on 14 October 1872 (Aoki *et al.* 2000:7–9; Semmens 2000:1–3).[4]

A significant point about the new line, certainly in terms of the history of the shinkansen, was its gauge (the distance between the two rails). Different countries use different gauges. The 'standard' gauge, that is, the one used in most European countries, is 1435 mm. This 'standard' is derived from the width of two side-by-side horses, which in the early days of railways were used to haul the carriages (often coal trucks), and was the gauge used by Stephenson, one of the pioneers of railways in Britain (Dugan 2003; Haraguchi 2003a:56). However, it was decided that a gauge of 1067 mm would be used in Japan.[5] There still appears to be some debate about who made this decision and why. It is commonly said that the reason derived from the experience of Morel, or one of his engineers, in British colonies which adopted this gauge, and that he considered it the most appropriate for Japan's topography as it required less space, was easier to construct, and consequently was also cheaper (Aoki *et al.* 2000:8; Ericsson 1996:33, 396ff19; Haraguchi 2003a:56–7). Yamanouchi (2000:12) suggests that it was the topography that was key, and that the decision was taken by the Japanese government. However, I have come across some Japanese that have referred to the decision to use the 1067 mm gauge as a mistake. I suspect the fact that it has been suggested that one of the factors that led to the decision was the idea that Japan was a third-world country and so there would not be demand for railway transport (Haraguchi 2003a:57; Yamanouchi 2000:12) probably fuels some resentment. Yet, I would argue that if it were not for this 'mistake', it is likely that the shinkansen, at least the system as it exists now, would never have been created and that Japan would possibly have now been suffering from many of the railway problems that are faced by some European countries.

It is worth noting that while the gauge adopted for most railways in Japan (hereafter referred to as conventional lines) was narrower than the 'standard', the loading gauge, which takes account of how big the locomotives and carriages can be, is greater than most 'standard' gauge railways. For this reason, trains on conventional lines in Japan are often wider than British trains, for example, which run on a 'standard gauge'.[6] The significance of the loading gauge will be addressed in further sections.

Following the opening of the first railway, further lines were soon developed throughout Japan. The development of Japanese railways was turbulent, largely due to the financial restraints that arose at various times. Private investment was encouraged, allowing the network to expand rapidly. The original line itself became part of the Tōkaidō Line. The final part of this line, which linked Tōkyō with Kōbe, was completed on 1 July 1889 (Aoki *et al.* 2000:15). Although this connected Tōkyō, Nagoya, Kyōto, Ōsaka and Kōbe using a route close to the old Tōkaidō 'road' that had been used for centuries by those travelling between these cities, the original proposal was that Tōkyō would be linked with these cities via the Nakasendō.[7] This route goes via the central part of Japan and would have allowed greater access to Yokohama port for the farmers of silk, which was becoming Japan's greatest export at that time. It was also initially the preferred route since it would not be at risk from attack from the sea. However, due to the particularly severe nature of the topography along the Nakasendō it was decided to develop the eastern seaboard route first, although this decision was not confirmed until after construction of the lines had already begun (Ericson 1996:44). Although no route existed between Kōbe and Fukuoka in 1890, with the rapid development of other lines, a map of Japan's railway network then (see e.g. Aoki *et al.* 2000:14), closely resembles the shinkansen network some 100 years later.

In the five years to 1890, the length of the government's network grew from 269.9 km to 885.9 km. In the same period, the length of private railways mushroomed from 299.3 km to 1,365.3 km. By 1905, 68% of the 7,696 km network was private (Aoki *et al.* 2000:206). Yet this soon changed. In 1904, on the back of a new found confidence following victory over China in 1895 and having signed the Anglo-Japanese Alliance in 1902, Japan went to war with Russia. Consequently, the government proposed to nationalize the railways to 'smooth domestic transport... and standardize and integrate railway infrastructure' (Aoki *et al.* 2000:40). Nationalization was completed between 1906 and 1907. By 1910, five years after Japan's victory over Russia, the proportion of the railways (not including trams) that were run by the government was 90% of the 8,660 km total (Aoki *et al.* 2000:206).

In the early twentieth century those who saw the need for standard gauge railways became more vocal. There was even the possibility of there being a private standard-gauge line between Tōkyō and Ōsaka, funded by the financier Yasuda Zenjirō and a new company, the Japan Electric Railway Company (Yamanouchi 2000:17). However this plan did not materialize as the government was already considering its own plans for such a line thanks to the work of Gotō Shimpei. Gotō had been a governor in Taiwan following the Sino-Japanese War and went on to become the first President of the South Manchuria Railway in 1906 and later the first Director-General of the Railway Agency (Aoki *et al.* 2000:41). Although Gotō continued to promote the plan for the adoption of the standard gauge and it being debated a number times in the Diet, political instability made progress difficult. Research was even conducted on the Yokohama Line near Machida that demonstrated the viability of using extra rails alongside existing rails to speed up the conversion from the narrow gauge to 'standard' (Haraguchi 2003a:59; Yamanouchi 2000:18), but the proposal to convert existing lines to standard gauge or to build a new standard gauge line between Tōkyō and Ōsaka never materialized. The main network, however, continued to expand, spurred on by the introduction of the 1922 Railway Construction Law, whereby politicians sought to have railways brought to their constituencies (Aoki *et al.* 2000:100; Ericson 1996:191–242; Kasai 2003:11) and promoted the policy of 'build first, convert [to standard gauge] later' (Yamanouchi 2000:18). However it also led to a situation where some lines 'zigzagged from one town to another or looped in a huge half-circle through several towns' rather than merely linking larger cities (Hosokawa 1997:49).

Dangan ressha – the original bullet train

Due to various domestic and international developments and pressures, Japan by the start of the 1930s was becoming a very different country. Although its economy had suffered as a result of the Great Kantō Earthquake on 1 September 1923 and the global impact of the Wall Street Crash, Japan remained one of the largest economies in the world. Meanwhile, partly a result of these factors, but also due to the nature of international relations and diplomacy in that period, Japan was politically shifting to the right, the army was exerting greater influence, and so colonial expansion became almost inevitable. This expansion had started over 50 years previously with the colonization of the Ryūkyū islands (modern day Okinawa Prefecture) and of Hokkaidō. Taiwan and Korea followed thereafter. By the 1930s, Japan was ready to extend its Empire further. Although it is the

8 December 1941 that is the day that has gone down in infamy,[8] the Pacific War had effectively begun about 10 years prior to that (see Storry 1987:182–213).

Railways were to play a significant role in the expansion. Manchuria, an area rich in minerals and agricultural land, remained underdeveloped, although Japan had first gained the territory in 1905 as a result of the Treaty of Portsmouth that ended the Russo-Japanese War. In the 1930s, Japanese were encouraged to relocate to Manchuria so that the Empire could better tap its resources and cement its hold on the territory. In order to persuade ordinary Japanese to take this step, Manchuria was portrayed as an Asian-mainland utopia. Central to this image were the railways, and specifically the *Pashina*-class express train, *Ajia* ['*Asia*'] (Charrier 2003; Hoshikawa 2003:30–1; Yamanouchi 2000:25). Charrier (2003) notes that the railways 'ordered space that was messy', and 'while not emphasizing the railways, the railways were used as a reference point'. On top of this, globally the 1930s was a period of speed. It was an age when countries, companies and individuals were competing to develop the fastest ocean liners, boats, planes, cars and trains.[9] Speed was important. Charrier (2003) argues that the advantage for the Japanese authorities trying to encourage the relocation to Manchuria was that 'speed kills or empties the landscape as it goes', an idea that had existed in Europe since at least the 1840s also (see Ericson 1996:69). Although I would argue that speed can add an extra dimension, it is undeniable that the imagery of the fast trains was important in the Imperial expansion.

But it was more than a matter of imagery. Asia is a huge landmass. Even the four main islands of Japan stretch for about 3,000 km. Air travel was in its infancy, and was not able to carry large numbers of personnel or carry large weights. As the Japanese Empire expanded further into Asia, Tōkyō became more isolated on the Eastern edge of this Empire. A means was needed to address this. Two proposals were put forward. The first was a new railway using the existing gauge. The second proposal was for a standard gauge railway. Despite being over 18 times the cost, the latter option was accepted (MLIT 2004a:1). The new trunk line, or 'shinkansen', would connect Tōkyō with Shimonoseki at the Western tip of Honshū. The train would travel at up to 200 km/h, cutting the travelling time for the approximately 1,000 km journey to a mere nine hours (Hoshikawa 2003:32–3; Yamanouchi 2000:19). This shinkansen was just part of an ambitious plan to create an Asian loop line that would have had tunnels connecting Honshū with Hokkaidō, as well as Japan to the Asian continent via Korea and via a northern route (Oka 1985:324–6). Priority was given to the route to Korea. One proposal suggested that the link between Japan and Korea could made via Tsushima using a tube standing on stilts on the surface of the seabed rather than using a conventional tunnel beneath the seabed. However, this plan was rejected due to its vulnerability to torpedo attack (Hoshikawa 2003:33).

Although the Shinkansen Construction Standards allowed for the train to be either electric or steam, due to concerns about the vulnerability to bombing and the time needed to repair an infrastructure using electricity, just twenty days after the Standards were adopted it was decided to use steam (Oikawa and Morokawa 1996:6–7). The train was referred to as the *dangan ressha* ('bullet train'). Land was gradually purchased and construction began on some of the major tunnels. However, the 'bullet train' never became a reality. With Japan struggling to cope with the advance of the American forces, resources became scarce, and the plan was abandoned.

Genesis of the shinkansen

Although Gotō Keita, president of Tōkyū Railway, did try to resurrect the *dangan ressha* plan after the war, Japan's priority was reconstruction and the organization was not in place to deal with such an immense project (Yamanouchi 2000:19). It was a difficult time for most, characterized by labour strife, political upheaval, uncertainty, and harsh living conditions. Japan's recovery was helped by the outbreak of war on the Korean peninsula, which led to significant supply orders from the Americans (Horsley and Buckley 1990:52; Storry 1987:243). This kick-started the Japanese economy. By the end of the 1950s, Japan was well on the road to recovery. However, much of Japan's infrastructure remained under-developed and was struggling to cope with the demands being placed upon it. The consequence was as inevitable as it was sad.

In the 1950s and early 1960s JNR was linked with the loss of over 1,800 lives. On 24 April 1951, 106 were killed in a train fire at Sakuragichō. On 26 September 1954, the-then second worst peace-time ferry disaster occurred with the sinking of the *Tōya-Maru* train ferry between Honshū and Hokkaidō, with the loss of about 1,200 lives. On 25 March 1955, 168 lost their lives when the JNR ferries *Shiun-Maru* and *Washu-Maru* collided. On 15 October that year, a further 40 were killed near Rokken station when a stop sign was violated (Aoki *et al*. 2000:134–6; Yamanouchi 2000:149). Although many of these lives were lost at sea, where poor weather played a part, there appeared to be a lack of a safety culture within JNR. Consequently, it was probable that further accidents would occur. This was particularly likely where there were capacity problems, which had become a major problem with some trains carrying 300–400% above their capacity, spawning the term '*kōtsu jigoku*' ('traffic hell') (Noguchi 1990:26). On the Tōkaidō Line between Tōkyō and Ōsaka, there was the problem of trying to fit on more trains than it could sensibly handle. Although the line accounted for only 3% of the country's network, it was carrying 24% of all JNR passengers and 23% of JNR freight (Semmens 2000:7). Something had to be done.

Three plans were considered. The first was to build another narrow gauge line alongside the existing Tōkaidō Line. The second was for a shorter narrow gauge line to be built between Tōkyō and Ōsaka. The final option was for a standard gauge line to be built (Kasai 2003:6). The first of these was rejected as by the 1950s much of the land lying next to the Tōkaidō Line had been developed, making it prohibitively expensive to purchase land for the new line. The second option had the advantage that it could be connected to the existing route, so each new section could become operational as it was completed. However, this option did not overcome the problem that all trains have physical limits on their top speed. The lower the gauge of a railway, particularly when a relatively large loading gauge is being used, the lower the maximum top speed is. Furthermore, at the time it was stipulated that trains on conventional lines must be able to stop within 600 m due to the existence of level crossings, for example.[10] This naturally further limits the top speed as braking a train, as I have found out from my own experience of trying to stop a train at JR East's General Education Centre in Shirakawa, is one of the hardest tasks in train operations. So the second option, while improving capacity of the Tōkaidō route, would only have limited impact upon journey times between Tōkyō and Ōsaka. It was partly for this reason that the final option, the shinkansen option, was chosen. However, although in hindsight the decision to build the shinkansen can be seen as the right one, at the time there was great concern about it and it was by no means straight forward.

Luck often plays a part in the development of great achievements. However, it is often overlooked or forgotten, perhaps due to a need for humans to feel as though they are in control of their destiny. The development of the shinkansen also benefited from a large element of luck – namely the existence of certain individuals who happened to have different and complementary skills, and who united together on this one project. Two of these men, Shima Hideo and Sogō Shinji, have become synonymous with the shinkansen, and without their cooperation, the shinkansen 'may well not have become a reality' (Kasai 2003:9).

Shima is often the man credited with building the shinkansen. He was the son of Shima Yasujirō, who had developed the plans for the shinkansen during the Pacific War as well as successfully carried out the previously mentioned tests on the Yokohama Line. Shima Hideo was also involved in the development of the *dangan ressha*, though seeing that Japan would probably lose the war had already began working on the creation of Electric Multiple Units (EMUs) rather than working on the high-speed steam locomotive as instructed (Oikawa and Morokawa 1996:7). Sogō was the President of JNR, and it was he who managed to persuade politicians to back the plan.

Other significant people were Miki Tadanao, Matsudaira Tadashi, and Kawanabe Hajime based at the Railway Technology Research Institute (RTRI), part of JNR. They were responsible for much of the technical development of the shinkansen (Oikawa and Morokawa 1996:9), and their story and involvement in the project was shown as part of NHK's popular *Project X* television series (NHK 2001). They had all worked on aircraft design during the war, but had decided that they would prefer to work at JNR than to join what became the Self Defence Forces.

Although it may seem hard to believe now, those who were prepared to publicly support the shinkansen in the late 1950s were few in number. The plan was first made public at a convention at the Yamaha Hall in Tokyo on 25 May 1957 by the three RTRI engineers (NHK 2001; Suda 2000:26; Yamanouchi 2000:101–2). As the project became a reality, primarily thanks to the skilful handling of politicians by Sogō, the central team involved in it were referred to as the 'daydream team' (Yamanouchi 2000:12) or 'crazy gang' and were 'sneered at' by others in JNR (Kasai 2003:8). Even in 1963, only a year before opening, the director-general of the Construction Department of JNR, stated to new JNR employees:

> The Tōkaidō Shinkansen is the height of madness. As the gauge of the Tōkaidō Shinkansen is different from existing lines, track sharing is not possible. Even if the journey time between Tōkyō and Ōsaka is shortened,

passengers have to change trains at Ōsaka in order to travel further west. A railway system which lacks smooth connections and networks with other lines is meaningless and destined to fail.

(quoted in Kasai 2003:7)

Many compared the construction of the Tōkaidō Shinkansen to the Great Wall of China and the Battleship *Yamato* as being a white elephant, and it was a considerable political risk to back the project (Hosokawa 1997:5; Kasai 2003:8, 14; Yamanouchi 2002:30). Sogō had always worked on this assumption and devised a plan to make it almost impossible for the government to withdraw its support, once given. Central to Sogō's strategy was the use of a loan from the International Bank of Reconstruction and Development ('World Bank'). This was apparently an idea put to him by future Prime Minister and then Minister of Finance, Satō Eisaku, who had previously worked under Sogō in the Railway Ministry (Kasai 2003:15). With the successful application for a $80 million loan (estimated to be no more than 15% of the cost of the line (Yamanouchi 2000:111)) in place, it ensured that the Japanese government had to remain committed to the project. At the same time Sogō, who had deliberately kept estimated cost figures of the Tōkaidō Shinkansen low for fear that if they were too high neither the Japanese government nor the World Bank would have supported the proposal, began to divert money from other JNR projects to the construction of the shinkansen. This was possible as once JNR's total budget had been approved by the Diet, the JNR President had 'discretionary authority' over how to spend it (Kasai 2003:15). It was not only the costing figures that Sogō kept down. Concerned that a proposal for a 250 km/h train, as was being designed, would be met with disbelief and distain, the maximum proposed speed for the shinkansen was kept to 200 km/h (Semmens 2000:7–8). Sogō also managed to circumnavigate opponents within JNR who planned to use the Council of Railway Construction, an advisory body to the Minister of Transport to determine the approval of *new* lines, by 'arguing successfully that the Tōkaidō Shinkansen was *not a new line*, but the *expansion of the existing* Tōkaidō Line to quadruple-track' (Kasai 2003:15, emphasis added).[11]

On 20 April 1959 the ground-breaking ceremony took place. Pictures of the event reveal the wartime history of the project as clearly visible in the background is the perfectly formed entrance to Shin-Tanna tunnel, the construction of which had been abandoned after ten months in January 1943 (Umehara 2002:14, 16). So significant was the tunnel's location that Kannami town in Shizuoka prefecture had been paying around ¥10 million per year for the maintenance of that end of the tunnel. There had even been plans to complete the tunnel and allow cars to use it before the shinkansen project was resurrected (Hosokawa 1997:192; Umehara 2002:14).[12]

The route of the Tōkaidō Shinkansen was similar to that of the *dangan ressha* plan, and approximately 18% of the 11.8 million square metres of land that needed to be purchased had already been done so (Hoshikawa 2003:33; Hosokawa 1997:193). However, there were areas where alternative routes were either suggested or implemented. For example, while the *dangan ressha* had a proposed stop at Yokohama station (Hoshikawa 2003:33; Suda 2000:24), and it was initially suggested that the Tōkaidō Shinkansen go via central Yokohama using Higashi-Kanagawa station, eventually Yokohama was effectively by-passed by creating a new station, Shin-Yokohama, on the outskirts of the city where the shinkansen crossed the Yokohama Line connecting Yokohama and Hachiōji (Umehara 2002:26–8), solving the significant problem of land-purchase within the densely populated central area of Yokohama. The other advantage of this route was that it was shorter and so would further cut the travelling time between Tōkyō and Ōsaka. At the time the location of the Shin-Yokohama station must have seemed strange. Although hard to believe now when one sees the huge office blocks and the 2002 World Cup Final stadium surrounding the station, the area in the 1960s was almost entirely rice-fields and a few tree-covered hills (see photograph in Semmens 2000:110).

Another variation from the *dangan ressha* plan was the location of the terminal in Ōsaka. Due to the rivers within Ōsaka and the need to easily continue the line beyond Ōsaka towards Kōbe, it was always envisioned that Ōsaka's shinkansen station would be on the outskirts of the city at a point where the line intersected the Tōkaidō Line. However, while the *dangan ressha* plan had the terminal within Suita City, Ōsaka City opposed this when the Tōkaidō Shinkansen route was discussed. The eventual location was 1.5 km closer to Ōsaka at a new station named Shin-Ōsaka. This change in location meant that the station was not only at the intersection of the Tōkaidō Line but also the planned Midōsuji underground line. Despite this move, its location is undoubtedly more inconvenient than many shinkansen stations and has been a source of some disquiet for the people of Ōsaka, in particular, who feel as though they were treated poorly in comparison to Tōkyō (Umehara 2002:29–30). This frustration was probably fuelled by the debate over the location of the Tōkyō terminal of the Tōkaidō Shinkansen. For although 14 locations in and around the important Yamanote Line, including Shinagawa station, were considered, it was the most expensive option of extending the Yaesu side of Tōkyō station that was selected (Umehara 2002:22–4).

However, without doubt the greatest difference between the Tōkaidō Shinkansen and *dangan ressha* routes was between Nagoya and Kyōto. The reasons for this difference will be analysed in Chapter 4, as the debate that ensued about this change became an issue that has plagued debates about other shinkansen lines. The result of this change meant that the only significant tunnel construction that became necessary was the completion of the Shin-Tanna Tunnel and a new conventional line tunnel at Nihonzaka, a point where the mountains reach the coastline along the Tōkaidō route. The 2.2 km Nihonzaka Shinkansen tunnel had been completed in September 1944 and following the war, was used by the conventional railway from May 1949 until September 1962, by which time a new tunnel for the conventional line had been constructed and so allowing shinkansen rails to be laid in the tunnel as had been intended almost 20 years previously (Naitō 2003; Umehara 2002:18–19).

Although the route and funding problems had been addressed, there were still technical matters that needed to be addressed. These problems related closely to the issue of safety, which remained a problem at JNR. That this was the case was graphically confirmed while construction of the Tōkaidō Shinkansen was underway. On 3 May 1962, 160 people were killed in an accident at Mikawashima. Another 161 lives were lost on 9 November 1963 near Tsurumi following a collision with a derailed freight train. The high death tolls owed much to the fact that other trains became involved due to not having enough time to stop (Aoki et al. 2000:134–136; JR East General Education Centre Accident Museum; Kasai 2003:17; Takahashi 2003a:68). Sogō resigned shortly after the Mikawashima accident, apparently to take responsibility for that and the escalating shinkansen costs (Kasai 2003:11; Takahashi 2003a:68–9). Sogō was replaced by Ishida Reisuke, who was not enthusiastic about the Tōkaidō Shinkansen (Kasai 2003:12). The accidents had indicated to many that JNR continued to have a significant safety problem, and there was a logical concern that faster moving trains, in the form of the shinkansen, would increase the probability that there would be a catastrophic number of deaths in the event of collision or derailment. Indeed, so certain were some that there would be such an accident involving a shinkansen, that some newspapers had apparently even prepared the headlines and background to such stories (Hosokawa 1997:204; Kasai 2003:12).

Part of the solution to these safety problems was provided by two of the former wartime engineers. Matsudaira was convinced that, based on the work he had done on the development of the Navy's *Zero* fighter, the cause of many derailments was due to vibrations created by the train rather than due to the rails, as was thought by many at that time (NHK 2001). Having been sidelined for many years due to his unconventional thinking, Matsudaira was eventually called upon to provide the design of the shinkansen's bogies and used many of those who had worked under him during the war to help him. A test track was built at Kamonomiya, which was easily accessible from RTRI's headquarters in Kunitachi and was on the route of the shinkansen so that the track could become part of the Tōkaidō Shinkansen itself in due course. Although during one speed test the train began to vibrate as it approached 250 km/h, the train did not derail and Matsudaira's work was confirmed as being correct (NHK 2001). Kawanabe had an even harder route to the implementation of his ideas. Having been accused of developing signal devices that helped the Imperial Navy during the war, he was fired from RTRI during the Occupation, though was later reinstated. Throughout he continued to work on an idea of stopping a train by using a low-frequency signal. This system became Automatic Train Control (ATC) (NHK 2001). ATC is the primary reason why the shinkansen can operate safely at high speeds without any fear of collision (see Chapter 6).

Naturally there were many other engineering problems that had to be overcome. Yet, it is without doubt that had it not been for the experiences and abilities of Miki, Matsudaira and Kawanabe in providing solutions to the key issues, the shinkansen would never have been born. Although Shima and Sogō tend to be credited with the creation of the shinkansen, one should not forget the significant work done at RTRI. Even today RTRI consider their headquarters to be 'the birth place of the shinkansen' (RTRI interview January 2001). When I visited RTRI in 2001 I was able to see one of the original test trains, which is clearly recognizable as its nose cone lights up. Although over forty years old, and clearly showing its age, it is housed in a building so it can both be preserved and be used to run various tests to help with the development of current trains. While being shown around the train I was aware of not only the great affection that RTRI staff have for it, but also the significant part it has played in the development of the shinkansen and also arguably, as will be discussed in Chapter 5, of Japan.

Tōkaidō Shinkansen

The first of October 1964 is a date that has become legendary in Japanese history. At 06:00, following a short ceremony, the first shinkansen left Tōkyō towards Shin-Ōsaka station. At the same time another shinkansen left Shin-Ōsaka towards Tōkyō (Hoshikawa 2003:46; Suda 2000:29, 45). Expectations were high. In August, a test train from Tōkyō was followed all the way to Ōsaka in a live four hour television broadcast on NHK (Hoshikawa 2003:46; Yamanouchi 2000:115). Yamanouchi (2000:115), who was on that train, describes the looks of excitement of the large number of people who had come to be by the line as the train passed and how it over-took some of the helicopters that were covering the event. Yet, despite his significant role in the development of the Tōkaidō Shinkansen, one of those who was not present at the ceremonies on the first day was Sogō. This was not only due his recent resignation, but also a sign of the significant anti-shinkansen feelings that were held by some influential politicians and JNR staff at the time. Although not on the platform on that first day, his impact and the justification for the project was later recognized with the establishment of a plaque in his honour at Tōkyō station (see Figure 2.1b).

Much of the world's media was already in Japan in preparation for the Tōkyō Olympics, and so they were given the opportunity to experience Japan's new technological and engineering feat. Professor Geoffrey Bownas, who was helping the BBC, reflected that

> Although most shinkansen journeys have now become blurred together due to their predictability, one thing I remember from the inaugural run of the shinkansen was when the train accelerated around Tamagawa [after leaving the residential area of Tokyo], I felt a surge in my stomach and I was looking around for a seatbelt.
>
> (Bownas interview September 2002)

Speed is an issue that will analyzed elsewhere. However, it is worth noting here the significant difference that the opening of the shinkansen had upon the journey time between Tōkyō and Ōsaka (see Figure 2.2). When the Tōkaidō Line began operations in 1889 the journey time was about 16 hours 30 minutes (Suda 1998:52). This was already a vast improvement upon the some two to three weeks

28 *From bullet train to low flying plane*

Figure 2.1 (a) Commemorative plaque at Tōkyō Station; (b) a 100-series shinkansen entering the platform at Tōkyō next to the plaque of Sogō; (c) Tōkyō station in the evening; (d) cleaners (circled) waiting for a shinkansen to arrive at the platform; (e) the fortieth anniversary shinkansen departs Tōkyō station; (f) a display – including a model of the N700-series shinkansen at the front – celebrating 40 years of the Tōkaidō Shinkansen.

that it had taken by foot or palanquin by the end of the Tokugawa Period (1603–1868), which was cut to a week with the introduction of the horse-drawn omnibus in 1881 and the few days taken by steamships after that (Ericson 1996:68; Mito 2002:24). While various enhancements, such as the opening of the Tanna Tunnel and electrification had managed to cut this to 6 hours 30 minutes by 1960, the limits of the gauge of the railway, let alone the capacity problems, meant that further cuts were practically impossible. The journey times on the Tōkaidō Shinkansen have been cut further since 1964 (with a reduction of 50 minutes taking place from November 1965). While the initial time saving of 2 hours 30 minutes, or about 38%, is impressive, it is the impact of this change that is note wothy. When the shinkansen began operations, a day trip between Tōkyō and Ōsaka became possible. Although some had done this before, it had been hard. To reach Ōsaka for an early afternoon meeting meant leaving Tōkyō at 07:00 (Sugiura 2001:44), and many workers were restricted to taking the slower trains that took some 10 hours (Tokorozawa 2003a). When one takes account of the time needed to prepare oneself and get to the station, even those on the faster

From bullet train to low flying plane 29

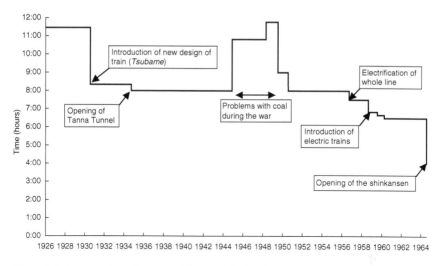

Figure 2.2 Journey times between Tōkyō and Ōsaka (1926–1964).
Source: Based upon data in Sugiura (2001:44) and Suda (1998:52).

services would have been lucky to return home much before midnight. The shinkansen saved five hours on this round trip in its first year alone. As will be discussed elsewhere, there are also losers with such an 'improvement', however.

The Tōkaidō Shinkansen was an instant success, helped by, and reflecting, events that occurred around it (though its influence upon some of these should not be discounted). It opened during the time of Prime Minister Ikeda's successful 'income-doubling policy' when the Japanese economy was experiencing unprecedented economic growth. As airline transport was only just taking off, the Tōkaidō Shinkansen provided the best means of travel between Tōkyō and Ōsaka. With face-to-face contact being considered so important in Japan (see Hendry 2003:249–50), the Tōkaidō Shinkansen was able to rely on an increasingly dependent business customer base. The Tōkaidō Shinkansen was given a further boost with Ōsaka's hosting of EXPO 70.[13] The extra tourist traffic that this generated saw the length of the shinkansen trains extended from 12 to 16 carriages (Suda 2000:30). Total passengers since the opening reached 300 million, about treble the Japanese population, by July 1970 (Aoki *et al.* 2000:144).

Since 1964, the Tōkaidō Shinkansen has seen several changes. The number of stations along the route has increased from 12 to 17. A further station, Biwako-Rittō,[14] is planned to open in about 2010 between Maibara and Kyōto (Hoshikawa 2003:27–8), now the longest stretch of the Tōkaidō Shinkansen without a station. However, as far JR Tōkai are concerned (based on interviews), there is little prospect of further stations being built in Shizuoka Prefecture under the new Shizuoka Airport or in Kanagawa Prefecture between Shin-Yokohama and Odawara, both of which some have campaigned for. For many passengers the most noticeable differences are changes in the design of trains (see following section), the addition of an extra service (see Chapter 5) and the changes brought about as a result of privatization.

Expanding the shinkansen network

With the success of the Tōkaidō Shinkansen, the opinions of politicians, JNR officials, bureaucrats and the public soon shifted to one of eagerness to have further lines. Writing some 40 years on from its introduction, and having not experienced it first hand, it is hard to gain a full appreciation of just what excitement the shinkansen created. However, there can be no doubt that for such a shift from concern and ridicule to enthusiasm within a short period of time was quite remarkable. The impression that I have from the reading that I have done and interviews that I have conducted is that Japan was gripped by some kind of fever, indeed Suda (2000:30) even refers to it as 'Shinkansen fever', albeit, perhaps, not an unhealthy one. Seen in this context, it is of little surprise that proposals soon came forward for further shinkansen lines to be constructed. These proposals culminated in the approval, just six years after the opening of the Tōkaidō Shinkansen, of the Nationwide Shinkansen Development Law (*Zenkoku Shinkansen Tetsudō Seibi Hō*). This law proposed a shinkansen network of about 7,000 km to be constructed by 1985 (Aoki *et al.* 2000:144, 146; Semmens 2000:28). This was incredibly ambitious considering at the time only 515 km was in operation and construction had only begun on the extension of the line beyond Shin-Ōsaka to Okayama three years previously.

With the worsening finances of JNR, as well as the uncertainties in the Japanese economy that occurred in the early 1970s following the Nixon Shocks and first Oil Shock,[15] the shinkansen entered its own 'dark valley' (see following sections) as it came under greater scrutiny and the bubble of enthusiasm that had surrounded it burst. A new plan, the *Seibi Keikaku*, was passed in November 1973. This plan placed a priority on certain routes (known as the *Seibi Shinkansen*), while the other lines (now referred to as the *Kihon Keikaku Rosen* – 'Principal Planned Routes') in the National Shinkansen Development Law were put on hold. Table 2.1 compares the situation between the Nationwide Shinkansen Development Law and the *Seibi-Shinkansen*. There now seems little prospect for any of the non-*Seibi-Shinkansen* lines ever being built. Once the *Seibi-Shinkansen* lines are completed it is more likely for further improvements to be made to those lines (MLIT interview January 2004). The only possible exception to this is the Chūō Shinkansen. This line and the *Seibi-Shinkansen* lines are introduced in the following sections. The case of the Narita Shinkansen will be addressed elsewhere.

Sanyō Shinkansen

The Sanyō Shinkansen is in many respects an extension of the Tōkaidō Shinkansen. Indeed, some trains continue from Tōkyō beyond Shin-Ōsaka along the Sanyō Shinkansen, and vice versa. During the days of JNR this continuation would have been even more seamless than it is today. For since 1987, the two lines have been operated by different companies. As a consequence, personnel change at Shin-Ōsaka, which accounts for the slightly longer stop that trains make there (about two minutes rather than 50 seconds).

The Sanyō Shinkansen was opened in two stages due to the need to secure the funding, as well as the time taken to construct the 18.7 km Shin-Kanmon Tunnel between Honshū and Kyūshū. The first section to Okayama opened in 1972, with the second section to Hakata (in Fukuoka) opening in 1975. Being built after the Tōkaidō Shinkansen, it was possible to introduce improvements into the design of the infrastructure. The most significant of these was the increase in the minimum radius of curves from 2,500 to 4,000 m. This change, combined with other alterations in the way the line was built (e.g. the greater use of elevated sections and laying the tracks on concrete rather than ballast) means that even today the maximum speed on the Sanyō Shinkansen is 30 km/h higher than that of the Tōkaidō Shinkansen (equivalent to about 15 minutes for a train travelling between Shin-Ōsaka and Hakata). The Sanyō Shinkansen is also different from the Tōkaidō Shinkansen in other respects. The most significant of these are that the cities along its route are much smaller and it does not have the advantage of serving the capital as the Tōkaidō Shinkansen does. As a consequence it is possible to see much shorter trains operating on the Sanyō Shinkansen as JR West attempts to maintain a regular service rather than an infrequent service using 16-carriage-trains.

Tōhoku Shinkansen

In April 1971, JNR and the Japan Railway Construction Public Corporation (JRCC, now JRTT) were given permission to construct the Tōhoku and Jōetsu Shinkansen. The Tōhoku Shinkansen has been built in stages. The first stage connected Ōmiya with Morioka. As there was no direct connection by shinkansen to Tōkyō, special 'relay' services on the conventional line were provided. Naturally, the aim was to have the line start in Tōkyō, but this was fraught with difficulties. The initial plan was to have the route south of Ōmiya underground, but it was found that the soil quality was so poor that such construction (around 30 km) would be extremely difficult and expensive (Yamanouchi 2002:95).

Therefore it was decided that elevated tracks would be used until close to Ueno (in Tōkyō), where the track goes deep underground. The plan was vehemently opposed by many living along the proposed route and a new terminal at Ueno was not opened until 1985, three years after the Tōhoku Shinkansen began operations. The tunnelling was hard and costly, with one collapse leading to fatalities. Although an improvement in many respects, Ueno's shinkansen platforms are so deep underground that the connection remained inconvenient for many passengers, with 96% of passengers saying they wanted the terminal to be Tōkyō (Mito 2002:213). Work to extend the line the remaining 3.6 km was slow, difficult and expensive. It was not until 1991 that services began from Tōkyō.

Table 2.1 The shinkansen network

Shinkansen line	Route	Distance (km)	Nationwide shinkansen development law	Seibi-Shinkansen	Construction approved
Tōkaidō	Tōkyō↔Shin-Ōsaka	515	(already constructed)		
Sanyō	Shin-Ōsaka↔Hakata	554	(already under construction)		
Tōhoku	Tōkyō↔Morioka	497	Yes	Yes	1971
	Morioka↔Numakunai	31	Yes	Yes	1995
	Numakunai↔Hachinohe	66	Yes	Yes	1991
	Hachinohe↔Shin-Aomori	82	Yes	Yes	1998
Jōetsu	Ōmiya[a]↔Niigata	270	Yes	Yes	1971
Narita	Tōkyō↔Narita Airport	65	Yes	Yes	1971
Hokuriku	Takasaki↔Nagano	117	Yes	Yes	1989
	Nagano↔Jōetsu	59	Yes	Yes	1998
	Jōetsu↔Toyama	111	Yes	Yes	2001[b]
	Toyama↔Isurugi	35	Yes	Yes	2004
	Isurugi↔Kanazawa	24	Yes	Yes	1992[c]
	Kanazawa↔Ōsaka	254	Yes	Yes	—
Kyūshū (Kagoshima)	Hakata↔Shin-Yatsushiro	130	Yes	Yes	2001[d]
	Shin-Yatsushiro↔Nishi-Kagoshima[e]	127	Yes	Yes	1991
Hokkaidō	Shin-Aomori↔Shin-Hakodate	150[f]	Yes	Yes	2004
	Shin-Hakodate↔Sapporo	210	Yes	Yes	—
Kyūshū (Nagasaki)	Shin-Tosu↔Takeo-Onsen	51	Yes	Yes	—
	Takeo-Onsen↔Isahaya	45	Yes	Yes	2004
	Isahaya↔Nagasaki	25	Yes	Yes	—
Hokkaidō	Sapporo↔Asahikawa	130	Yes	—	—
Hokkaidō Minami-Mawari	Oshamanbe↔Sapporo	180	Yes	—	—
Uetsu	Toyama↔Aomori	560[g]	Yes	—	—
Ōu	Fukushima↔Akita	270	Yes	—	—
Chūō	Tōkyō↔Ōsaka	480	Yes	—	—
Hokuriku/Chūkyō	Tsuruga↔Nagoya	50	Yes	—	—
San'in	Ōsaka↔Shimonoseki	550	Yes	—	—
Chūgoku Ōdan	Okayama↔Matsue	150[h]	Yes	—	—
Shikoku	Ōsaka↔Ōita	480	Yes	—	—
Shikoku Ōdan	Okayama↔Kōchi	150	Yes	—	—
Higashi Kyūshū	Hakata↔Nishi-Kagoshima[e]	390[i]	Yes	—	—
Kyūshū Ōdan	Ōita↔Kumamoto	120	Yes	—	—
Total length (including already constructed)			6,926	3,416	

Source: Kagoshima-ken Kyūshū Shinkansen Kensetsu Sokushin Kyōryoku Kai 2002:20; http://www.mifuru.to/frdb/

Notes
a Intended terminal was Shinjuku in Tōkyō, but approval was never given.
b Construction for section between Itoigawa and Uozu approved in 1993.
c Originally approved as Super-*Tokkyū*.
d Construction initially approved for Funagoya↔Shin-Yatsushiro in 1998.
e Station renamed Kagoshima-Chūō in 2004.
f Includes the 53.9 km Seikan Tunnel.
g Not including distance where Hokuriku and Jōetsu Shinkansen lines are used.
h Not including distance where San'in Shinkansen is used.
i Not including distance where Sanyō Shinkansen is used.
All distances have been rounded to nearest km. Distances for those lines only included in the Nationwide Shinkansen Development Law are approximate as exact route had not been decided.

From bullet train to low flying plane 33

Figure 2.3 (a) A sign showing the distance from Tōkyō to Shin-Aomori with a diagram behind showing what the future station, which will completely transform the area, may look like; (b) a conventional train arrives at the current Shin-Aomori station; (c) a view along the current Shin-Aomori platform, which will cross underneath the platforms of the Tōhoku Shinkansen terminal in the future; (d) a van goes over a level crossing shortly after a *tokkyū* passes by along the single-track line at Shin-Aomori.

It has not just been at the southern end of the Tōhoku Shinkansen that construction work has been done in stages. The northern extension to Hachinohe was opened on 1 December 2002, about 20 years after the opening of the original Tōhoku Shinkansen. Construction on the remaining section to Shin-Aomori (see Figure 2.3) is continuing and is due to be completed in about 2012.

Jōetsu Shinkansen

In many respects the history of the Jōetsu Shinkansen is intertwined with that of the Tōhoku Shinkansen. Approval for its construction was given at the same time and it began operations only six months later. Although it also serves cities with relatively small populations, it is a very different line. The route is significantly shorter, yet the construction was a major challenge due to the mountains which the route passes through. Construction of tunnels proved to be a major challenge and expense. The greatest of these was the Dai-Shimizu Tunnel. Construction took just over seven years, and when completed was the longest land tunnel in the world at 22.2 km. Yet it was a shorter tunnel along the route, the Nakayama Tunnel, that caused the greatest problems when streams of high pressure water were discovered within the volcanic rock (Semmens 2000:29). In the end it was necessary to divert the route from the original plan, creating a kink in the route that forces trains to drop their speed to 160 km/h in this section (Umehara 2002:58–60). While it was possible to overcome most of the route's problems during construction, the route's other significant problem, snow, is annual (see Chapter 7).

Like the Tōhoku Shinkansen, the Jōetsu Shinkansen remains incomplete. However, unlike the Tōhoku Shinkansen it is extremely unlikely that the route will ever be completed. At present the start of the Jōetsu Shinkansen is officially at Ōmiya. In practice virtually all shinkansen start or terminate at either Ueno or Tōkyō, but for this section they are officially using the Tōhoku Shinkansen. The start of the Jōetsu Shinkansen was planned to be Shinjuku, the busiest station in the world, which lies on the western side of the Yamanote loop line and in the middle of one of Tōkyō's business districts. Due to the costs of land purchase and construction at a time when JNR finances were already in a severe state, and considering the problems building the Tōhoku Shinkansen in Saitama and Tōkyō, the plan to build this section of the Jōetsu Shinkansen was abandoned.

The Jōetsu Shinkansen has also become the most infamous of the shinkansen lines. For it is widely suggested that it was constructed due to the influence of politicians more than due to transportation reasons (see Chapter 4). In October 2004, the Jōetsu Shinkansen made the news worldwide when a shinkansen derailed following a devastating earthquake – the first time a passenger-carrying shinkansen had ever derailed (see Chapter 7).[16]

Hokuriku Shinkansen

While all of the lines mentioned earlier were completed, at least in part, during the JNR era, the network has continued to expand since privatization. The first new line to be undertaken was the Hokuriku Shinkansen. Construction began in 1989 after a freeze on shinkansen construction was lifted in 1988 (Umehara 2002:71). The initial stage was to link Takasaki (on the Jōetsu Shinkansen) to Karuizawa. The route would be a major challenge due to the mountains, but once opened would significantly cut journey time from Tokyo since the conventional track had to use the Usui Pass with its steep 66.7% (1:15) slope that greatly reduced speeds (Aoki *et al.* 2000:134). However, soon after construction began, Nagano was selected to host the Winter Olympics and JR East was asked to construct the line between Karuizawa and Nagano in time for the Olympics.[17] As the line currently only operates as far as Nagano, it is commonly referred to as the Nagano Shinkansen. Yet the route is still under construction, although it seems unlikely that it will be completed as far as its intended terminal in Ōsaka, and that it may only be completed as far as Kanazawa.

Kyūshū Shinkansen

While the expansion of the shinkansen network had largely appeared like lines spreading out from Tōkyō, the opening of the Kyūshū Shinkansen in 2004 broke this pattern, as it is not even directly connected to the rest of the shinkansen network yet. Although the eventual route will link Fukuoka and Kagoshima, it was decided that initially only the southern section between Yatsushiro and Kagoshima would be constructed. The reason for this can be clearly seen when one looks at the impact the two different sections of the Kyūshū Shinkansen have upon travel time (see Table 2.2).

Building a shinkansen line disconnected from the main network, particularly where many passengers are likely to be using the trains north of Shin-Yatsushiro, has meant that a conventional line platform has also been constructed at the station (see Figure 2.4). Although this platform will be used by shinkansen in the future, until then conventional trains will use it so that passengers can walk across to the other side of the platform to easily connect with the shinkansen. While the main Kyūshū Shinkansen route has already been partially opened and the rest is under construction, another line on the island has yet to be fully approved. The Nagasaki Shinkansen, although officially starting from Shin-Tosu, would connect Nagasaki with Fukuoka (see Chapter 5).

To the final frontier – the Hokkaidō Shinkansen

The tragic loss of the *Tōya-Maru* prompted the construction of the Seikan Tunnel between Honshū and Hokkaidō. Construction began in 1964, shortly before the opening of the Tōkaidō Shinkansen, and was completed some 24 years later (Oka 1985; Semmens 2000:43). Despite the growing competition from air travel, in 1972 it was decided to upgrade the tunnel so that it would be capable of taking

Table 2.2 Kyūshū Shinkansen construction

	Conventional line		Shinkansen		Reduction	
	Distance	Time	Distance	Time	Distance	Time
Hakata ↔ Kumamoto	118.4	1:15	98.2	0:35	20.2 (17.1%)	0:40 (53.3%)
Kumamoto ↔ Shin-Yatsushiro[a]	35.7	0:20	31.8	0:10	3.9 (10.9%)	0:10 (50.0%)
Shin-Yatsushiro[a] ↔ Kagoshima-Chūō[b]	165.7	2:05	126.8	0:35	38.9 (23.5%)	1:30 (72.0%)
Total (Hakata ↔ Kagoshima-Chūō[b])	317.1	3:40	256.8	1:20	60.3 (19.0%)	2:20 (63.6%)

Source: Nihon Tetsudō Kensetsu Kōdan 2002a.

Notes
a Distance/time calculated to Yatsushiro for conventional line data.
b Nishi-Kagoshima prior to opening of Kyūshū Shinkansen.
Shinkansen journey time based on original operating speed of 260 km/h.

Figure 2.4 (a) On the left is the space where the shinkansen rail will be installed by JR Kyūshū at Shin-Yatsuhiro in the future – showing the difference in gauge between the conventional line and the shinkansen line; (b) end of the line at Kagoshima-Chūō with Sakurajima in the distance; (c) a viaduct – to be used by the Kyūshū Shinkansen in the future – near Hakata-Minami station (shinkansen access to which is visible below the viaduct); (d) model cranes on the conventional line platform at Izumi and origami-style cranes in the roof of the shinkansen station; (e) on the right the conventional line curves away from Shin-Yatsushiro to go to join the original Kagoshima Line, on the left the route which shinkansen will take in the future when the Kyūshū Shinkansen is completed; (f) a 0-series-renewal waiting at the narrow Hakata-Minami station – the space to the left will become part of the Kyūshū Shinkansen in the future; (g) the sail-inspired design of Shin-Minamata station.

shinkansen (Aoki et al. 2000:146; Oka 1985:327). Yet today there are still no shinkansen operating in Hokkaidō.

Indeed, the changed climate surrounding the shinkansen has meant that in 2003 it was decided that the proposal for the Hokkaidō Shinkansen to link Shin-Aomori and Sapporo should be put to one side, and that initially a proposal to only link Shin-Aomori and Hakodate would be considered (interviews with JR Hokkaidō staff and Hokkaidō Prefectural government officials, 2004). This proposal was finally accepted in late 2004 with the plan for the line to open when the final section of the Tōhoku Shinkansen is completed, thus avoiding for any additional construction being needed at Shin-Aomori for passengers wanting to continue their journey.[18]

Table 2.3 Top air routes in the world

Rank	City-pair	Region	Passengers in 2002 (million)[a]
1	Tōkyō Haneda–Sapporo	Japan Domestic	9.51
2	Tōkyō Haneda–Fukuoka	Japan Domestic	8.28
3	Tōkyō Haneda–Ōsaka[b]	Japan Domestic	7.32
4	Hong Kong–Taipei	Asia International	5.46
5	Seoul–Busan	Korea Domestic	5.39
6	Sydney–Melbourne	Australia Domestic	5.35
7	Seoul–Jeju Island	Korea Domestic	4.91
8	Tōkyō Haneda–Okinawa	Japan Domestic	4.33
9	London–New York	Europe–N. America International	3.90
10	Sydney–Brisbane	Australia Domestic	3.50
11	Taipei–Kaohsiung	Taiwan Domestic	3.46

Source: Japan Airlines 2004.

Notes
a Some data is from slightly different time periods.
b Includes Itami and Kansai International airports.
Generally transfer passengers are not included; unless stated otherwise city names includes all airports within that city.

Hokkaidō is not a train-friendly environment, not only because of the harsh winter conditions, but also as most journeys are done by car. Those that have to travel greater distances tend to use planes. Indeed the Sapporo (Chitose Airport) – Tōkyō (Haneda Airport) route is the busiest in the world (see Table 2.3). According to officials at Hokkaidō Prefectural Government (interview January 2004), 'Most Hokkaidō people have never used the shinkansen. They have no feel for what it is.' The hope of those promoting the Hokkaidō Shinkansen is that a line to Hakodate will help to develop an interest in the shinkansen amongst Hokkaidō people, so that demand for the completion of the line to Sapporo will build. Also, due to the decision taken to change the specifications of the Seikan Tunnel, much of the proposed route has effectively already been constructed, although a means by which non-shinkansen traffic (particularly freight trains) can still use the tunnel has to be addressed (JR Hokkaidō interview January 2004).

It is worth noting that Sapporo is not as far away from Tōkyō as many may believe. Hokkaidō is often regarded as Japan's wild frontier and that Sapporo is a distant outpost in the frozen north, at least from the prospective of those living in Tōkyō and further west. Hokkaidō was not one of the original eight islands of Japan,[19] and it only formerly became part of Japan in the nineteenth century. It is now Japan's largest prefecture by area and seventh biggest by population, with Sapporo being Japan's fifth biggest city. However, the average per capita income of Hokkaidō is ranked only twenty-third of the 47 prefectures and is below Japan's national average (Asahi Shimbun 2004:32, 44–5). Yet Figure 2.5a shows that Sapporo is not significantly further away from Tōkyō than Fukuoka (based on the conventional railway distance). Indeed, if the shinkansen is completed, it

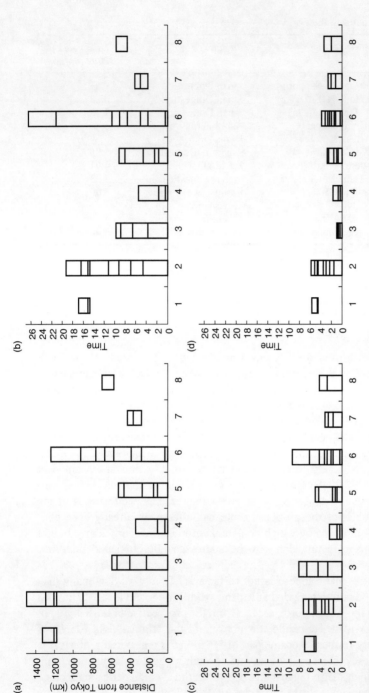

Figure 2.5 (a) Distance of lines, (b) travelling times (1963), (c) travelling times (2004), (d) travelling times (all shinkansen lines completed).

Source: Timetables and Kawashima (2002).

Notes

Due to space only certain key stations have been represented. Lines not starting at '0' indicates that no direct service exists or existed between that route and Tōkyō. Distances are those by conventional railway – the shinkansen route is shorter in many cases, with some station locations also changing. Key (stations listed in order from bottom of column to top (i.e. closest to Tōkyō to furthest away): 1 = Nagasaki route (Hakata, Tosu, Nagasaki); 2 = Tōkaidō, Sanyō and Kyūshū route (Tōkyō, Nagoya, Ōsaka, Hiroshima, Hakata, Tosu, Kumamoto, Kagoshima); 3 = Chūō/Nakasendō route (Tōkyō, Shiojiri, Nagoya, Nara, Ōsaka); 4 = Jōetsu route (Tōkyō Ōmiya, Takasaki, Nīgata); 5 = Hokuriku route (Tōkyō, Takasaki, Karuizawa, Nagano, Toyama, Kanazawa); 6 = Tōhoku and Hokkaidō route (Tōkyō, Ōmiya, Fukuyama, Sendai, Morika, Hachinohe, Aomori, Hakodate, Sapporo); 6 = Ou Route/Yamagata Shinkansen (Fukushima, Yamagata Shinjo); 7 = Tazawako & Ou route/Akita Shinkansen (Morioka, Akita).

will be closer. Yet Figures 2.5b and 2.5c show the relative disadvantage in using train travel to Hokkaidō both historically and currently, while Figure 2.5d shows the situation should the shinkansen network becompleted. While Hokkaidō and Sapporo have enjoyed a somewhat poetic and idyllic status in comparison to the cramped urban confines of the cities in Kantō and Kansai, there is no doubt that Hokkaidō has not always been included in the benefits of modernization that the rest of Japan has enjoyed.

The linear shinkansen

While the lines mentioned above are all *Seibi-Shinkansen*, there is one further line that was not included as part of this plan that may yet be constructed. Due to its unique characteristics, work on its possible introduction has been progressing for a number of years. The Chūō Shinkansen would link Tōkyō and Ōsaka. However, rather than using standard shinkansen, the trains would be 'linear motor cars' using magnetic levitation ('maglev'), meaning the train effectively flies. By eliminating the problems caused by friction, for example, it means that much greater speeds are attainable. Research into 'linear' or 'maglev' shinkansen started in Japan in 1962. A test track was set up in Miyazaki and research was conducted for a number of years until it was realized that JNR's worsening finances meant that its implementation was unlikely. The plan for a linear shinkansen was not resurrected until after the reform of JNR. In 1990 the government gave JR Tōkai permission to start research and a test track was built in Yamanashi Prefecture. The target is for the line to be able to connect Tōkyō and Ōsaka in just one hour, requiring speeds in excess of 500 km/h.

It should be noted that in 1974 Japan Airlines (JAL) began developing its own shinkansen rival in the form of the High Speed Surface Transport (HSST) as a means to connect city centres with airports quicker. The HSST uses on-board electromagnets rather than being a track-based system. Although JAL's interests in the research stopped for a few years around its privatization, the HSST research has continued and demonstration lines have been used at various EXPOs and other locations over the years (JAL interview 2004). However, the HSST system would never be used for the Chūō Shinkansen route as JR Tōkai and RTRI continue to develop their track-based electromagnetic system, and those developing HSST would not have access to the funds to enable such a huge project and would have problems gaining approval from MLIT due to the separation of turf between the railway and airline sections of the bureaucracy and government.

Mini-shinkansen

Looking at a map of the shinkansen network, one will often see two additional lines that have not yet been mentioned. These are the Yamagata and Akita Shinkansen. Although they are often referred to as being shinkansen, they are what is known as 'mini-shinkansen' (officially '*shinkansen chokutsūsen*'). The crux of the difference goes back to the origins of the shinkansen itself. The

40 *From bullet train to low flying plane*

Yamagata and Akita Shinkansen, while having the same gauge as the main shinkansen, have the loading gauge of conventional lines. The result is that the shinkansen that operate upon them are significantly narrower than other shinkansen, as can be clearly seen upon arrival at Tōkyō station, for example, when a footplate automatically rises to bridge the gap between the train and the platform. As the 'mini-shinkansen' have not been upgraded to 'full-shinkansen', with the most noticeable difference between the two being the existence of level crossings on the former, the top speed along these lines is only 130 km/h. However, once on the main shinkansen lines, the trains are capable of travelling at normal shinkansen operating speeds.

The Yamagata Shinkansen was the first to be built. The first stage of the line was between Fukushima and Yamagata by widening the Ōu Line. The shinkansen stop at major stations, whereas standard-gauge conventional trains stop at all stations. Following the success of the Yamagata Shinkansen, the Akita Shinkansen was developed on the same basis using the Tazawako Line between Morioka and Ōmagari (see Figure 2.6) and the Ōu Line between Ōmagari and Akita. The Yamagata Shinkansen was later extended from Yamagata to Shinjō.

Figure 2.6 (a) An E3-series shinkansen battles through the snow near Tazawako; (b) An E3-series shinkansen in the 'snow country'; (c) an E3-series shinkansen preparing to join up to an E2-series shinkansen at Morioka station; (d) an E3-series shinkansen joined to an E2-series shinkansen travelling next to the Saikyō Line (on left) south of Ōmiya.

The trains

It is worth noting here some of the basic issues relating to the trains themselves. Perhaps the most significant point to emphasize about the shinkansen is that they are EMUs. They do not use locomotives, meaning that the shinkansen has passenger seating areas at the end carriages also. The motors of the shinkansen are on the underneath of a number, sometimes all, of the carriages. This means that weight is more evenly distributed along the train, reducing the wear and tear on the track. It also has the advantage that it is possible to run the train even if some of the motors are defective. Although some countries have continued to pursue locomotive technology, the shinkansen has clearly shown the superiority of the EMU system (Kasai 2003:6).

Today there are many series of shinkansen in operation. The first, with its familiar round 'nose', is now known as the 0-series (see Figure 2.7). Although it was initially planned that the first shinkansen introduced would soon be replaced by a revised version. It was not until the opening of the Tōhoku Shinkansen, that the original shinkansen was officially classified as the 0-series (Oikawa and Morokawa 1996:84), although it had evolved over 30 times in its design (Yamanouchi 2000:136). Naturally each series of shinkansen is a development upon previous ones. However, this evolutionary process has been complicated by

Figure 2.7 A 0-series shinkansen near Higashi-Hiroshima. Note how although this service is only 6 carriages long rather than 16 as in the past, it has more pantographs (three) than are found on newer shinkansen.

the changes since privatization. For this reason, it is easiest to introduce the trains according to the lines they operate on rather than the order by which they appeared.

The 0-series was the first to be introduced on the Tōkaidō and Sanyō Shinkansen. In 1985, the 100-series was introduced. The difference between the two was largely cosmetic, with the 100-series having a sharper, more *Concorde*- and less 'Jumbo'-like appearance. In 1993, JR Tōkai introduced the 300-series. This train not only looked significantly different, but it was also a huge technological advance using AC motors and lighter materials. Although it suffered from a few technical problems at first, its introduction meant operating speeds were raised to 270 km/h. The train was later also adopted by JR West. JR West, however, had also been doing their own research and in 1997 introduced the 500-series, the fastest train in the world with a top speed of 300 km/h (see Chapter 5). Although initially only used on the Sanyō Shinkansen, it was later used on the Tōkaidō Shinkansen also, although JR Tōkai have not bought any. In 1999, JR Tōkai introduced its 700-series. Although it has a lower top speed than the 500-series, it is significantly cheaper, while having many improvements upon the 300-series. The most notable development was in the area of noise pollution and aerodynamics. The end result was that while the 700-series has a very sleek profile, it has gained the nickname of 'duckbill-platypus' due to its appearance when seen from the front. Again the 700-series has also been used by JR West, both in the standard 16-carriage formation, but also for their own 8-carriage version that solely operates on the Sanyō Shinkansen. With these various developments, inevitably, older trains have been retired. It is no longer possible to see either the 0-series or the 100-series on the Tōkaidō Shinkansen. On the Sanyō Shinkansen, 0-series and 100-series shinkansen do operate, though in much shorter formations than in the past, and some have been completely renewed and repainted. Although not currently linked to the Sanyō Shinkansen, the shinkansen on the Kyūshū Shinkansen, the 800-series, is based largely upon the 700-series.

On the Tōhoku and Jōetsu Shinkansen, the first shinkansen to be introduced was the 200-series. It looked much like the 0-series, but had various alterations to help combat the problems caused by snow. In time another version of the 200-series was introduced which resembled the 100-series, and was given the designation of 200-2000 rather that its own series number. With the development of the Yamagata Shinkansen, a shinkansen that was able to travel on both the main shinkansen and the mini-shinkansen line with its smaller loading gauge was needed. The result was the 400-series. After this shinkansen, JR East changed the numbering system of its shinkansen by using the prefix 'E' (for East). The first of these was the E1-series in 1994. Its most notable feature is that it is completely double-decker, reflecting the change in the type of passenger that was using the JR East shinkansen. The E2-series, although externally different (especially if seen from the front), has much in common with the 700-series, and has become JR East's main shinkansen as the company replaces its ageing 200-series shinkansen. The E3-series was introduced for use on the Akita Shinkansen and took advantage of further developments made since the 400-series was introduced on the Yamagata Shinkansen. The E3-series has subsequently been used on both lines. The E4-series, like the E1-series, is essentially double-decker, but has again taken advantage of various developments made in recent years.

There are other shinkansen also. There have been various test trains over the years, with further trains being planned. The N700-series, for example, will be tested extensively before being introduced on the Tōkaidō Shinkansen as the first tilting-shinkansen aimed to cope with the 39.5% of the Tōkaidō Shinkansen which is curves (Central Japan Railway Company 2003d:12). JR East has also announced plans for new test trains as it aims to introduce shinkansen capable of over 360 km/h in time for the opening of the final stage of the Tōhoku Shinkansen. There are also shinkansen, as will be discussed in Chapter 6, that are used for ensuring that the line remains safe.

Conclusion

It is possible to break the history of the shinkansen into six periods. The first three of these are those prior to the commencing of passenger operations. The first period, which I refer to as 'Imperial Expansion', runs from 1938, with the original plans for the *dangan ressha*, to 1944 when construction was abandoned. The period that follows, 'The Wilderness Years', was a time when the priority was reconstruction and development of the conventional lines rather than constructing a shinkansen. Then, in 1957, with the announcement of the plan for a high speed railway, the period of 'Genesis' of the shinkansen began. With the start of passenger operations on 1 October 1964, the increasing passenger usage and discussion about a nationwide shinkansen network, I refer to this period as the 'Boom Years'. However, the boom came to an end in November 1973, with the announcement of the *Seibi Keikaku*. This period, which still saw some improvements, is probably best remembered for the continuing problems of JNR. This period was the shinkansen's 'Dark Valley'.[20] The following period, 'A New Hope', began with the privatization of JNR. In fact, some elements of this period, such as improvements to services on the Tōkaidō Shinkansen, had already begun to take place, but the date of privatization is perhaps the most symbolic. This period was relatively short, but provides a bridge to the one that began in 1992 with the introduction of the 300-series. Since then the shinkansen has entered a period which I am sure will looked back upon as its 'Golden Age'. It is a period of renewed construction of lines, technological improvements, new series of shinkansen and the first export to a foreign country. Ironically it overlaps with the period often referred to as Japan's 'Lost Decade'. That significant, but incremental, changes have occurred, as with the shinkansen, while economic progress was not as spectacular as that of the 1980s, during this time reveal just what an empty term 'Lost Decade' is.

Having briefly introduced the history of Japanese railways and the shinkansen, it is now possible to analyse the main issues relating to the shinkansen.

3 Ambassador of Japan

> (T)he bullet train speeding past Mount Fuji... is an icon of modernity or, more precisely, a meeting of two icons representing modernity and tradition, the new Japan and the old; it brings us no closer to the lives of those who are ranged inside the streamlined carriages. The pure lines of Mount Fuji rise upwards and the pure lines of the bullet train intersect its base, composing themselves into a perfect aesthetic image.
>
> (Littlewood 1996:62)

The image described is one that has become familiar to many around the world. The shinkansen and Mount Fuji almost seem to be inseparable at times. Littlewood's comments refer to an illustration in the Fodor travel guide to Japan. That the image should appear in a travel guide is perhaps of little surprise. But, as Littlewood (1996:62) points out, the Fodor guide has 'Japan refined to its mythical elements, purged of modern Japanese and of all the other unwelcome realities of the present day – modern buildings, modern clothes, modern cars, modern technology', this illustration is the only concession that the book is set in the present day. Why has this image become so well-used? Littlewood's comments suggest a probable answer to this. But this suggestion leads to further questions. Does the shinkansen truly reflect Japanese modernity? Why has the shinkansen become an icon? Does this icon also influence the society of which it is a part? It is these questions that this and subsequent chapters will address.

Name values

Before discussing the imagery, let us consider the name. The word 'shinkansen' in Japanese can mean the line, the train, or even several lines or trains since there is no singular or plural variants for nouns. Yet, the origin of the word is very specific. Considering the *kanji* (Chinese characters) that make up the word, it is clear that the word originally refers only to the railway line. Literally the meaning of 'shinkansen' is 'new trunk line'. The wartime plan for the project clearly had a defined name for the train that would use the line – the *dangan ressha*. So why the confusion now?

When operations began in 1964 it was referred to as '*yume no chōtokkyū*' ('super express of dreams') or just '*chōtokkyū*' ('super express'). The former phrase was used in promotional materials and reflected the incredible nature of the train considering the problems that had been overcome to make it reality and its pioneering nature in comparison to trains around the world. However, the term '*chōtokkyū*' was merely its operational classification. Referring back to Figure 1.3, it is possible to see how there are many types of services within Japan. '*Futsū*' ('ordinary' or 'local') services stop at all stations, '*kyūkō*' at significant stations, and '*tokkyū*' at main stations only. That there is now no '*chōtokkyū*' classification and the top service is 'shinkansen' is evidence of how the term never came to be fully accepted, at least by the public.

Part of the problem may be due to the ambiguity of the way in which the word 'shinkansen' is used. For example, on board the Tōkaidō Shinkansen, the Japanese passenger is thanked for using the 'shinkansen',[1] with details then given about the specific service. Yet the English announcement welcomes the passenger to the 'shinkansen', without any mention of the line name, and then speaks of the service as being a 'super express'.[2] There is no equivalent word used for 'super express' in the Japanese announcement. Given that in 1964 the train itself also had no official designation in terms of a series number, it is perhaps unsurprising that the word 'shinkansen' came to be used to refer to the train as well as the line itself.

Aoki *et al.* (2000:181) suggest that the shinkansen's success meant that the word 'became synonymous with high-speed train in the English language'. I doubt this. This may be true of those that work in the railway industry, but it has certainly not entered the English language, to which this study is limited, in the same way as the words hara-kiri, tsunami or karaoke have. The few Japanese words that have entered the English language are often mispronounced,[3] and the longer they are, the greater the problem appears to become. It is hardly surprising, therefore, that 'bullet train' became the standard term. I would like to suggest that there are two further reasons for this. At the start of the Tōkaidō Shinkansen, the term '*dangan ressha*' was being used in some rail magazines when discussing the shinkansen. It is possible, therefore, that this term was picked up by journalists and then translated into English to describe the shinkansen. Furthermore, 'bullet train' neatly reflected two of the key features of the train – the shape of the front of the train (see Figure 3.1) and being considerably faster speed than other railway trains in the world at that time.

Recently, I have begun to detect a shift, albeit a minor shift. I have come across dictionaries that only provide the word 'shinkansen' as a translation of the Japanese term. For example, one dictionary I consulted contained an entry for '*chōtokkyū*' for which not only is 'super express' offered as a translation, but also 'bullet train'. That it offers 'bullet train' as a translation of '*chōtokkyū*' but not 'shinkansen' appears to reflect the problems of terminology back in 1964 and how the term 'shinkansen' may be becoming more widely known.[4] I believe there are two main reasons for this shift. First, there are significantly larger numbers of people visiting Japan today than there were 40 years ago, due to tourism, business

Figure 3.1 The familiar rounded 'bullet' shape front of the original shinkansen – here in its 'renewal' design – at Fukuyama station, with Fukuyama Castle in the background.

and schemes such as the JET Programme.[5] Those who go to Japan discover that the term 'bullet train' is not widely used and soon become familiar with the term 'shinkansen'. Even the revised version of the Japan Rail Pass, used by many foreign visitors to Japan, now uses the term 'shinkansen' rather than 'bullet train'.

Another reason why the word 'shinkansen' may be becoming more widely used is that Japan's position in the world, particularly in relation to railways, has also shifted in the past 40 years. Japan is not the exotic, mystical country that it once was. It has become a country, and its railways a system, to which many have looked to learn from. This situation can inevitably lead to both confidence on the part of the Japanese to use the words which they find most appropriate, as well a desire by those looking at Japan to use the correct terminology. Although I have come across those that have suggested that 'bullet train' is a term that can be used to apply to any high speed train in any country (e.g. Soloman 2001), and this in itself could be interpreted as a compliment in relation to the shinkansen's global influence, as far as being used to mean the shinkansen itself, 'bullet train' now seems past its sell by date. However, it is likely that those who speak of the 'shinkansen' to a general audience are always likely to need to initially follow up the word's first usage with the 'translation' of 'bullet train'.

The difference between different types of services on conventional lines was described above. When the term '*chōtokkyū*' was introduced, it was as if a new, faster type of service had been introduced. However, the reality was more complicated than this. For even on the shinkansen a system of different types of services was introduced from the beginning. On the Tōkaidō Shinkansen, there were two types of service – one that stopped at all stations and one that only stopped at major cities. Strictly speaking, until 1972, it was only the latter of these two services that was '*chōtokkyū*', while the former were '*tokkyū*' (Suda 2000:30; Umehara 2002:86). To avoid confusion, and to follow the convention used by JNR for its other '*kyūkō*' and '*tokkyū*' services, the two types of services were named. As this is a system that is not widely used in Europe, and as the names have a significant symbolic value, it is worth expanding on this point further.

Originally it was planned that individual train services would be numbered in much the same manner as planes are given specific flight numbers, but the change was made due to popular demand for named services (Umehara 2002:35). The popular element is significant as, since the initial shinkansen services were introduced, all but one name has been chosen by popular vote.[6] Perhaps inevitably it was the first ever shinkansen naming competition that gained the greatest interest, with over 550,000 votes being cast in the voting period of about two weeks (Umehara 2002:35). This is over four times the number that were cast for the naming of the Kyūshū Shinkansen in 2003 (Kyūshū Railway Company 2003). Although the public is given the opportunity to vote, with prizes being given to some of those that suggested the name that is eventually selected, it is not necessarily the most popular name that is selected. The selection is done by the railway company.

Looking at the names that have been selected, there appear to be two main patterns. First, there are the shinkansen services that take names that were formerly used by other trains that used to run along the equivalent conventional line. Such a method ensures a degree of continuity between the old and the new. The other pattern, which often overlaps with the first, are service names that reflect the region which they serve. Interestingly, if one looks at the shinkansen service names across Japan, one sees that it is the first pattern that is of greater significance for the lines west of Tōkyō, and the second pattern that is of greater significance for those running north from Tōkyō.

West of Tōkyō there are only four main service names in operation. The first two to be adopted, for the Tōkaidō Shinkansen, were '*Hikari*' and '*Kodama*'. Both of these were service names already being used. '*Kodama*' was used for *tokkyū* services between Tōkyō and Ōsaka just prior to the opening of the shinkansen and '*Hikari*' was used on *tokkyū* services in Kyūshū between Hakata and Nishi-Kagoshima (Umehara 2002:35, 36). Although '*Hikari*' gained the highest number of votes (19,845), '*Kodama*' was only tenth (Umehara 2002:36). These two were selected as their meanings ('light' and 'echo') were felt to make a good pairing (Umehara 2002:36). With the opening of the Sanyō Shinkansen, it was decided to use the same service names since many services continued past Shin-Ōsaka.

The other name to be selected by public vote in the region west of Tōkyō was *'Tsubame'* for the Kyūshū Shinkansen. JR Kyūshū had said that they would always go for a very symbolic name (JR Kyūshū interviews). Despite the high placing of *Sakura* ('cherry blossom'),[7] *Tsubame* ('swallow') was selected due to the high speed imagery linked with the name. In so doing it became the fifth service to carry the name, with the then service being the *tokkyū* on the route which the Kyūshū Shinkansen partly replaced. Previously it had been the name of the original *tokkyū* service between Tōkyō and Ōsaka and was also the name of the JNR in-house magazine. It also means that effectively *'Hikari'* and *'Tsubame'* have swapped the locations they serve compared to when they were originally conventional train service names. Other suggested names included *Kibiaiyanse* (a Kagoshima dialect word for *gambare* ('try hard')), *Wazzeka* (Kagoshima dialect word for 'very'), *Monosugoi* ('extremely'), *Pegasasu* ('Pegasus'), and *Maguma* ('Magma' – Kagoshima is famous for its volcanoes) (Kyūshū Railway Company 2003). These more unusual suggestions reflect some of those made for the Tōkaidō Shinkansen: for example, *'mahha'* (Mach), *Kennedy* (JFK had recently been assassinated at the time of the vote) and Sogō (Umehara 2002:36). Nearly 40% of the votes for the Kyūshū Shinkansen service name came from people living outside Kyūshū (Kyūshū Railway Company 2003), and although some may originally have been from Kyūshū, it does suggest that railway enthusiasts and others also consider it important to vote.

The other name on the lines West of Tōkyō is *'Nozomi'*. This was introduced without public vote by JR Tōkai. The word means 'hope', and this has allowed JR Tōkai to use puns in various marketing campaigns over the years. For example, in 1997, when the Japanese economy was entering a period of instability and public morale seemed particularly low, the campaign was *'Nippon ni, Nozomi ari'*, which could be translated as either 'In Japan, we have hope' or 'In Japan, we have the *Nozomi'*. That the Tōkaidō Shinkansen's revenue tends to be closely correlated to the state of the Japanese economy added extra poignancy to the slogan. It is interesting to note that *'Nozomi'*, like *'Hikari'*, had also be used as service names by the Japanese between Korea and Manchuria during the Pacific War (Pearce 2004:40–1).

On the shinkansen lines that are now operated by JR East, the situation regarding names has been more complicated. The original pairs of names for the Tōhoku Shinkansen were *'Yamabiko'* and *'Aoba'*, with *'Asahi'* and *'Toki'* on the Jōetsu Shinkansen. However, these names were not popular with all. Although *'Yamabiko'* has a similar meaning to that of *'Kodama'*, the service was the equivalent to the Tōkaidō Shinkansen's *'Hikari'* service, and so some felt it inappropriate. On the Jōetsu Shinkansen, there were some that did not like the name *'Asahi'* since it is part of the name of many companies (Umehara 2002:56). The public vote itself did not muster the same level of enthusiasm as the votes for the Tōkaidō Shinkansen, with a total of 148,950 being cast for the two lines (Umehara 2002:55–6). With the expansion of the Tōhoku Shinkansen, a new name, *'Hayate'*, which had done well in many other votes, was introduced.

At the same time as the introduction of the '*Hayate*' service, the '*Asahi*' name was retired. The official reason was to help avoid any confusion at Tōkyō, in particular, as two of the three *hiragana* characters that make up the name are the same as the Hokuriku Shinkansen service – '*Asama*' (name of a volcano on the route of the Hokuriku Shinkansen). However, I have heard from some of those that were involved in the decision making process that this was just an excuse used in order to get their way in having the name changed, and that the real reason was merely to have a desire to have '*Toki*' become the main service name. '*Toki*' had been the slower service name for many years before being withdrawn in 1997. However, the '*Toki*' (ibis) is said to be a significant symbol for the people of Niigata and so there were many that wished to see its reintroduction. Some of the trains have subsequently been painted in accordance with the name change (anonymous interview; Umehara 2002:57) (see Figure 3.2). The name that had replaced '*Toki*', '*Tanigawa*' (a mountain on the route of the Jōetsu Shinkansen) has been maintained.

On the Tōhoku Shinkansen, the '*Aoba*' (a park in Sendai) name was withdrawn in 1997. Its replacement, although they both existed at the same time for a period of about two years, was '*Nasuno*' (name of an area on the route). The need for service names is demonstrated by the fact that on some sections of the

Figure 3.2 (a) An E1-series shinkansen in its original colours at Gāla-Yuzawa towards the end of the skiing season; (b) an E1-series-renewal shinkansen in its livery based on the colours of the ibis ('*toki*') which provides the name of the service the train primarily operates under.

Tōhoku Shinkansen, stations are served by five main services. Ensuring passengers catch the right train is extremely important, and service names are signed to help in this respect, although the large number of different patterns on some lines (see Appendix 2) means that merely using the service name is no guarantee that the train is going to the desired station. In addition to the three Tōhoku Shinkansen services ('*Hayate*', '*Yamabiko*' and '*Nasuno*'), there are also the Yamagata and Akita Shinkansen services. Both of these names were selected by public vote. '*Tsubasa*' (wing) was selected for the Yamagata Shinkansen and '*Komachi*' for the Akita Shinkansen. '*Komachi*' has various meanings, including the name of a local rice, but it was on the basis of the meaning of the legendary beautiful women of the area which the service name was selected. Clearly the names of services can have a strong link to regional identity. This seems to be particularly the case for the lines that go to more rural areas. Although the names are decided upon by those in Tōkyō, there appears to be a strong desire to bring an essence of the identity of the area at the 'other end' of the line to the capital.

While there is some local flavour in service names, the names of stations are much more directly linked to their locality. As obvious as this sounds, one has to remember that the location of shinkansen stations has not always been straight forward. Many stations are located away from the town or city they serve, with some being away from the main conventional station for that town or city. In most cases, particularly along the Tōkaidō and Sanyō Shinkansen, the solution to this problem was to add the prefix 'Shin' ('new') to the name of the town or city, for example Shin-Ōsaka, Shin-Kōbe, and Shin-Iwakuni. When Mikawa-Anjō was opened in 1988, it was not possible to use the name Shin-Anjō, however, as this name was already used by the Meitetsu private railway line. Shin-Fuji, which opened at the same time, probably benefited from the privatization of JNR, as there was already a station called Shin-Fuji in Hokkaidō, but as this was now operated by JR Hokkaidō rather than the JNR, as well as being hundreds of kilometres from the location of the shinkansen station, it was not an issue for JR Tōkai (Tokorozawa 2003b:165).

There have been a few interesting situations relating to names. Although not a new station for the shinkansen, Maibara is an odd name as the town which it served was Maihara. The station name dates back to the Meiji period and reflects the more 'standard' reading for the two *kanji* that make up the town's name. This top-down-Tōkyō imposed naming has apparently caused a problem for the town, for I have observed that it tends to use *hiragana* in much of its advertising material.[8] In 2005, the town merged with some other local habitations to become Maibara city, thus ending the confusion. While many stations on the Tōkaidō and Sanyō Shinkansen have the prefix 'Shin', those on lines now operated by JR East have not used this prefix as much and there has been a greater emphasis on the regional identity or the tourist resort located near to that station. Perhaps the best example is Nasu-Shiobara. Given the location of the station, it may have been expected that this station would have been called Shin-Kuroiso. However, it was felt that this station would be better used if it appealed to tourists, who may visit the nearby Nasu plateau, rather than using the name of a relatively small city

(Tokorozawa 2003b:165) despite the idea being unpopular with many in the city and with some politicians (Yamanouchi 2002:41–2). It should also be noted that the Imperial Family have a residence in the area, which may have been a further reason for wanting both the location of the station and its name to be divorced from Kuroiso. In the end the name Nasu-Shiobara was pushed through, but due to the trouble caused and the 'unpleasant aftertaste', Yamanouchi resigned from his position at JNR (Yamanouchi 2000:197). According to staff at one hotel I visited in the area most visitors come by car, and that neither using the prefix 'Nasu' for the station nor the service name '*Nasuno*' appears to have any major impact on rail usage. However, their existence may effectively work as a form of free advertising for the area as people are reminded of the resort when seeing the station name or service name when travelling on other JR services or looking at timetables, for example. So while JR East may not benefit greatly from the policy, local business related to the tourist trade may do.

That adding a prefix that helps to locate the station may be a benefit can be illustrated by contrasting two other cases. Shinjō is the terminal of the Yamagata Shinkansen. Yet, trains that go there are always indicated as going to Yamagata *and* Shinjō. JR East feel that many passengers would be unfamiliar with the name Shinjō and so feel it necessary to add 'Yamagata' (JR East interview 2003), which is the name of the prefecture and the prefectural capital. Numakunai, on the other hand, was a station on the Tōhoku mainline. When this station became connected to the Tōhoku Shinkansen in 2002, it was decided to rename it to Iwate-Numakunai. Iwate is the name of the prefecture as well as the town in which Numakunai is located. With the Iwate prefix in place people are more likely to be aware of its location. Whether the area will benefit from extra business as a consequence, however, remains to be seen.

When the Sanyō Shinkansen was being built, a station was planned in the town of Ogōri in Yamaguchi prefecture. This station would provide a convenient change-over for those going to Yamaguchi city, the prefectural capital, as well as to the historically important city of Hagi. Yamaguchi government campaigned for the station to be called Shin-Yamaguchi (Umehara 2002:50). This would have been similar to the way in which Nagatoichinomiya, also in Yamaguchi prefecture, was renamed as Shin-Shimonoseki (Umehara 2002:52) or Tamashima was renamed to Shin-Kurashiki (Tokorozawa 2003b:165). However, whereas Nagatoichinomiya was just a minor station within Shimonoseki city, some locals in Ogōri did not want to lose their station and the name was confirmed following a local mayoral election in which candidates campaigned on their position regarding the station name (Umehara 2002:50). However, for many, such as some businesses, in Yamaguchi, not having a station with a more symbolic name remained a problem. When *Nozomi* services were introduced, they did not stop at Ogōri. However, with the significant timetable alterations introduced in 2003, it was decided to not only make the station a *Nozomi* stop, but also to rename the station to Shin-Yamaguchi. Another factor that helped bring about this change is that it is at this station where tourists can catch the steam train *Yamaguchi*, also operated by JR West, at certain times of the year. That the station is only a few hundred metres from being in Yamaguchi city boundaries, although about 12 km from the centre of Yamaguchi city, adds a certain irony to the debate. Given that the station is in Yamaguchi prefecture, an original name of Yamaguchi-Ogōri, a system that had previously been used for Gifu-Hashima, could possibly have solved a lot of problems. While this problem took nearly 30 years to resolve, a similar geographic problem elsewhere did not raise such problems. Despite Nishi-Gōmura being only slightly smaller than Ogōri, its shinkansen station is called Shin-Shirakawa, even though it is just outside the boundary of Shirakawa city. Although some locals did wish for their town's name to be used, there was no significant protest (Yamanouchi 2002:41).

Naming can be a problem and in one case has led to an apparent compromise. The station Tsubame-Sanjō on the Jōetsu Shinkansen serves the cities of Tsubame and Sanjō. Both cities have conventional railway stations. However, due to cost reduction and the infeasibility of having the shinkansen serve both cities, the Jōetsu Shinkansen passes in between the two cities. The station is also located next to the Hokuriku Expressway. Yet while the station, which strictly speaking is within Sanjō city, was named Tsubame-Sanjō, the expressway junction, which is mostly within Tsubame city, was called Sanjō-Tsubame.

Another station that has been renamed is Nishi-Kagoshima. It has always been bigger and a more significant station than Kagoshima station. With the opening of the Kyūshū Shinkansen, it was decided to rename it. This was a decision taken at the local level, and one that JR Kyūshū apparently had no involvement (JR Kyūshū staff and Kagoshima Prefectural government interviews). The apparent reason for a change was that 'Nishi-Kagoshima' (literally 'West Kagoshima') did not give the impression that the station was part of the city. So, rather than confusing local people by renaming Kagoshima station to Higashi-Kagoshima (East Kagoshima), and then Nishi-Kagoshima to Kagoshima, for example, it was decided to create a new name, Kagoshima-Chūō ('Kagoshima Central').

Station names are clearly important, partly as the station is seen in Japan as being the 'point of reference' (Noguchi 1990:47) or *genkan* (term used to describe the entrance way of a Japanese house where the shoes are removed) of a city or prefecture. Given this, there are two shinkansen station names which stand out as being somewhat odd. These are Hakata and Kokura which are in Fukuoka and Kitakyūshū respectively. These names date from the historical names of the cities. For example, Kokura was one of five cities that merged to form Kitakyūshū. Yet, there appears to be no move to have the station renamed as Kitakyūshū. Given the debate surrounding Shin-Yamaguchi this may seem surprising. However, Kokura and Hakata are names with which many Japanese would be familiar due to their historical importance. City mergers are an increasing phenomenon in Japan as an attempt to help rationalize administrative work and reduce costs. Some of these mergers, for example the one that created Saitama city, has meant that some shinkansen station's names, for example, Ōmiya, are different to that of the city it represents. In another example, the city name, Satsuma-Sendai, sounds more like a shinkansen station name than the station, Sendai, itself. Changes in shinkansen station names due to mergers seem unlikely in the short-term. For example, given its proximity to Tōkyō and consequently it being relatively well known, as well as the history connected to Ōmiya within the railway industry,[9] it is unlikely that it would ever be renamed. One could imagine a change from Tokuyama to Shūnan at some point in the future, particularly as it would help to cement the new name in the public consciousness. However, using names that hark back to a region's past, for example many conventional line stations use a prefix relating to the old domain name before modern prefectures were established in the nineteenth century, continues to be common-place in many areas of Japan.

The image of the shinkansen

In considering the image of the shinkansen, it is worth considering the thoughts of one of the original designers. Miki wanted the shinkansen to be 'beautiful'. Yet, how is beauty defined? Miki's target was a form that had the minimum air resistance as this would improve performance (NHK 2001). So, one can say that the shinkansen's shape and beauty evolved out of mechanical demands. But, there were many prototype models made and although it is probable, given that specialist engineers were working on the project, they would have had limited air resistance, it appears that Miki smashed these models and complained of their ugliness (NHK 2001). It would seem that it was more than just pure mechanical demands.

It is often said that 'beauty is in the eye of the beholder', and so it was necessary for the engineers to find out what Miki's definition of beauty was. The answer to this was the *Ginga*, a Japanese navy dive bomber. Its streamlined design, which Miki had designed, became the key influence (NHK 2001). So was it beautiful because it solved the technical problems that high speed performance raised? What is it that makes people conclude whether something is beautiful or not? There are likely to be a number of influences. Some of these may be hidden or genetic, while there are also the experiences of the individual. For some, past experiences may lead to something being considered 'beautiful' as it is familiar and awakens some kind of nostalgic emotion. For others, it may be that what is being looked at is different to the point of being exotic and that this produces the response of 'beauty'. Current influences cannot be overlooked. Tastes change, both an individual's and a society's. As fads, fashions, products and personalities come and go, so their influence upon society changes, though this is by no means a one-way process since it is also society's tastes that would have enabled the success. Naturally not all are successful due to their beauty, but all are likely to influence society, at least some parts of it, and the way they perceive beauty.

Although the shinkansen's design was said to be derived from the *Ginga*, there may have been other influences that Miki was either not aware of or chose not to comment on. In the railway world at that time, Odakyū's Super Express was considered to be beautiful by many, and this train was responsible for many technical developments that were incorporated in the shinkansen (Haraguchi 2003a:35–6). Early design sketches of the shinkansen were also clearly influenced by the Odakyū train (see Haraguchi 2003a:39). Internationally, the *Dakota* aeroplane's design was also highly regarded, and this may also have had some influence, directly or indirectly, upon Miki and his fellow designers. However, I would like to suggest that there is another level at which the shinkansen's beauty can be assessed. That is to use the 'criterion of beauty in Japan' (*nihon ni okeru bi no hyōjun*) of Yanagi Sōetsu.

Yanagi was 'a creator of *Mingei* theory and a leader of the *Mingei* (Folkcrafts) movement' (Kikuchi 2004:xv). The ideas espoused by Yanagi appear to have had some widespread influence, and his 'criterion of beauty' may capture *a* Japanese sense of what beauty is. Kikuchi (2004:53) summarizes them under the following twelve headings of beauty (*bi*): (1) 'handcrafts' (*Shukōgei*), (2) 'intimacy' (*Shitashisa*), (3) 'use/function' (*Yō/Kinō*), (4) 'health' (*Kenkō*), (5) 'naturalness' (*Shizen*), (6) 'simplicity' (*Tanjun*), (7) 'tradition' (*Dentō*), (8) 'irregularity' (*Kisū*), (9) 'inexpensiveness' (*Ren*), (10) 'plurality' (*Ta*), (11) 'sincerity and honest toil' (*Seijitsu na Rōdō*), (12) 'selflesness and anonymity' (*Mushin/Mumei*). Yanagi himself was 'generally negative about machines' (Kikuchi 2004:54) particularly the idea of whether a machine could ever make anything beautiful. In this respect it is worth noting the difference in attitude I encountered when visiting two of the companies that construct shinkansen. While Hitachi are apparently striving towards greater automation, the view at Nippon Sharyō is much more in line with Yanagi's way of thinking. However, the two companies, and the others making shinkansen, produce identical products, in terms of external appearance and their performance, which can only be identified by their construction plates at the end of carriages.

The shinkansen achieves beauty on many of the criterion on Yanagi's list. As the shinkansen has a quality of 'commonness and familiarity', it meets the criterion of 'intimacy' as Yanagi defined it (Kikuchi 2004:55). In terms of 'use/function', Yanagi believed that 'beauty and use inevitably coincide and create one total quality' (Kikuchi 2004:55), and the shinkansen is without doubt useful. Regarding 'health', Yanagi's concern appears to have been whether the object, including its production, was appropriate to its use (Kikuchi 2004:55–6). In this respect, the shinkansen would appear to meet the criterion. As for 'naturalness', rather than whether something is made of natural materials, Yanagi appeared to be more concerned with its interaction with nature (Kikuchi 2004:56). In this respect, the shinkansen's streamlined shape would suggest that it is 'natural', although the shape of the original 0-series shinkansen actually proved to be a problem (see Chapter 7). 'Beauty of simplicity' refers to there being no excessive colouring or patterns (Kikuchi 2004:57), which generally the shinkansen does not have. Although a modern design, the shinkansen owed many elements of its design and style to traditional ideas and methods of work and would meet the criterion of 'beauty of tradition'. Despite the great cost involved in its production, the shinkansen itself, in terms of the cost of a ticket, is something that is available to most, which is how Yanagi appeared to define 'inexpensive' (Kikuchi 2004:58). As the shinkansen is mass produced in quantities sufficient to meet mass demand, it meets Yanagi's definition of 'plurality' (Kikuchi 2004:58).

There are some items where the shinkansen does not meet Yanagi's criterion. The shinkansen does not appear to fully possess the qualities of being *shibui* (austere), *wabi* (plain) or *sabi* (tranquil) or being 'grotesque' that were of concern to Yanagi in his criterion of 'beauty of irregularity' (Kikuchi 2004:58). Similarly, although the shinkansen is often made by unknown craftsmen (even allowing for the increase in automation as mentioned above), many of whom may have pride in their work, the underlying reason for the work is money and this runs counter to Yanagi's ideas of 'beauty of sincerity and honest toil' and 'beauty of selflessness and anonymity' (Kikuchi 2004:58–9).

On balance, the shinkansen does well on Yanagi's 'criterion of beauty'. Yanagi's criterion are not the only ways to judge beauty, of course, but the shinkansen does seem to be 'beautiful'. Yet, one needs to consider the differences between series, which raises subjectivity as using Yanagi's criterion will not allow any significant differentiation. While the 0-series is often seen as being the 'classic' design, the 500-series has tended to be a favourite amongst rail enthusiasts. On the other hand, the 700-series has a smart profile, but is more well-known for its 'duckbill-platypus' (*kamonohashi*) front profile (see Figure 3.3). Therefore some may argue that the shape of the 700-series, and perhaps the E4-series even more due to its similar front and its extra height, are bordering on the 'grotesque' in comparison

Figure 3.3 (a) A 700-series shinkansen passing a speed-boat racing (one the few forms of legalized gambling in Japan) stadium near Hamanako; (b) a 700-series shinkansen passing an exhibition hall in Shizuoka; (c) inside the cab of the 700-series; (d) the smart profile of the 700-series.

to some of the other shinkansen. That beauty can be an issue is revealed by both JR Kyūshū's and the Taiwan High Speed Railway Corporation's desire to have the 700-series modified so as to lose its distinctive front, leading to the 800-series and 700T respectively (JR Kyūshū and THSRC interviews). Even JR Tōkai's planned N700-series has a smarter front and JR Tōkai was quick to suggest that its front is similar to that of a Japanese white eagle (JR Tōkai interview).

Since 1958 and 1961 there have been two awards in Japan that could also be said to be a barometer of Japanese train beauty. The first, 'The Blue Ribbon Award' is awarded each year by railway enthusiasts to the best looking train introduced during the previous year. The 0-series shinkansen picked up the award in 1965. A shinkansen did not win the award again until 1998, when it was won by the 500-series. Although it brings no significant reward, many at JR Kyūshū, a company that has won the award three times since 1988, set out with the aim of having the 800-series win this award (interviews JR Kyūshū), which it did. The other award, the Laurel Award, is similar but emphasizes the technology and design of the new train. This was not first won by a shinkansen until 1983 (by the 200-series) as in the early years the award tended to be only given to commuter trains, which tended not to win the Blue Ribbon Award. The 100-series won it in 1986, the 300-series in 1993, and the 700-series in 2000. For both awards there have been years when there has been no winner, but only for the Laurel Award are there sometimes multiple winners (Japan Railfan Club 2004).

The diversity in designs has also led to signs at stations, which use a simplistic picture, or what could be termed as an 'abstract symbol', of the front of shinkansen to help guide passengers in the correct direction, varying. Whether everyone would understand the picture as being of a shinkansen, particularly some of the newer designs, were the word not written next to it in Japanese and in Romanized letters is questionable. Across Japan, and even within individual stations, different pictures are used to represent the shinkansen – often reflecting when a station was opened. As van Leeuwen (2001:107) suggests 'In the age of the logo, abstract symbols may yet again become increasingly important'.

Symbols are beautiful. Although they may not always be considered as such at first, for example, the Eiffel Tower, their symbolic appeal helps them to become seen as such. In turn this may help to define what is considered as beautiful in a society. The shinkansen may have had the advantage of being beautiful when created, and this may have helped it to become such a significant symbol. Yet, it may have been further aided by the image of it passing Mt Fuji (see Figure 3.4). Indeed Sogō said 'I want the shinkansen to complement the beauty of Mount Fuji' (Hosokawa 1997:203). This image brings together the modernity of the shinkansen with the naturalness of Japan's picturesque highest peak and long-term icon.

There are various means to study images and symbols. Iconography uses textual analysis, which include 'pointers' as to how the image should be interpreted, and contextual analysis, which relies upon the image having an established meaning (van Leeuwen 2001:107–8). A popular method of study is content analysis. However, as Bell (2001:13) points out 'Content analysis alone is seldom

Ambassador of Japan 57

able to support statements about the significance, effects or interpreted meaning of a domain of representation'. Barthian visual semiotics suggests that there are two layers in pictures – denotation, of 'what, or who, is being depicted' and connotation of 'what ideas and values are expressed through what is represented, and through the way in which it is represented' – in relation to a picture's meaning (van Leeuwen 2001:94). In the case of the image of the shinkansen and Mt Fuji, a possible meaning is suggested by Littlewood at the start of this chapter – it is the combination of tradition with modernity. But does the shinkansen's position at the front of the picture suggest modernity is of greater significance, or is tradition of greater significance as suggested by the height of Mt Fuji?

The significance of both of these icons of Japan can be seen by looking at how and where they are used. That the image is often used in materials relating to tourism, particularly those coming from abroad, is hardly a surprise. This is particularly so when one considers that many planned short tours for foreigners to Japan include a trip on the shinkansen past Mt Fuji, and others will often use a Japan Rail Pass, which until 2004 used a graphic of a 0-series shinkansen passing Mt Fuji despite it not operating on this route since 1999,[10] that allows almost unlimited use of the shinkansen as well as many other JR trains. While it is relatively uncommon to see Mt Fuji from the shinkansen (see Chapter 7), the shinkansen has become part of the Japan experience for a large number of those

Figure 3.4 A 300-series shinkansen passing Mount Fuji.

that visit Japan. The two icons have become so interlinked that one has to question whether non-Japanese would recognize one in the absence of the other. One survey I conducted (Hood 2002), found that when asked to identify which trains were shinkansen from a selection of photographs, about two-thirds correctly identified the 0-series. Other series which were correctly identified by at least half of the respondents were the 200-series, 300-series and 100-series. However, more people incorrectly identified a TGV as being a shinkansen than correctly identified the 400-series, 700-series, E1-series, E2-series, E4-series, or E3-series as being shinkansen. It is notable that many of these are JR-East shinkansen that tend not to appear in pictures – let alone those that include Mt Fuji – or be used by tourists as much as JR Tōkai shinkansen. That JR East's 200-series was the second most correctly identified probably owes much to its 0-series-style front. Interestingly not many correctly identified the 500-series, despite its image being used in a number of publicity materials for Japan 2001, a range of events that took place in Britain between 2001 and 2002 of which the survey was a part.

Rather than how the image of the shinkansen is used for those travelling to Japan, of greater significance is its more general use. In this respect, its usage in the media is particularly important. For the media, one of the greatest sources of images is the Getty Image Bank, a searchable database, which can even suggest pictures when words relating to emotions are entered. This in itself is an interesting phenomenon as the database is likely to be culturally loaded, as interpretations of one image by people of different cultural backgrounds would not necessarily produce similar, let alone the same, response (van Leeuwen interview 2003). In February 2003, I conducted a search of the Getty Image Bank to see what images were available in relation to Japan (i.e. images that were presented when the keyword 'Japan' was used). In total there were 2,068 images. Many of these were not obviously related to Japan and were probably presented due to the photographer being Japanese, or that the image of the monkey, for example, was in Japan although this may not immediately be identifiable as such. There were 22 images that included a shinkansen.[11] Of these 22, four of them are of a shinkansen in front of Mt Fuji and another one is a view of Mt Fuji from inside the shinkansen. Doing a search for 'Mt Fuji' brought up 58 images of which 5 included the shinkansen in some form. When the moving image database was searched, shinkansen clips accounted for 19 of 344 Japan-related clips, five of which were taken in front of Mt Fuji (which accounted for 25% of the 'Mt Fuji' clips). Generally the images and clips were of the Tōkaidō Shinkansen. So although the shinkansen would not appear to be a dominant image statistically, the Tōkaidō Shinkansen remains of greater significance and the link with Mt Fuji is relatively strong. On top of this, the shinkansen is significant amongst the images of high speed trains globally.[12] In a nod to the significance of the shinkansen and Mt Fuji image, it is worth noting that JR Kyūshū uses the image of a 800-series shinkansen imposed in front of Sakurajima, the active volcano in Kagoshima bay, in its marketing campaigns despite the fact that such a view would never be possible.

Such searches of the Getty Image Bank (the supplier) do not reveal the details of which images are actually used or who they are used by (the consumer). Finding this information is not straight forward, as even after an image is purchased from Getty, there is no certainty that it will be actually used or used in a public domain, and so would require an extensive search of various media to see the context in which they were used. Instead, let us look at some areas where the image of the shinkansen has been used; in non-Japanese films and in (non-tourism related) advertisements outside Japan.

Considering films, I have had to keep my study to English-language films. The most apparent part of this study is how few films are actually set either in totality or in part in contemporary Japan, although there has been a recent increase. Various films have included the shinkansen, either after the film has already been clearly located in Japan (e.g. *The Yakuza* (1975) which included both external and internal scenes, *Mr. Baseball* (1992) where a shinkansen not only appears but blows its horn in a location when this would not normally happen in an apparent attempt to draw the audience's attention to it, and *Lost in Translation* (2003) which includes views of both the outside of the shinkansen and views of Mt Fuji from it[13]) or to help locate the film in Japan (e.g. *Gung Ho* (1986), showing the American's 'first encounter with Japan: neon lights, a capsule hotel, the bullet train, strange food juxtaposed with a flash of McDonald's, confused bowing, and finally an image of the hero lost in the middle of a rice-field' (Littlewood 1996: 53)). Notable films that have not included the shinkansen are *Black Rain* (1989) and *You Only Live Twice* (1967), which is slightly surprising given the adventures that James Bond had on board trains over the next decade.[14]

Clearly the shinkansen does have some symbolic value internationally. Perhaps unsurprisingly there was a shinkansen character in the Andrew Lloyd Webber musical *Starlight Express*. Yet shinkansen have even appeared in adverts internationally. For example, in Britain, a 300-series shinkansen appeared at the beginning of an advert for a headache pill that claims to relieve the consumer from the stresses of modern fast-pace life.[15] Also, Fullers brewery in their campaign 'Worth travelling for' promoting London Pride beer used an image of a 100-series shinkansen passing the iconic Battersea Power Station in South London. Fullers had sought imagery representatives from 'almost the four corners of the globe'. The image of the shinkansen was chosen as it was considered to be 'visually stunning'. The 100-series itself was selected as it was felt that it would be more recognizable as Japanese than some of the more modern shinkansen. To reinforce this, the word 'shinkansen', written in *kanji*, was added to the front window by the advertising agency as there was a concern that at a glance some people may confuse it with Eurostar which also passes Battersea, and they wanted it to appear a 'a little more "Japanese" without losing any of the beauty of the picture' (Fullers 2003).

Mention should also be given to travel diaries. These books, sometimes with accompanying television programmes, are different from travel guides as they do not act primarily as a means for tourists to help find their way around another country. Rather, these accounts are a portrayal of experiences of the writer that, depending on the experiences and the nature of the reader, may serve to either entice the reader to visit that place or deter them altogether. Often these travels are meant to be entertaining and as a consequence one has to be careful about drawing too many conclusions about how representative the accounts are. The most well-known travel writer in Britain that has been to Japan is Michael Palin, formerly of *Monty Python* fame. His global travels, which have been documented as BBC TV series and books, have taken him to Japan twice. During his first round-the-world trip, Palin made a rather odd journey so he could take the shinkansen. Having arrived at Yokohama port, and needing to get to Tōkyō, he took the more timely and inconvenient option of going to Shin-Yokohama and taking the shinkansen. Although not explained in the book or programme, this peculiar routing had been caused by missing his intended ferry from China which would have taken him to Kōbe and allowed a greater journey on the shinkansen. But due to the iconic importance of the shinkansen, particularly in a programme that featured many modes of transport, the detour was considered necessary (Powers 2004). On his second visit, he apparently used the shinkansen a number of times while in Japan during his tour of the Pacific rim. In one entry, he wrote 'Leave Tokyo on the Hikari super express, heading south-west at a furious but almost imperceptible pace on a specially constructed high-speed track' (Palin 1997:62). The wording is interesting as he has used neither 'shinkansen' nor 'bullet train'. Some of Palin's other comments appear in a later chapter. Another such traveller to Japan has been Fergal Keane. Although the BBC programme was entitled 'Great Railway Journeys: Tokyo to Kagoshima', he made many detours from this route. In terms of imagery, his visit to the linear shinkansen test centre is most relevant to this chapter; 'I'm no train spotter, but the Maglev was a work of art' (BBC TV 1999a). Clearly there is something about shinkansen, whether the original designs or the futuristic designs, that many find appealing.

Another 'traveller', although not necessarily an international traveller, are special and commemorative postage stamps.[16] The importance of their design has long been recognized due to the potential to portray a certain image. Indeed, this role has been recognized as being so significant that the Japanese government, though not alone in this respect, has used them to 'define Japan's place in the world' (Frewer 2002:12). With the recognition of this role, their issuance rose significantly in the 1960s, from under ten series a year made up of about 15 stamps to over twenty series with 30–40 stamps (Frewer 2002:12). Yet, as Dobson (2002:22) points out 'Unfortunately, in the study of the creation of identities and public opinion, the importance of what is commemorated, emphasized, ignored or forgotten in these everyday images [stamps] has generally failed to receive attention.' This is despite the fact that 'stamps are products or "windows" of the state that illustrate how it wishes to be seen by its own citizens and those beyond its boundaries' (Brunn in Dobson 2002:22). The shinkansen has appeared on a large number of stamps. Indeed, its image has even been used when the issue was 'devoted to the economic development of the country' (Frewer 2002: 15) rather than any special event in relation to the shinkansen itself.

So how and why does the shinkansen appear on these stamps? Regarding 'how', it is necessary to look at the process by which stamps are chosen. The Post Bureau seeks ideas from each of the ministries and agencies (Dobson 2002:26). At the final decision-making stage at least three possibilities are presented to the director-general after the ideas have been worked on by both internal and external designers (Dobson 2002:26). Chapter 4 will look at the issue of *seiji eki* (political stations), but the existence of the terms '*seiji kitte*' (political stamps) and '*daijin kitte*' (ministerial stamps) reveal the apparent intervention of politicians, that supposedly began with Tanaka Kakuei (Dobson 2002:30), in what designs are selected. This intervention occurs despite it being 'difficult to quantify whether this form of propaganda was effective or profitable' (Dobson 2002: 30). As to 'why' the shinkansen's image was used on stamps, part of the reason will appear in the following section.

The postage stamp, although it can have a 'global reach', is largely domestic as the majority of postage items remain within the issuing country. In continuing this progression from the international image of the shinkansen to the domestic, it is worth concluding this section by looking at a selection of the material which is primarily for the domestic market. Although with greater internationalization, some of these materials are now available outside of Japan, the global market was not the initial target. Those that are outside of Japan who seek to obtain these materials are likely to do so due a specialist interest or because they are ex-patriot Japanese seeking to maintain links with popular cultural back in Japan.

Konno (1984:76) claims that 'although the Shinkansen is the epitome of the modern railroad, seldom is it celebrated in song or made a setting for a film'. This comment appears to be as true today as it was twenty years ago. Even JR Tōkai's promotional song *Ambitious Japan!* (see Chapter 5) does not explicitly mention the shinkansen. Konno suggests that there are two types of popularity in relation to railways, and that while the shinkansen is popular in terms of 'everyone' wishing to have it serve their district,

> The other kind of popularity, which is virtually absent in the case of the Shinkansen, can be observed in the passion to photograph trains, put their image on calendars, record the sound of their passing, and use them in songs and films... The Shinkansen, at least to date, has shown no sign of possessing this 'something'. While without doubt this train is a valued asset of civilization, does it somehow lack a cultural dimension?
>
> (Konno 1984:76)

This statement appears less valid today, certainly in terms the use of photographs of the shinkansen. Semmens (2000:iv) even appears to question whether it was ever true in his statement that 'Although the development of the Shinkansen system is basically a technical matter, the "Bullet Train" quickly caught the imagination of artistic photographers'. That Konno's argument has less validity today may be due to the shinkansen having entered a 'Golden Age', with there being many more designs of trains and greater choice of location, compared to the situation that existed for much of the first twenty years of the shinkansen's history. Also, due to its four decades of history, the shinkansen is now better placed within the mindset of the Japanese people. For as Konno (1984:77) suggests 'culture is born at the point where nostalgia is satisfied'. Noguchi (1990:49) also suggests that railway nostalgia is particularly 'imbedded' in Japan due to the large proportion of the population that use the railways. Japanese of virtually all ages can now remember either the first day of the shinkansen or their first trip on a shinkansen. Those that have experienced a trip on a steam train, which has tended to hold a romantic image in many countries, are becoming increasingly few in number.

That the shinkansen has become popular can be seen by Sanrio's launch of a shinkansen range of goods in 1999 (Sanrio 2002:18). Sanrio is probably best known globally for Hello Kitty, but in the domestic market it has 24 principal characters (Sanrio 2002:17–8). Sanrio is one Japan's leading companies, with net sales in excess of ¥100 billion in 2002, of which over 90% came from the sale of items such as gifts (Sanrio 2002:Appendix). The shinkansen range itself remains domestic due to the terms of their license with the JR companies (Sanrio interview 2003). At present the Sanrio shinkansen characters are based on JR East, JR Tōkai and JR West shinkansen only. A character based on JR Kyūshū's 800-series may be introduced in the future 'as the children like to have a big choice of characters' (Sanrio interview 2003), to go with the toy of Hello Kitty dressed as one of the on-board staff. Although had JNR continued to exist, Sanrio does not believe that launching a range would have been impossible, privatization certainly made it easier, particularly as since 1987 different types of shinkansen have been introduced (Sanrio interview 2003). Sanrio chose to launch its shinkansen range following market research that indicated that of the characters that it could use,[17] the shinkansen was the one that would be most popular amongst young boys due to it not only being something familiar but also more special than conventional trains (Sanrio interview 2003). The demand for such products apparently came from mothers wanting the company to have a range that would appeal to their sons as it was 'a problem trying to keep their sons amused when shopping at Sanrio with their daughters' (Sanrio interview 2003). That it was not launched until 1999 was largely due to the fact that such 'cute' (*kawaii*) goods had previously not been popular with boys (Sanrio interview 2003).

The shinkansen range of goods is extensive and includes drawing books, stationery, clothing, toys, and food products. Although originally targeted at children aged from three to eight years old, a baby range has also now been introduced, but the range is not extensive as that for Hello Kitty which includes products for people of all ages (Sanrio interview 2003). The shinkansen range has been very successful for Sanrio, and has proved to be popular amongst girls, including Sanrio's female staff (Sanrio interview 2003). Showing how iconic the first shinkansen remains, 'the "classic" stock of all modern high-speed trains throughout the world' (Oikawa and Morokawa 1996:89), the 0-series character is the main one used and most popular, although all of the Tōkaidō Shinkansen are used a lot as they tend to be the ones seen the most and are known abroad

(Sanrio interview 2003). The success of the shinkansen range can be seen by walking into any of their shops in Japan where it appears to be occupying nearly as much shelving space as Kitty, or by observing the goods being used by elementary school children. Although many slogans appear on Sanrio's shinkansen goods, perhaps the most revealing is the one proclaiming 'Shinkansen is the only way to go'.[18] Ironically, the shinkansen characters produced by Sanrio all have broad smiles, whereas the iconic Kitty is not even drawn with a mouth.[19] Of course for trains to be given faces is nothing new. Thomas the Tank Engine (and friends) is as popular in Japan as in its native Britain. A difference is that Thomas the Tank Engine began with stories and other related products followed later. With Sanrio's shinkansen it has been the other way round. In fact, Sanrio have yet to make stories based upon the shinkansen characters, though this may happen in the future (Sanrio interview 2003), although Sanrio's main product, Hello Kitty, appears to have been successful due to the lack of stories rather than despite them (Belson and Bremner 2004:120–3). Yet in marketing, Sanrio's approach is not unique as I have observed JR East posters, for example, giving personalities to some of their shinkansen, albeit by giving them a human form (see Figure 3.5).

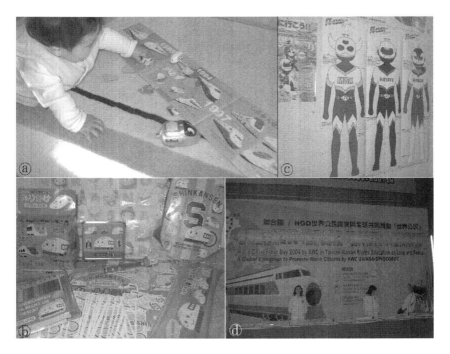

Figure 3.5 (a) A child plays with a range of shinkansen toys; (b) examples of Sanrio's range of shinkansen goods; (c) JR East's shinkansen characters representing the E4-series, E2-series and E3-series shinkansen – note that the *Komachi* figure is feminine in keeping with the origins of its name; (d) the image of the shinkansen being used to promote 'World Citizenship Day' in Taiwan.

The shinkansen has also featured on television programmes. It has been the central feature of the story in some television dramas in Japan, for example, *Shinkansen Kōankan* ('Shinkansen Railway Police', 1977) and *Shinkansen '97 Koimonogatari* ('Shinkansen '97 Love Story', 1997). Both of these series were fictional rather than 'fly-on-the-wall' documentary-type-soaps, as have plagued British television in recent years. In April 2004 NTV broadcast the first of two planned special shinkansen trivia programmes, while Tōkyō TV, in celebrating its own 40th anniversary in 2004, did a two-hour 'documentary-drama' on the creation of the shinkansen. To have such programmes would suggest that the shinkansen has achieved a certain cult status. It is also likely to indicate that there is no shortage of railway enthusiasts or wannabe railway enthusiasts in Japan.

This latter point would appear to be supported by a recent increase in the number of books written about the shinkansen. Kawashima Ryōzō is one author who makes a living out of writing about railways in Japan, producing five to six books a year (Kawashima interview 2004). However, books on the shinkansen are not limited to non-fictional books. For every line and every series of shinkansen there appears to be at least one novel. Most of these are disaster novels featuring train disappearances or murders, while in others, despite the title, the shinkansen only has a relatively minor part. Although not as well established as Hercule Poirot, policeman Miyanohara created by Kitani Kyōsuke, has featured in many of these books and has had to contend with crimes on conventional railways and famous locations around Japan as well. While *Murder on the Orient Express* is the arguably the most well known English language train-based story, others do exist, but I suspect most would find it hard to name many. Similarly, in Japan, these novels also remain relatively unknown, yet there appears to be a market sufficient enough to warrant the output of new stories.

As with *Murder on the Orient Express*, there is one story that is more well known, partly, like its English counterpart, as it has been made into a film too. *Shinkansen Daibakuha*, written by Katō Arei, became a box office hit in 1975 when the film, starring Takakura Ken, was released. The book itself has been translated into English (Rance and Katō 1980), though under the title of 'Bullet Train' rather than the literal translation of 'Shinkansen Big Explosion'.[20] As the more literal translation suggests, the story revolves around a bomb on the shinkansen. Indeed, it appears that the story may have influenced some outside Japan for the film *Speed* (1994) seems to have borrowed the idea of a bomb being armed once the vehicle goes above a certain speed and detonating should the vehicle go slower than 80 km/h (50 mph).

So far this section has been concerned with the image of the shinkansen and how it has been used, both domestically and internationally. However, the shinkansen itself has been exported too. Discussion about the export of the shinkansen to Taiwan, the bid for exports to other countries, including China, where its adoption has been referred to as being a potential 'symbol of good relations with Japan' (Yomiuri Shimbun Chūbu Shakaibu 2002:110), will be looked at later. Another shinkansen has already been exported. A 0-series shinkansen can be found at the National Railway Museum (NRM) in York. I was present for the unveiling ceremony in 2001, and it was possible to feel the sense of excitement amongst the staff at the NRM as well as the other guests gathered there that day. It is a unique display as it is the only foreign train at the NRM, in the sense that it was neither manufactured nor operated in Britain, as well as the largest display there. The shinkansen's long journey to the NRM was documented in *York's Oriental Express* (BBC TV 2001).[21] Richard Gibbon (Head of Engineering, NRM) makes is clear why the NRM wanted to have this exhibit; 'It's an icon that has changed the world... They produced this wonderful great white bird, the bullet train... the Japanese bullet train has revolutionized high speed rail travel in the world forever' (BBC TV 2001). Gibbon also commented on the image of the shinkansen:

> I think there are certain shapes, both engineering and architectural, that are forever, that are icons... the *Dakota*, the *Spitfire*, the E-type Jaguar. All of these shapes that man has created look right. When you look at the *Spitfire* aircraft, when you look at this [shinkansen], it just looks right and feels right.
> (BBC TV 2001)

In an apparent confirmation of a point made above, Gibbon sometimes used the word 'shinkansen' when talking to JR West, who supplied the retired train, but always 'bullet train' when speaking to the BBC programme makers. JR West was obviously well aware of the importance of this export and in the speeches it was stated that it was 'sad to part with this vehicle because we cherished it very much... Like sending a daughter to a wedding in another country'. Accordingly JR West completely renewed the train, whereas normally NRM receives trains at the end of their life and are not in such good condition (BBC TV 2001).

Machines can develop a character, or what Clarkson (2004b) refers to as a 'soul', in the same way that any animal or person can. According to Clarkson, for a machine to have 'soul', while looking good is important, there appear to be two further qualities needed. Computer technology should be limited in the machine, with it being relatively 'mechanical'. Further, and seemingly more significant in many of the examples that Clarkson presents, there must be a 'flaw'. In this respect, while the 500-series is clearly seen as beautiful by many, the more mechanized 0-series, which also had its flaws, and the flawed, in the sense that it just not look quite right for a high-speed train, E4-series arguably have more 'soul'. Yet Yamanouchi (2000:121–2) suggests that the shinkansen perhaps does not have 'character' in the same way as the named trains like 'Japan's *Tsubame* or *Hato*, Britain's *Flying Scotsman*, or France's *Le Mistral*' which have their 'own name and features found on no other trains'.

The role of a symbol

Images and symbols have their uses, but also their limitations. They can be created in an attempt to create a fresh idea, but they can also reflect an outdated state. They are, naturally, central in the development of stereotypes, which in themselves need not be negative as they can provide a basis on which to build greater understanding. Unfortunately this second step is not always taken. Indeed Littlewood contends that

> One by one, the time-honoured images turn out to be true. But in doing so, they obscure all the other things that are true – which is why they are so dangerous. They teach us what to look for, and that is what we find; everything else becomes a background blur. We are left with a reality selected for us by our stereotypes.
>
> (1996:xiii)

Stereotypes are often developed when we are still children. My own experience of seeing a picture of a shinkansen passing Mt Fuji may have been the seed that turned into this study. Looking at some children's books and toys, it is interesting to reflect upon the imagery used when selecting national symbols. Typical examples for selected countries are; a panda for China; kangaroos, koalas, and the Sydney Opera House for Australia; the Statue of Liberty and an eagle for the United States; a moose for Canada; football for Brazil; the Eiffel Tower and French bread for France; Big Ben for Britain and the shinkansen and a woman or girl wearing kimono for Japan. The fact that such items seem appropriate may say more about us and what we look for in a particular country, even our own country, than what it really tells us is truly representative of that country.

Japan is a particularly interesting country to study in respect to symbols. As Buruma (2001:111) points out 'What meets the eye in Japan is often all there is. Japan is, after all, as Roland Barthes observed, the empire of signs, the land of the empty gesture, the symbol, the detail that stands for the whole.' This point is supported by Hendry (1993:35), who suggests that in countries where the language is rife with ambiguities, as Japanese is, there is 'greater use of other means of communication, such as through material objects'.

There are many different types of symbols, but what concerns us here are 'public symbols' (see Hendry 1999:82). According to the *Concise Oxford Dictionary*, a symbol is something that is 'regarded by general consent as naturally typifying or representing or recalling something by possession of analogous qualities or by association in fact or in thought'. As Hendry (1999:83) notes, the use of the word 'naturally' is significant as 'it is often the case that people within one society remain blissfully unaware of the relativity of their symbols'. Yet we need to be careful in discussing the significance of symbols. Indeed, Cohen (1986:7) questions whether the symbols are significant for the 'people we are supposedly describing. Or ourselves?'

One of the most significant symbols has been a flag. In Japan noble families (*daimyō*) had their own crests (*mon*) that would be used on these flags. In due course, Japan also adopted a national flag, although it together with the national anthem was not legally recognised until 1999 (Hood 2001:70). These types of symbols have evolved in the modern era in the form of company logos. The significance of them can clearly be seen by the frequency that logos are changed

Ambassador of Japan 67

when a company seeks to change its direction, or the prominence of logos when companies sponsor some activity in the hope that it will improve the company's profile and business. People can also be symbols – monarchs and emperors being a prime example of this. Yet, symbols need not be visual. Indeed the Japanese Emperor historically remained virtually unseen. National anthems also symbolize the nation. Befu (2001) describes how a 'symbolic vacuum' was created because of the controversy surrounding Japan's flag and anthem, which were linked to Japan's imperial expansion and the role of the Emperor. The shinkansen helped to fill the vacuum as part of Japan's growth in 'cultural nationalism'.

'Symbolism' is not necessarily a positive term. A symbolic gesture, in particular, can have negative overtones. It may not always be possible to know whether what is observed is a true reflection of what is intended and felt, or whether it is merely a token action. This can be equated with the Japanese terms '*honne*' and '*tatemae*'. Furthermore, how does one know whether what is observed is symbolic of the 'norm' or more exceptional? For example, were the scenes in London following the death of Princess Diana in 1997, symbolic of the British people's love for Diana and the Royal Family? It could be argued that although the popularity of the Royal Family did appear to increase following the accident, initially what was observed was exceptional and more a reflection of their guilt for creating a situation where demand for stories about Diana may have led to the accident. Also in Britain, fireworks displays are held on 5 November to remember Guy Fawkes and the importance of not bowing to terrorism, although the desire to have fun has probably meant that the original meaning of the day has all but been lost to many. Many in Britain will wear a poppy around 11 November to symbolize their remembrance of those who lost their lives fighting for the country in wars. The wearing of the poppy, as with stickers and bows for charities, rather that just donating money, would appear to be an important symbolic act.

Events can also be significant. For example, weddings are particularly symbolic, being designed as a gathering so that the couple can proclaim their love for each other in front of others, with rings often being exchanged as a symbol of this. Large international events, such as the Olympics and World Cup, are highly symbolic. So significant are these events that some countries have chosen not to send teams due to the message it may send (for example, the United States not sending a team to Moscow in 1980). The 1964 Tōkyō Olympics was such a symbolic event. It signified Japan's return to the international arena and economic recovery. That the shinkansen was deliberately completed in time for these Olympics further ties these two symbols together.

Since industrialization, machines, particularly means of transport and other products have become symbols too. For example, as discussed elsewhere, steam engines and the construction of lines to connect towns with cities were symbolic of the development of a particular country, and that modernization was reaching a region within that country. The twentieth century was full with these symbols of modernism and development which often aimed to portray some form of superiority over others. Many of these appeared due to the conflicts in that century. For example, in the air there were the *Lancaster*, *Spitfire*, *Zero*, the *Hindenburg* airship, V1 and V2 flying bombs, the *Dakota*, the *Flying Fortress*, *Harrier* jump-jet and B52; at sea there were *HMS Ark Royal*, *HMS Hood*, *Nimitz*, *Missouri*, *Yamato*, *Bismarck* and the U-boats. During peacetime, symbols that had a closer link to the people and consumers appeared. In the early part of the century examples would include the *Titanic*, the *Mallard* and *Flying Scotsman* steam locomotives, and the Model T Ford. As the century progressed so symbols that were used for more peaceful means, although some of them were clearly linked to the Cold War, appeared. For example, the hovercraft (especially Dover-Calais), Sputnik, the Apollo space missions, the shinkansen, *Concorde*, the 'Jumbo' jet, the Space Shuttle and Formula One motor racing cars. One of the features of many on this list is that they were a world first, a pioneering piece of technology or feat. Indeed, Konno (1984:74) argues that one of the reasons the shinkansen is popular 'is that it satisfies the Japanese penchant for world firsts. This constant search for "the best in the world" or "the best in Japan" stems in part from the Japanese complex about catching up with other advanced nations.' The shinkansen was the symbol that Japan had not only caught up, but, in terms of railway technology at least, in 1964 had overtaken the rest of the world.

Konno (1984:77) is less than enthusiastic about the shinkansen's accomplishments. He questions whether it can been seen as 'symbolic of the history of civilization in an era that is supported by industry and technology'. Konno's concern is that

> Every means of transport that human beings have used to date has made a major contribution to cultural dissemination and exchange. The result has been the evolution of culture and the unfolding of history. The Shinkansen, too, is helping create new culture and will be highly evaluated by later generations, though just what sort of culture it will bring into being is yet unknown.
>
> (1984:80)

However, it is important to stress that the shinkansen is not only involved in the creation of new culture, but is also reflecting the evolution of Japan's society and culture.

Returning to the list of peacetime symbols I presented, most of them are 'national' symbols. They represent the achievements of a nation or constituents of that nation. Of them the hovercraft, the 'Jumbo' jet and the shinkansen, are perhaps special in that they were more accessible to the 'common person' than many of the other symbols which were used by companies or the nation to state their progress. Indeed, I would argue that part of the strength and the reason why they are so significant was their commonness. Consequently the shinkansen is a 'face' of Japan (*Nikkei Shimbun*, 13 April 2002), a face that has continued to change as Japan has also changed.

Although these symbols I have presented are 'national', many of them are also 'spatial' as they represent and reflect a particular time in that nation's development. Although there have been further improvements and developments to some over the years, many of them are no longer in use. These 'spatial' symbols, as with 'national' symbols, are not restricted to modes of transport. Common 'spatial' symbols are people, music, films, and other forms of popular culture. There are also the products of the modern world, for example, the mobile phone could be considered a symbol of the current times just as the personal stereo was of the early 1980s.

Another significant point about the above examples is that they gained international renown. All countries have their own symbols, the majority of which are not known beyond their local region. If I were to list the various symbols of Japan here, they would be meaningless for many readers. In some cases, even if the image is known, the word may not be. Some symbols are also likely to have limited impact in some parts of the globe. Furthermore, the nature of symbols also changes over time. Given some of the examples that I have presented, one has to speculate as to why *ThrustSSC* and the linear railway in Shanghai have not become famous symbols. *ThrustSSC*, like the *Bluebird* land-speed and water-speed vehicles before it, was a record breaker. Not only is it the fastest land-speed car it is also the only one to have broken the speed of sound. Yet, even in its native Britain, it has not the enjoyed significant adulation it is probably due and the project was nearly bankrupt on several occasions (Noble and Tremayne 1998). As for China's high speed railway, it is also suffering from financial challenges. Perhaps this example is less symbolic due to its relatively short operating distance and the fact that it is a German development exported to China rather than home-grown. The examples of *ThrustSSC* and the Chinese linear railway also suggest Konno's assertion (1984:77) that the popularity of the shinkansen is derived from its size and speed, as 'It satisfied humankind's materialistic desires and responded to the demand for the biggest and the fastest' may not be valid. Either humankind has reached a point where these two values are no longer important or they were never as significant as Konno suggests. There is another possibility I would like to suggest. Many of the examples I presented not only looked fantastic in their own right, but also enjoyed a great backdrop – whether it be Mt Fuji, the white cliffs of Dover or space itself. Even the greatest portraits have an attractive setting.

Symbols can generate different reactions in different countries. Although the shinkansen remains largely a Japanese symbol, *Concorde* was one that crossed boundaries, for it was a result of Anglo-French cooperation. Yet, when the loss-making plane was retired, the event apparently passed without significant attention in France. Although this may in part have been due to the crash at Paris that was a contributing factor in *Concorde's* eventual withdrawal, it is also appears accurate to say that *Concorde* never captured the French imagination in the same way as it did for the British. For *Concorde* was special and symbolic for many British people and, as Clarkson (2004b:14) suggests, when it crashed in Paris, 'we were actually mourning the loss of the machine itself'. The final commercial flights of *Concorde* were covered by special live broadcasts by all of Britain's leading news networks. The BBC described *Concorde* as being at the 'cutting edge of modernity although its design is around 40 years old', a phrase that could easily be applied to the 0-series shinkansen, and described the atmosphere on the day of the final flights as being 'like a national wake – a mixture of celebration and mourning' (BBC News special 24 October 2003).

In conclusion, we can see how the shinkansen can and has played an important role in the development of Japan's image domestically and internationally. As Littlewood (1996:xi) states, 'Popular impressions of Japan owe little to historians or social scientists' but are more a result of being 'scrambled together' from various popular images and products. Much of the rest of this study will analyse the symbolism of the shinkansen and how it reflects different aspects of Japanese society.

4 Whose line is it anyway?

> There is nothing more deceptive than an obvious fact.

Having established that the shinkansen can be seen as symbolic, while acknowledging that we need to be aware of possible limitations of doing this, it is possible to use it as a tool to study Japan and its society. This chapter will concentrate on an area that has become perhaps the most infamous issue in relation to the shinkansen. The standard terminology in English used to describe this issue is 'pork-barrel politics'. Given the sentiment expressed in the statement above, and his belief that the more obvious something is, the harder it is to prove,[1] Sherlock Holmes would no doubt enjoyed investigating the supposed link between political intervention and the shinkansen.

Pork-barrel politics is not necessarily the same as corruption. Pork-barrel politics implies an attempt to provide better services and employment opportunities to your local electorate, something that is both logical and at one level desirable. However, the problem is that it can lead to an overly local focus at the expense of the national issues. This may lead to inefficiency and greater overall costs for the majority. This appears to have been the case in Japan, with the creation of the *dokken kokka* ('construction state'), whereby 10% of Japan's workforce was engaged in construction, supported by *dangō* (bid-rigging) (Kingston 2001:25), and accounted for over 20% of Japan's GDP compared to 10.4% in the United States and 7.6% in Britain (Kingston 2004:124). Despite this, it is important to remember that not all pork-barrel politics is undesirable and corrupt, nor is all corruption in relation to construction a form of pork-barrel politics.

The situation in relation to corruption is perhaps confused further in Japan by *giri* – obligation – and the associated culture of gift-giving, at which politicians 'have to be past-masters to be effective' (Buruma). This can lead to a situation where

> It is not always easy to distinguish this custom from real bribery, particularly when cash changes hands. It is a common practice, for instance, for mothers to pay substantial fees to teachers in return for a helping hand in securing a place for their children in prestigious schools.
>
> (Buruma 2001:151)

In other words, when we consider the practices that go on in Japan, we need to be careful not to judge everything by 'Western standards'. Japan does not necessarily operate by these standards and these standards may neither be appropriate in Japan, nor necessarily as inherently desirable or 'right' as many in the 'West' appear to assume.

Creation of the *seiji eki*

Although it is the Jōetsu Shinkansen that is normally mentioned first in relation to pork-barrel politics and the shinkansen, the suggested link between the influence of politicians and the shinkansen dates back to the Tōkaidō Shinkansen. On 18 November 1959, with the route of the Tōkaidō Shinkansen having been announced, the press proclaimed the existence of a 'political station' (*seiji eki*) (Umehara 2002:20). The station in question was Gifu-Hashima. The reason for the press to be suspicious is clear to see. For the opening of the Tōkaidō Shinkansen there were only twelve stations being proposed. These stations were all to serve major cities, tourist destinations or interchanges with conventional lines. The average distance between stations was over 60 km. Yet, Gifu-Hashima was to be located in Hashima, which had a population of just 40,000 (Umehara 2002:20). There were no connections to other conventional lines – JNR or private – at the time. There are no significant tourist attractions in the area. The station was also to be located only 25 km from Nagoya and 41 km from Maibara stations. A further issue was that the original route of the shinkansen, that is, that for the *dangan ressha*, which the Tōkaidō Shinkansen largely followed, was significantly different in this area. The conclusion that the press came to was that the local politician, Ono Banboku, was involved. So great was the criticism from the press that an investigation, which lasted until January 1961, was launched (Umehara 2002:20, 34).

The investigation found that there was nothing improper about the location of the proposed station. Cynics might suggest that this is not a surprise. Yet analysis of the facts would suggest that the conclusion of the investigation was probably correct. The first issue that has to be considered is the variance from the original *dangan ressha* route. This route would have been significantly shorter than the proposed Tōkaidō Shinkansen route. This would clearly have saved time on the journey itself, and would have helped in achieving one of the goals of the project – to connect Tōkyō and Ōsaka in three hours (NHK 2001). However, there was another time consideration in the construction of the line. The project had a strict deadline. It had to be open in time for the Olympics in October 1964. This meant that the line would have to be constructed by around mid-1964 so that safety checks, training, etc. could be carried out. This gave less than five and a half years to complete the work on over 450 km of infrastructure. This was going to be a major challenge whatever route was selected. However, any route that needed major tunnelling work would make the challenge an impossibility. As it was Shin-Tanna tunnel, on which some work had already been completed, was only finished a few months before the final rail was laid. The *dangan ressha* route would have meant a tunnel would have been needed under Mount Suzuka, as well as other tunnels nearby. According to JNR, this would have made it impossible to complete the construction work on time (Umehara 2002:21). Clearly an alternative route was needed.

The solution to the problem came partly from history. Rather than following the traditional Tōkaidō route via Mount Suzuka, the shinkansen would take a more northerly route and go via Sekigahara (see Figure 8.1). Sekigahara was the location in 1600 of a major battle, the result of which was to effectively unite Japan for the first time and led to a period of unprecedented peace and stability for over 250 years. The reason for the location of this battle was in large part due to it being a bottle-neck on the East–West route across Japan avoiding major mountains. The Tōkaidō conventional line had also followed this route. However, while the Tōkaidō Line goes via the city of Gifu, the shinkansen was to keep further south. The reason for this was that it would reduce the total route distance, save time, and also save significant construction costs as building in cities was naturally more expensive than in open areas. Indeed, not going via Gifu was expected to save 15 minutes travelling time and ¥10 billion (¥58.3 billion in current prices) in construction costs (Umehara 2002:33)

Although this explains why the route did not go via Gifu, it does not explain why there was a station or why it was located in Hashima. The answer to the first question was due to meteorological concerns. Despite not having the snow-fall problems that the Jōetsu, Tōhoku and Hokuriku Shinkansen lines later had to contend with, the area around Sekigahara and Mount Ibuki, was felt to be a potential problem. So great was this problem likely to be, that there was a concern that trains may have had to be stopped. Clearly this would mean that a station would be needed so that passengers could alight. As a consequence stations near the problem areas were planned. Maibara was an obvious location to the West as it is also a major interchange with other lines. To the East, a station within Gifu prefecture would be needed. That the two stations were designed for this purpose can be seen by the construction of an extra platform at Maibara and two extra platforms at Gifu-Hashima which would make it easier for more than one train to be terminated and emptied at a time. The justification for this decision was demonstrated on a number of occasions in early years before other methods for dealing with the snow were developed (see Chapter 7).

Having established that a station within Gifu would be necessary, the only question that remained was its exact location. The most logical was Ōgaki as this was a city of 94,000. Ōgaki was also a significant station on the Tōkaidō conventional line. However, due to the local conditions and the significant amount of space that is need to construct a shinkansen station, Ōgaki was not considered practical. Another potential site within Ōgaki city, at Minoyanagi station on the Kintetsu network, was also ruled out due to the limitations in construction techniques at the time. As a consequence, despite not being connected to any other railway network at the time, the decision was taken to locate the station at Hashima (Umehara 2002:34). The use of the prefix 'Gifu' is perhaps an indication of how minor a city Hashima was – reflecting the fact that many may not

74 Whose line is it anyway?

have known where it was without the prefix, although I suspect that Ono would not have been opposed to this extra publicity of his prefecture's name.

Yet, it was via Gifu *city* that Ono wanted the shinkansen to go. Clearly this did not happen, so why did the issue of *seiji eki* arise and, to some degree, continue to be associated with Gifu-Hashima station? The answer to this is perhaps in part due to the statue outside of the station with Ono and his wife pointing towards the station and Tōkyō (see Figure 4.1). Although Ono may not have been a decisive factor in the location of the station, he clearly felt that it was in his interest to make people think that he was. Given the nature of the line, there were potentially votes to be won by being associated with having one of the few stations located within his prefecture. Effectively the locating of the station became free publicity for Ono.

The next station that was said to be a *seiji eki* was Shin-Iwakuni as it was within the constituency of Satō Eisaku, then Prime Minister of Japan. However, such an idea overlooked some significant points. First, stations were being constructed on average 36 km apart on the Sanyō Shinkansen. This allows faster services to pass slower ones, as well as provide a sooner stop should technical problems develop on any train. Shin-Iwakuni is 44 km from Hiroshima and 38 km from Tokuyama, so above the average distance between stations. Iwakuni is also a major tourist

Figure 4.1 (a) Statue of Ono Banboku and his wife outside Gifu-Hashima station, where a 300-series shinkansen is waiting for a faster service to pass; (b) statue of Tanaka Kakuei outside Urasa station.

destination. Iwakuni station was also a stop for express trains on the conventional line. So there would appear to be need for a shinkansen station. However, due to the nature of the local topography and the problems that would have been in attempting to purchase land in the area, it would not have been practical to build the shinkansen station at the site of the conventional station (Umehara 2002:51–2).[2]

Tanaka's shinkansen

Although it is the term *seiji eki* that has gained greatest usage, in terms of appropriateness, I would suggest '*seiji sen*' ('political line') would be more apt. For although some still speak of Gifu-Hashima and a lesser extent Shin-Iwakuni, it is the Jōetsu Shinkansen that has become the prime example of pork-barrel politics in Japan, let alone in relation to railways. The reason for this is that the man who is most associated with the line is Tanaka Kakuei, then arguably LDP's most powerful and influential politician. His involvement in the Lockheed bribery scandal, of which he was found guilty in 1983, may be what he is best remembered for now, but while he did not appeal against the suspended jail sentence, he continued to be a successful and influential politician within the LDP, and was instrumental in enabling Nakasone to become Prime Minister in 1982 (Hood 2001:31). Although Buruma (2001:163) argues that 'politics are by definition polluted by the calculating ways of society', as a consequence of Tanaka's case and other similar cases, Horsley and Buckley (1990:131) conclude that 'in Japan corruption can pay'. Tanaka symbolized what many disliked about LDP-style politics, although nationally it was not until the 1993 general election that the LDP lost power for the first time since 1955 due, at least in part, to the public dissatisfaction with the apparent corruption within the party. Yet Tanaka remained popular, at least within his own constituency, until his death in 1993.

On the back of the shinkansen 'Boom', as Minister of International Trade and Industry, Tanaka proposed that the whole Japanese archipelago should be connected by a network of shinkansen lines. This was just one part of a proposal to redesign Japan in an attempt to better distribute production, resources and people. His book (Tanaka 1972) was a best seller. Yet questions remain about how much of the book was actually the work of Tanaka, with ghost writers from within MITI and other agencies thought to have been used (Babb 2000:74). Indeed, it was joked that when Tanaka 'heard that his book had become a best seller, he said "I guess if it's so popular, I'll have to buy a copy and read it too" ' (Passin 1975:281). Whether the ideas were originally Tanaka's or not, the project certainly became identified with him. The section on the shinkansen network, which is accounts for less than 3% of the book, is what people tend to associate with the plan – to the extent that I expect many think that that was all the book was about. There is no doubt that 'The idea caught the public's imagination, but it was spoiled by the greed of speculators' as construction and real-estate companies bought land along proposed routes (Horsley and Buckley 1990:111).

Prior to turning his interest to the shinkansen, Tanaka was clearly a man who was frustrated by the lack of progress being made by JNR in the construction of new lines to rural areas, for he led a group that set up the Japan Railway Construction Public Corporation in 1964 following Sogō's resignation (Kasai 2003:11, 34; Matsuda 2002:13). Tanaka also promised his constituents that he would get the conventional line to Niigata double-tracked. By 1967 this had been done (Babb 2000:66). Although undoubtedly Tanaka had great influence within politics and the bureaucracy, it is questionable that this improvement was only due to Tanaka.

The area that Tanaka represented was Niigata. Niigata prefecture is, in area, Japan's fifth largest.[3] When Japan began to modernize in the late nineteenth century, Niigata was the largest prefecture in terms of population (Hiraishi 2002:55).[4] However, by 1970, although its population had risen by about 750,000, the Japanese population had been rising at a much greater rate. The 1960s had also seen the beginnings of huge migration from rural areas to the large cities, as well as the impact of baby booms within the metropolitan areas. Tanaka's plan was to reduce the need for people to move from Niigata to Kantō by providing jobs in the local area. This meant bringing new investments to the area and also improving the area's link to the capital so that both existing and new businesses could conveniently do business with companies in Tōkyō. While the logic of what Tanaka was wanting to achieve is irrefutable, the questions that remain are whether it was an appropriate use of JNR (i.e. government) funds, whether the choice of route was due to any political influences, and whether Tanaka was responsible, as has been suggested, for the construction of the line.

It is easy to get carried away in criticizing the fervour for wanting to build shinkansen lines such as the Jōetsu Shinkansen and to jump to conclusions about possible corruption and involvement of politicians. Thirty years on, and with the benefit of hindsight of what occurred during the 1970s in relation to the economy, it is easy to be sceptical about the reasons for building the line. But that overlooks the reality of the situation in the second half of the 1960s and early years of the 1970s when the Tōkaidō Shinkansen was seen as a shining example to the world of Japanese technology and a solution to some of the country's transport problems. Undoubtedly, upon reflection, some of the suggested shinkansen lines make little sense, as can be seen by the lack of use of even the conventional lines even today, but one cannot question the desire and reasons for them and the greater logic behind the lines which were included as *Seibi-Shinkansen* (see Table 2.1). It may have been jumping on a bandwagon, but it was a fast moving and popular one.

When the decision was taken to build the Jōetsu Shinkansen, the shinkansen was enjoying a boom from the early success of the Tōkaidō Shinkansen – particularly following 'EXPO 70' – and construction of the Sanyō Shinkansen was on-going. There was much hope and enthusiasm associated with the shinkansen. When one considers all of the lines that were included within the Nationwide Shinkansen Development Law, the Tōhoku, Hokkaidō, Kyūshū, Hokuriku and Jōetsu Shinkansen lines (which became many of the *Seibi-Shinkansen*) were arguably the most needed. Of these, the Tōhoku and Jōetsu Shinkansen were always likely to be given preference. The Hokkaidō Shinkansen would be effectively an extension of the Tōhoku Shinkansen and with speeds at the time limited to 210 km/h, it was unlikely that the shinkansen would be able to greatly reduce the growing demand for air transport between Tōkyō and Sapporo. On top of this, the work on the Seikan Tunnel had only recently been started and was not expected to be completed for many years. Similarly the Kyūshū Shinkansen would effectively be an extension of the Sanyō Shinkansen and would also require significant tunnelling and inclines. While in terms of route topography, the Kyūshū Shinkansen and Jōetsu Shinkansen are comparable, the Jōetsu Shinkansen has the advantage of the huge Kantō market.

It was also unlikely that the Hokuriku Shinkansen would be given preference. The Hokuriku Shinkansen uses the same route as the Jōetsu Shinkansen as far as Takasaki, although one proposal was for it to branch off at Nagaoka, thus avoiding Nagano and the mountains in that prefecture.[5] The Takasaki–Nagano route would require a challenging incline to Karuizawa,[6] and dealing with the change in electricity frequency from 50 to 60 Hz.[7] Indeed when JR East first proposed to build the Hokuriku Shinkansen, it was only to be as far as Karuizawa as this would have addressed the problems faced by the conventional line in dealing with the incline. That Nagano became the terminal due to the Olympics is perhaps an indication of the relatively lack of demand anticipated for that route initially. To build the Hokuriku Shinkansen via Nagano would have been unthinkable. Therefore, the Jōetsu Shinkansen was the most likely line to be built together with the Tōhoku Shinkansen.

Considering the Jōetsu Shinkansen route in more detail, the conventional line from Tōkyō to Takasaki was already at capacity, and so it was clear that this section would benefit from a shinkansen line. Given this situation, there would appear to have been two options. First, to only build as far as Takasaki. Second, to build the complete Jōetsu Shinkansen line. That the second option was selected should not be seen as surprising. Although costly due to the tunnel work needed, there was, at the time the decision was taken, no logical reason not to build the whole line. The economy was still growing strongly. Given what happened in the years following the start of construction, rather than it being a surprise that the decision was taken to build this line, it is perhaps more of a wonder that work on the line beyond Takasaki was not abandoned, as it was on the Narita Shinkansen.

Given that the whole line to Niigata was to be built, it was then a matter of deciding the exact route. On the whole, the route was as would be expected. However, the Jōetsu Shinkansen also has an apparent *seiji eki*, namely Urasa. Like Gifu-Hashima, outside the station there is a statue of a politician. The statue is of Tanaka Kakuei (see Figure 4.1).[8] Statues are not particularly rare in Japan. Although there appears to an abundance of mostly anonymous female nudes in this apparently conservative country, *tanuki* (similar to racoons), and characters or products of local significance,[9] most of the statues of people tend to be of political figures that have achieved something of *national* significance, for example Saigō Takamori's statues in Kagoshima and Ueno Park (Tōkyō), Sakamoto Ryōma in Kōchi and Ōtomo no Yakamochi in Takaoka.[10] Indeed, much of the text under the statue of Tanaka relates to his national and international achievements. Furthermore the statue's existence could be more a sign of the wealth of Tanaka's *kōenkai* (support association) than Tanaka's role in the location of the station.

So was Urasa built at the request of Tanaka? There is no doubt that this choice of location for a station seems odd. Urasa was not a stop for either *tokkyū* or *kyūkō* trains on the conventional line. JNR had even been looking to reduce the number of personnel working at the station (Umehara 2002:64). The likely stops along that part of the route would have been at either Muikamachi or Koide as they were stops for faster trains on the conventional line and provided connections to other lines. However, Urasa had the advantage that it would be about halfway between Echigo-Yuzawa and Nagaoka on the shinkansen. On top of this, a stop at Urasa avoided the need for a significant curve, which would have reduced the speed of the shinkansen, further tunnelling work, or a station unconnected with the conventional line (Umehara 2002:69). I have heard it suggested that Tanaka was keen to ensure the shinkansen station was not built in Ojiya, which had not voted for him on a number of occasions, but its location is so close to Nagaoka that a station there was unlikely and this story may just be another part of the myth surrounding Urasa. As with Gifu-Hashima on the Tōkaidō Shinkansen, therefore, there were clear practical reasons for the location of the station at Urasa.

So even if Urasa was not a *seiji eki*, is the Jōetsu Shinkansen a *seiji sen*? It is certainly the case that Tanaka was involved in pushing for the line to be built. In an interview with Nakasone Yasuhiro, whose Takasaki home and constituency is also on the Jōetsu Shinkansen, he commented that he had worked together with Tanaka to see the fruition of the line (Nakasone interview 2001). Yet, that these politicians were working to have the line built does not equate to it being built solely due to their efforts. Indeed, Nakasone (interview 2004) believes that due to the importance of Niigata and a link to the Japan Sea coast, the line would have been built even without Tanaka's involvement. As with Ono, however, these politicians had seemingly little to lose and much to gain from being associated with the decision to have the line built, which, as already discussed, appeared to be justifiable at the time construction started. It is more probable that their influence may have been needed to ensure that the construction work, at least beyond Takasaki, was not stopped when the economic problems associated with the oil shocks struck. With the economic downturn in the 1970s while plans for new lines were not pursued, other shinkansen construction was not abandoned. Indeed, Babb (2000:92) points out that Tanaka pushed ahead with the spending on the shinkansen and 'other public works to buoy up the domestic economy and avoid recession.' Furthermore, it is likely that construction of the Jōetsu Shinkansen had reached a point whereby abandonment would have seemed a waste. I expect there was also a certain degree of national pride involved as the Dai-Shimizu Tunnel was, when opened, the longest land tunnel in the world. In many respects, the decision not to complete the construction work would probably have been harder than the decision to begin the construction work initially.

So why is the image of Jōetsu Shinkansen so bad then? Part of the answer is that the line became a victim of the times. As a result of the oil shocks, rising costs due to environmental issues and construction problems, the cost of the construction rose significantly. I suspect that many have either forgotten or are unaware of these issues. What is known and remembered is the high cost of the line (see Chapter 5). There is no doubt the fact that there appears to be over-capacity in the design of the line – with larger stations than perhaps necessary being built, for example. On top of this, although within the context of what had been planned for Japan when construction began, which I suspect many are now unfamiliar with, the line was not extraordinary, for many years, in comparison to the other lines (Tōkaidō, Sanyō and Tōhoku), the Jōetsu Shinkansen probably seemed like an extravagant branch line to many (e.g. Babb 2000:82). When the issue of Tanaka, in particular, is added to the equation, it is little wonder that the Jōetsu Shinkansen has become one of the most used examples of pork-barrel politics in Japan. Yet, the example does not appear to be as valid as is usually assumed, being based on supposition rather than facts and logic.

Political interference after JNR

With JNR's debts reaching catastrophic proportions, all new shinkansen line construction was halted following the opening of the Tōhoku Shinkansen between Ueno and Ōmiya, which had been the only section under construction since 1982. Following the break-up and privatization of JNR, it was expected that it would be possible to 'bid farewell to the political meddling that came with being a special public corporation' (Matsuda 2002:3). With the shift to being private companies, investment in new lines should only occur if there were economic grounds to do so. Even if government provided the money for the initial construction, the private company would be expected to operate the line. The private company's burden would obviously be greater if they had to pay any money towards the cost of the construction. So, with this in mind, and the background of the horrific state that JNR finances reached fresh in the memories of Japanese people, it was expected that there would be no possibility of the government burdening any of the new private companies with new loss-making lines. Although some have suggested that 'unlike in the JNR days when demand for shinkansen services was directly reflected in political pressure on JNR, at least the JRs are now free from such interference' (Aoki et al. 2000:199), there appears to be evidence to suggest that political interference has survived privatization.

By 1991, JR East had already begun construction on several new sections of shinkansen line. In fact most of these were constructed and initially funded by JRCC (see Chapter 5). The first new section of line to be completed was the extension of the Tōhoku Shinkansen from Ueno to Tōkyō.[11] Although incredibly expensive, the need and logic to have Tōhoku and Jōetsu shinkansen services being able to access the capital's main station was clear. On top of this, plans for additional lines, such as the Yamagata Shinkansen, and the difficulty in increasing capacity at Ueno, where the platforms are deep underground, further supported this. The other construction in the JR East region that was continuing in 1991 was the Hokuriku Shinkansen between Takasaki and Karuizawa, which had begun in 1989. Despite the section being only about 42 km long, there were various technical challenges that this line would have to address. Two of these have already been mentioned above. The third was that, as some services were not expected to stop at Takasaki, the world's longest turnout (or 'point'), whereby trains change from one track to another, would be needed. Railway technology in Japan had reached a point where all of these challenges could be successfully met. The powerful E2-series shinkansen was to be developed to cope with the incline and would have special converters to cope with the change in frequency. As for the turnout, rather than the standard 1:18 design, a 1:38 turnout was developed, which is 135 m long on a lead curve radius of 4,200 m. To put it more graphic terms, this allows the shinkansen to change tracks at the speed of 160 km/h (RTRI & EJRCF 2001:91)

Although the line between Takasaki and Karuizawa would have only been a relatively short spur, it would have addressed the problems caused on the conventional line by the incline, which would have improved overall journey times to other towns and cities in Nagano, as well as make access to the popular tourist destination of Karuizawa easier. However, in 1991, JR East was effectively ordered to extend the construction to Nagano so that the line would be ready in time for the 1998 Winter Olympics. Despite all the hope that the new era would see an end of political meddling, it seemed that politics and the shinkansen could not be separated for long. Although the opening of this part of the Hokuriku Shinkansen allowed JR East to cease services on one of its loss making lines (see Chapter 5), it meant that it was also effectively having to take on a greater financial burden sooner than had been anticipated (Kasai 2003:203).

Following the successful completion of the line to Nagano, construction work began on the section to Jōetsu. Subsequently approval was also given to extend the line from Jōetsu to Toyama. In fact, approval for construction on some of this section was given in 1993. On another section, between Isurugi and Kanazawa, approval was given in 1992 for construction to *super-tokkyū* standards (see Chapter Five). Yet, approval for the section between Toyama and Isurugi was not given until 2004 when the whole route to Kanazawa to shinkansen standards was approved. There is still no approval, or even an agreed route, for the section of the Hokuriku Shinkansen beyond Kanazawa to Ōsaka. Why is it that the Hokuriku Shinkansen has progressed in this way? The answer appears to be that it has benefited from significant pressure from politicians from the region. So significant has this influence been that some, for example Kawashima (interview 2003) even consider this line to be more political than the infamous Jōetsu Shinkansen.

Hokuriku, like much of the Japan Sea-side of Japan, is an area that has benefited greatly from public works projects. Given that the public works projects have supposedly been an attempt to redistribute wealth from the metropolitan areas and reduce the migration to the big cities, it should perhaps not be a surprise that this has been the case, since the Japan Sea-side of Japan has many of the country's poorer prefectures. Indeed Tanaka's plan of a national shinkansen network was to 'serve as pump-priming investments to promote regional development to Hokkaidō, Tōhoku, Hokuriku, Kyūshū and elsewhere, and contribute to the closing of the gap between the Pacific coast and the Japan Sea coast, northern Japan and Southern Kyūshū' (Tanaka 1973:121). Hokuriku has 21 of Japan's 52 active nuclear power stations (Asahi Shimbun 2004:153). It also has six airports, one of which is Noto Airport, which opened in 2003 despite fears that it could never make any money, leading to claims that it may be a *seiji hikōjō* ('political airport') (Kawashima interview 2003). Once the shinkansen reaches Toyama, and then Kanazawa, the future for many of these airports is likely to be bleak. Yet, in the meantime, it is clear that Hokuriku, or at least the construction companies there, has been managed to benefit from these injections of capital. Part of the reason for this is likely to be the work of the *zoku*, or cliques, within the LDP in particular, which work to have their particular interests funded. That there is also a regional demand which can work together with both the railway and airport *zoku* has undoubtedly helped Hokuriku. However, the different *zoku* compete for resources, which appears to have led to a lack of national coordination of resources as to what transportation infrastructure was really needed. However, while the *zoku* continue these battles largely out of the public eye, certain politicians can become more symbolic of new initiatives. If Tanaka is the example used for the Jōetsu Shinkansen, then the equivalent for the Hokuriku Shinkansen may be Mori Yoshirō (Kawashima interview 2003), who comes from Kanazawa, and became Prime Minister in strange circumstances in 2000. Although his time as Prime Minister was short-lived due to his personal unpopularity and a number of faux pas, his influence within the LDP and more widely has been much more significant.

So where does this leave JR East? Part of the problem is that beyond Jōetsu the Hokuriku Shinkansen will enter JR West territory. As a consequence there will probably be a division of ownership (as happens at Shin-Ōsaka between JR Tōkai and JR West) rather than this part of the line being operated by JR East. Although there are other examples of companies owning and operating shinkansen lines within the areas of other companies (e.g. JR West and the Sanyō Shinkansen in Kyūshū, JR Tōkai and the Tōkaidō Shinkansen west of Maibara and east of Atami), as the Hokuriku Shinkansen could be extended all the way to Ōsaka, so more than half the line would be in JR West territory, it is unlikely that this will be an option. Between Nagano and Jōetsu there will be no major stops. Although JR East will receive a share of the income from fares of passengers travelling on the line beyond Jōetsu (or those from that area that come into JR East territory), JR East is likely to be burdened by the cost of the expensive section through the tunnels and mountains between Nagano and Jōetsu. Although, it is highly probable that the line to Kanazawa will be profitable, the continued extension of the Hokuriku Shinkansen may not necessarily be a good thing for JR East.

So far this chapter has concentrated on pork-barrel politics and began by pointing out that this did not necessarily equate to corruption. However, there have been some specific cases where corruption has been suggested. For example, according to Ozawa Kazuaki of the Japanese Communist Party (JCP), contractors involved in the construction of the Kyūshū Shinkansen donated ¥1 billion to nine Kyūshū-based LDP politicians and their local chapters over a seven year period from 1995 to 2001 (*Japan Today*, 28 February 2003). While Ozawa has suggested that these 'donations' were improper, that there have been cases where LDP politicians, in particular, have been arrested for corruption, it could be that other parties may be using this as an issue where they suspect the public is likely to believe that the accused has done something improper regardless of how 'normal' such activities are or were in Japanese society.

Perhaps the oddest form of potential political interference that has occurred in relation to the shinkansen has been the suggested actions of the Governor of Shizuoka Prefecture. Frustrated by the fact that *Nozomi* services do not stop in the prefecture, and following the announcement of the increase in *Nozomi* services and reduction in *Hikari* services to be introduced with the opening of Shinagawa station, he proposed a 'passing tax' to be levied on the tickets of passengers that use the *Nozomi* service. The proposed ¥100 tax would raise an estimated ¥3.1 billion a year (*Shizuoka Shimbun Online*, 10 December 2002). I suspect the fact that some cities smaller than Shizuoka and Hamamatsu are *Nozomi* stops on the Sanyō Shinkansen, made possible by the lesser frequency of trains along that line, probably frustrated the Governor even more. As yet this tax has not been implemented. If, as I suspect, it was intended as a threat, JR Tōkai has not given any indication that it is very concerned about it and is planning any changes to the stopping patterns of *Nozomi* services (interviews with various JR Tōkai employees). That the Governor was suggesting such a tax, however, is a clear indication of not just how important the shinkansen is viewed but also how critically the need to be able to connect to the capital and other large cities quickly – even if it means only a few minutes difference – is.

It is important not to underestimate the importance of local politics in Japan. Governors and other elected officials have much greater powers and prestige than their counterparts in Britain, for example. With respect to the shinkansen, not only is their involvement expected, it is also now a requirement. Pre-privatization of JNR, local government's role in railway construction was limited to contributions to stations for which they had sought JNR to construct. However, they are expected to share the cost of construction of new shinkansen lines (Aoki *et al.* 2000:199). On top of this, some may also have to take on the burden, at least in part, of local lines that would otherwise have closed due to the opening of new shinkansen lines (see Chapter 5). This change has meant that the JR companies, despite the example of Shizuoka earlier, are having to forge better relationships with local communities. This has perhaps been another of the positive results of the privatization process itself. Without such local involvement not only would the Kyūshū Shinkansen, for example, not have been built, but the Yamagata and Akita Shinkansen lines, which it should be remembered are not *Seibi-Shinkansen*, would never have been constructed.

It has already been pointed out that there was an expectation that politicians were expected to meddle less in the operations of the JR companies after the privatization of JNR, but that this hope was apparently misplaced. In fact, given the way in which privatization progressed, this should not have been a surprise. Following the proposal made by the Second Administrative Reform Council to privatize JNR, the JNR Reform Advisory Panel began to look at how this could be achieved. At this point, Kasai (2003:162) argues that 'politics started to intervene and it became clear that the reality of the reform process would be determined by mostly political motivated decisions.' An example of this, according to Kasai (2003:164), is the way in which it was decided that JR Tōkai should cross subsidize JR West and JR East from the huge profits made on the Tōkaidō Shinkansen (see Chapter 5). However, even Kasai (2003:185) believes that after privatization there was a 'reduction in political interference in management'. For example, management no longer needs to spend time at Diet sessions, which used to number about 200 days a year, to answer 'what were very often meaningless questions', and no longer could politicians intervene 'in key management issues, for example, investment plans, setting of fares and wages and other operational aspects.' Furthermore, Kasai (2003:204) points out that 'quick decision-making became possible because we did not need to consult anymore with the politicians and bureaucrats about key decisions. We judge and decide ourselves, and once we have decided we put this decision into practice very quickly.'

Despite his own concerns about government intervention, Kasai (2003:203) does appear to believe that the government should be involved in the operations of the railways. At the very least there needs to be 'some flexibility'. For example, he suggests that the government should maintain a stake in the privatized companies. His reasoning for this is that while the first ten years of privatization have been 'very successful', 'ten years down the line things could change dramatically'. Of the three Honshū JR companies, JR Tōkai is the only one which is not fully privatized. Similarly it is JR Tōkai that has been most actively voicing the need for government intervention in the construction of a new shinkansen line – the Chūō Shinkansen. This fact cannot be seen in isolation from Kasai's apparently favourable comments about government part-ownership of the privatized companies. The Chūō Shinkansen cannot and will not be built without government support. If it is ever built, not only will it be arguably the most exciting, expensive and ambitious national project Japan has ever seen, but it will probably become the prime example of a '*seiji sen*' ('political line'). That no major LDP politician's constituency is located on the route and that part of the proposed route goes through Nagano prefecture, which has elected its Governor, Tanaka Yasuo, on the basis of his campaign to end public works projects in the prefecture, would suggest that the line is unlikely to be constructed. It could be suggested that the best hopes for the Chūō Shinkansen died in 1996 with the death of LDP 'king-maker' Kanemaru Shin. Kanemaru was a highly influential politician within the LDP who had also be tainted by corruption scandal. The proposed Chūō Shinkansen does pass through his constituency. Indeed, it is in his constituency that the Yamanashi test centre and track are located.

The shinkansen is always bound to be linked to politics. Today this is in part due to the need for financing – whether it be from local government or from the national government. However, the symbolic value of railways, steam in the Meiji Period and shinkansen today, mean that there is always likely to be a group somewhere that is calling for support for a line to be constructed in its region.

84 *Whose line is it anyway?*

Graphic demonstrations of this can be seen travelling around Japan. Outside certain stations that would become shinkansen stations, in the middle of the countryside where the route of a proposed shinkansen line intersects a conventional line, and outside city or prefectural government buildings, huge advertising hoardings or signs can be seen calling for the quick construction of the particular line. Examples I have seen are at Nara and Nakatsugawa stations (Chūō Shinkansen), near Kameoka station on the San'in Line (Hokuriku Shinkansen), near Kikonai station on the Kaikyō Line (Hokkaidō Shinkansen), outside Aomori station (Tōhoku Shinkansen), and outside Hokkaidō Prefectural Government building (Hokkaidō Shinkansen) (see Figure 4.2). On top of this, at the Nagatachō underground station, which is located closest to the Diet and MLIT, I have seen posters calling for the construction of the Chūō Shinkansen and Nagasaki Shinkansen. One has to question what influence these posters will have – especially since the politicians tend to travel by car. When I asked a MLIT official about this, he responded that 'it is probably so that we do not forget about that line', but added that there was little chance of that since they take phone calls

Figure 4.2 (a) A sign outside Kikonai station calling for the construction of the Hokkaidō Shinkansen; (b) a sign outside Hokkaidō Prefectural Government calling for the construction of the Hokkaidō Shinkansen; (c) a sign on a platform of Kasumigaseki underground station calling for the construction of the Nagasaki Shinkansen; (d) a sign outside Aomori station calling for the quick completion of the Tōhoku Shinkansen.

from local governments calling for funds and approval of the line to their area on 'sometimes a daily basis' (MLIT interview 2004).

Following the reorganization of the Ministries and Agencies in 2001, the Ministry of Transport merged with the Ministry of Construction, the Hokkaidō Development Agency and the National Land Agency to become MLIT. It will remain to be see whether this merely stands for Ministry of Land, Infrastructure and Transport or whether it will be the Ministry for LDP Interests and Territories as its predecessors often appeared to be.

Opposition to the shinkansen

The impression given thus far has largely been that the shinkansen is something that local communities desire and that politicians will do anything within their power to ensure that they bring the shinkansen to their constituents in the expectation that this will bring them victory at the polls for years to come. As Konno (1984:75) commented about the situation in the days of JNR, 'From the standpoint of local communities, the best possible present they can receive from the state is a role in the state-financed Shinkansen network.' The 1922 Railway Construction Law and the creation of JRCC had enabled politicians and communities alike to work towards having their connection to Tōkyō. Although, as will be discussed in Chapter 5, with some lines it is inter-regional transportation that is probably of greater economic significance, the greater symbolic significance is the ability to be able to travel to Tōkyō. Trains 'go up' to Tōkyō. Although Tōkyō's main station is merely called Tōkyō and not Tōkyō Marunouchi[12] as Konno suggests had a European naming style been used,

> The popularity of the Shinkansen, we might say, is rooted in the fact that to people in the provinces, the train seems almost to be Marunouchi itself on the move. The Shinkansen makes people feel that they have a direct link with Tōkyō, that the gap between Tōkyō and the provinces has been bridged, that modernization has been achieved. With this perception of the Shinkansen as a foundation, a complex superstructure of political manoeuvring to select routes for and to finance the superexpress came into being. Quite unconsciously this "perceived Shinkansen" found a place in people's hearts.
>
> (Konno 1984:75)

However, the shinkansen is not universally popular. Opposition to its construction in some areas has been significant, and in one case, the Narita Shinkansen, led to the abandonment of construction altogether. Indeed, there was even opposition in some areas to the building of the Tōkaidō Shinkansen, leading to JNR taking on extra staff to deal with the problem (Hosokawa 1997:193). About 50,000 households were evicted for its construction (Kakumoto Ryōhei (1964) *Tōkaidō Shinkansen*, Tokyo: Chūōkōransha, 123 quoted in Groth 1996:215) and a market building near Kyōto station was demolished despite opposition from the shop owners (Nakabō 2000:20–1). Opposition to lines, largely due to environmental pollution, has also been an issue (see Chapter 7). Clearly opposition has not always been successful. One strategy was that local politicians would visit the protestors and attempt to change their minds. Those who persisted in their opposition were branded traitors and the new lines were built, even though it meant moving houses out of the way and digging tunnels under schools (Konno 1984:73). More often than not, however, before being moved the protestors would be paid compensation. Japan, like Britain, has a compulsory purchase law. Unlike Britain, however, it is rarely enforced in Japan.

The most famous example of protestors not wishing to move in order for new construction to occur has been at 'Narita Airport'. Officially, the airport is the New Tōkyō International Airport, but it is located some 60 km from Tōkyō city centre. So significant was the opposition at Narita Airport that its leaders became well-known figures (Konno 1984:73). Despite opening in 1978, the disputes continued, so its 'capacity has been pathetically limited' (Horsley and Buckley 1990:99), which has prevented it becoming a major transport hub in Asia as well as being a problem when extra capacity was needed during the 2002 World Cup, for example. Although most of the protestors have agreed to move, even now planes that taxi to Terminal 2 go past a house that is effectively an island, surrounded by huge fences and a security tower, in amongst the taxi-ways of the airport (see Figure 7.4c). The living conditions, given the noise and lack of view, cannot be good, but land is important in Japan and it is not given up easily. I have even seen a new dual carriageway under construction that appeared at the end of someone's back garden and then start again at the front of the house. The road remained useless until the owners of this house agreed to move and it could be demolished. Such resistance would be pointless in Britain, for example, as the residents would be forcibly removed if necessary. In Japan construction begins as soon as approval has been given. The result, particularly when bridges and elevated sections are needed, are pockets of completed construction work across the landscape surrounded by housing, factories and the like that have yet to be demolished to make way for the new infrastructure. One employee involved in the construction of railways in Japan admitted that this policy is done to try to pressure those that have not sold their land in to doing so. With the Compulsory Purchase Law as a last resort, there is little concern that land will not become available leading to a need for re-routing and wastage of the original construction work. It is unlikely that there will be a repeat of the situation that exists with Narita Airport, whereby anti-construction campaigners attacked those who were meant to sit on the prefectural committee that meets to decide applications relating to the Compulsory Purchase Law. Consequently the committee members resigned and as a consequence no applications can be heard in Chiba prefecture (JRTT interview 2004).

In the case of the shinkansen, while the Compulsory Purchase Law was used around Minamata for the construction of the Kyūshū Shinkansen (JRTT interview 2004), it has largely been money that has been used. Unsurprisingly, the greater the opposition, the higher the eventual compensation paid out had tended to be

(Konno 1984:76), so that, at least during the JNR days,

> The 'Shinkansen construction game' is the most enjoyable and profitable game around. The announcement of a new route immediately sparks an opposition movement that has far greater support than any other kind of antiestablishment campaign. The game can be viewed as a kind of ceremony in which the residents use their ingenuity to make money from the Shinkansen... With the residents and the authorities working at cross-purposes, construction costs soar to unnecessarily high levels.

Even during the building of the Tōkaidō Shinkansen, there were stories of landowners quickly building a house so that they would gain greater compensation, or demanding ¥200,000 for an orange tree that would have to be cut down (Hosokawa 1997:194). The 'game' has its risks, however. In the case of the Narita Shinkansen, the opposition was so great, even taking the form of sabotage (Horsley and Buckley 1990:99), that in the end the construction was abandoned. Part of the line was constructed – including the airport's stations, in which the space by the current conventional lines are testament to the fact that they were initially designed for trains of a larger gauge. For many years as a result of the abandonment of construction work there was no direct rail link between Tōkyō and Narita Airport, which may be another indication of the problems of *zoku* and lack of national co-ordination. Eventually a link was made, and in the case of the JR line,[13] it uses part of the construction, including the first part of a large viaduct in Narita city, that would have been used for the Narita Shinkansen. Other tracks of space by the conventional line are reminders of where space was made or land bought for the Narita Shinkansen. Whether the opposition was 'real' or whether the residents were playing the 'Shinkansen construction game' and had their bluff called is debatable. Questioning residents is likely to provide rational answers that point to the former, but it is unlikely that many would openly admit to the latter even if it were true. It is not always easy to find all the truth in these matters, just as it would also be hard to find those prepared to give evidence about pressure from the *Yakuza*,[14] who have strong links with the construction industry, in encouraging them to sell their land.

However, even Groth's significant study (1986 summarized in 1996) on the opposition to the construction of the shinkansen appears to support Konno's idea of a 'Shinkansen construction game' as he points out that

> Because of the immense importance of the Shinkansen to Japan, the citizens' movements did not oppose the Shinkansen per se. Rather they sought comprehensive solutions to the problems caused by the project. During 1980–1982 only the radical fringe of the movements in Tōkyō and Saitama could be described as 'anti-Shinkansen'.

(Groth 1996:216)

Groth studied 'Shinkansen Citizen Movements' in several locations; Tōkyō's Kita Ward, Toda, Urawa and Yono (all in Saitama Prefecture) – in relation to the building of the Tōhoku Shinkansen from Ōmiya to Ueno; and Nagoya – in relation to environmental protests (see Chapter 7). Regarding the Tōhoku Shinkansen, it is interesting to reflect upon the role of the media in the residents' campaigns. Konno (1984:76) has suggested that 'The mass media make heroes of the resisters, and Diet members come to pay tribute.' However Groth's work would suggest that this conclusion is too simplistic, as 'The mass media can distort messages and trivialize goals, perhaps at best providing superficial coverage of a movement's most flamboyant leaders' (Groth 1996:213). The problem was not helped by the media sometimes becoming misinformed by relying upon familiar sources of information regardless of whether the person was still representative of the Movement to which they were affiliated when the relationship began (Groth 1996:222, 239). This led to action being taken against the media organizations by the Shinkansen Citizen Movements. For example, 'The citizens' movements in Saitama used *oshikake* (perhaps best translated as "mob action at the doorsteps") against media organizations that had portrayed the movement unfavourably', and after one such action, the newspaper targeted apparently gave the Movement 'no major problems' (Groth 1996:224). In Kita Ward, the Movement threatened not to pay their NHK fees,[15] and as a result, the leader of the Movement was interviewed by NHK and their coverage became 'somewhat more balanced' (Groth 1996:224). In terms of direct action, the Kita Ward Movement also disrupted the ceremony marking the start of construction in the area, with around 100 members preventing the Mayor entering the site (Groth 1996:234). These various forms of action clearly show the determination of the Movements, suggesting it is more than a mere 'game', and they may have even been addressing some real imbalances or inaccuracies in the reporting. It also reveals that despite the image of Japanese being harmonious, desperate times can lead to significant protests, which can even become violent. However, this type of action appears to be something that is seen less than in pre-affluent- and pre-politically apathetic Japan of the 1980s onwards.

The other strategy developed by the Movements to counter the mass media which did not portray them in their preferred image of being 'powerful, united, committed to the public good, and having the support of many organizations' was to create 'mini-media' (Groth 1996:223).

> These mini-media were several kinds: newspapers, and pamphlets, and other publications; mobile street broadcasting from speaker trucks, including day-long 'caravans' through Kita Ward; *bira maki*, the distribution of printed statements at railroad and subway stations during rush hours, accompanied by speeches over loudspeaker systems; documentary films and slide shows about Shinkansen pollution and the activities of the citizens' movements; and even songs for 'Karaoke days'... all these mini-media, with the exception of 'Karaoke days', are standard tactics used by a wide variety of groups in Japan, including labour unions, student movements, and pacifist groups.
>
> (Groth 1996:228)

Although the use of mini-media is a 'basic media strategy of a protest movement' in Japan, it also means that the movement will require significant resources since their production requires 'money, personnel with media skills, significant time and effort from many participants, and access to expensive equipment' to print them (Groth 1996:235). As with some Movements' behaviour towards the mass-media, the mini-media also, in some cases, had its less pleasant features, which pointed to problems with the Movements as a whole. In Yono, the mini-media focused upon 'betrayers' – those leaders that allegedly took bribes from those supporting construction of the shinkansen or that negotiated their own compensation with JNR (Groth 1996:229). Although, the situation was different in Nagoya where similar actions by one leader led to his resignation being merely reported as being due to 'poor health and old age' (Groth 1996:229), such actions clearly reveal that some were playing the 'game' and that, as is so often the case, once a movement or group is created, internal politics of that organization can seemingly become of greater significance than whether the aims of the organization are actually met. This may be particularly the case in Japan, where group harmony is considered to be so important.

The Shinkansen Citizen Movements' situation was further complicated by the actions of those that may have also had different agendas to the members of the Movements. Lawyers, for example, were often used, but these lawyers often tried to control what was included in the mini-media and who dealt with the mass media, what was said and what other activities were carried out (Groth 1996:230, 233). For example, a lawyer for the Kita Ward movement recommended that the group's three-storey protest tower in Akabane be dismantled as it 'implicitly associated the citizens' movement with the violent protest against the Narita airport' which used a similar tower (Groth 1996:234). Such intervention by the lawyers further strained the workings of the Movements. For example, some members not only wanted to keep the tower as it attracted attention, but even wanted funds to renovate the tower as it had looked 'shabby' when it was featured on the television (Groth 1996:230, 234). In fact the members position may not have been so misplaced for 'JNR also treated the symbolism of the protest tower seriously' and as a consequence, a suit was filed demanding the tower's dismantlement on the grounds that it was on JNR land (Groth 1996:241). In Kita Ward, the JCP 'helped' with the distribution of *bira maki*, but did so while wearing party armbands, which meant that the Movement became closely associated with the JCP (Groth 1996:231). Given the JCP's low level of support, it is likely that this perceived closeness could have harmed the Movement. This was not the only problem the Movement had in the area, however. For in Akabane those giving out leaflets were having to compete with others that regularly distribute leaflets and the like at such stations – namely 'gangsters, pimps, and missionaries' (Groth 1996:235).

Although the Movements often tried to work with other neighbourhood associations and local government, this was not ultimately successful. For example, in both Kita Ward and Saitama Prefecture, although neighbourhood associations and local government were initially supportive, they gradually came to approve the construction of the Tōhoku Shinkansen between 1978 and 1980. By 1982, the

Shinkansen Citizen Movements 'had changed their own official positions, from absolute opposition to conditional approval', i.e. approval in return for 'generous compensation *and* alternate sites of land for residents who faced eviction' as well as a reduction in 'Shinkansen noise and vibration to tolerable levels' (Groth 1996:218).

While Groth (1996:215, 234) found that the groups covered a wide spectrum of Japanese society in terms of age, gender, educational background and voting patterns – although it tended to be the leftist members that were most vocal and active in sit-ins and other protests – the issue that united them was land. For in Japan land is scarce, especially in urban areas, and it is important. Although there may have been a 'game' to be played, for some it was undoubtedly more than a 'game' and the consequences of 'losing' have been severe.

> Residents evicted because of the Shinkansen have thus faced extreme difficulty in finding alternative sites of land. Some have been forced out of their homes, off their little pieces of land, and into apartments. These people have protested that compensation for their land and property was grossly inadequate. Owners of large tracts have been no less vocal, complaining about the high taxes they must pay on the money they receive.
>
> (Groth 1996:216)

These taxes in one case were estimated to have been 70–80% of the compensation, and even with this money there was little, if any, land to be bought in the same city (Groth 1996:237). A sad irony is that if one looks out of the shinkansen when passing Toda today one case see many spare plots of land near to the shinkansen line which became disused following the construction of the line and the repositioning of the Saikyō Line. Although a less than ideal location due to being in the shadow of the elevated section of the railway lines, it is a shame that there appears to have been such a wastage when land has been at a premium in the area.

Opposition to the construction of new shinkansen lines appears to be minimal now. This is primarily as construction is mostly in rural areas, and so avoiding areas which affect many people, or in tunnels. Where larger numbers are affected – there are some pockets of opposition in areas between Hakata and Kumamoto – the probability of a movement being created may have been reduced due to the relatively small populations and the lack of accessibility to the resources needed to mount a mini-media campaign. In the meantime, the shinkansen network continues to spread across Japan due to the need for such lines and due to the influence of national and local politicians, although those most associated with particular lines or stations are not always as significant as it may appear.

5 The bottom line

> The stationary is dull; the declining melancholy.
> (Adam Smith, *The Wealth of Nations*)

Having discussed the support and opposition for the construction of the shinkansen, it is now possible to consider whether the construction of these lines has been worthwhile financially. Yet to do so is not easy. There are many factors that need to be taken into consideration and one needs to keep in mind what the objectives of the building of the shinkansen were. One needs to remember that their purpose was not merely to be a profitable railway line, but to help regional development and national development – not just in economic terms, but also, as discussed in Chapter 3, in the development of pride through its use as a symbol of what can be achieved.

The cost of the shinkansen

First, let us consider the cost of both the shinkansen lines and the trains that operate upon them. To do this, it is necessary to take account of the fact the shinkansen's history is four decades long. On top of this, the nature of the lines varies a great deal, so direct comparison is not easy. Figure 5.1 compares how much of each line is made up embankments and cuttings, elevated sections, bridges and tunnels. Although in a country with a topography such as Japan's, the use of tunnels is to be expected, what Figure 5.1 hides is the increasing importance and use of tunnels. For the extension from Morioka to Hachinohe, 73% of the route is inside tunnels (Kōsoku Tetsudō Kenkyūkai 2003:85). Indeed one of these tunnels, Iwate Ichinohe Tunnel, is now the longest shinkansen tunnel, and the longest land tunnel in the world at 25.8 km in length.[1] As the Tōhoku Shinkansen continues northwards, so the need for tunnels will continue, as over 60% will be in tunnels and will include the 26.5 km Hakkōda Tunnel (Kōsoku Tetsudō Kenkyūkai 2003:217). Similarly on the Hokuriku Shinkansen from Nagano to Toyama, tunnels will make up 55% of the whole route and will include the 22.2 km Īyama Tunnel (Kōsoku Tetsudō Kenkyūkai 2003:219, 221). If the Chūō Shinkansen is ever approved, it is certain that most of this route will be in tunnels. This gives the

92 The bottom line

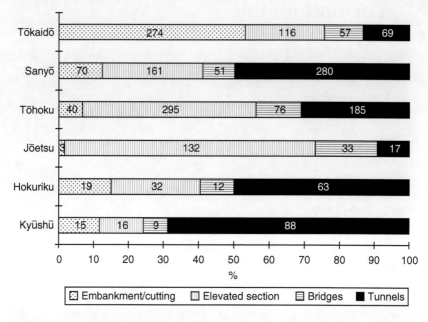

Figure 5.1 Construction of shinkansen lines.

Notes
Figures on the graphs show the distance in kilometres (rounded to the nearest km) for each type of construction work. Data for Tōhoku Shinkansen is for Tōkyō–Hachinohe; for the Jōetsu Shinkansen is for Ōmiya–Niigata; for the Hokuriku Shinkansen is for Takasaki–Nagano; for the Kyūshū Shinkansen is for Shin-Yatsushiro–Kagoshima-Chūō. Based upon data in Kōsoku Tetsudō Kenkyūkai 2003:85 and Nihon Tetsudō Kensetsu Kōdan 2002a.

impression that rather than building a high-speed railway network, Japan is building the world's fastest underground system. However, although most of the Kyūshū Shinkansen at present is in tunnels, when completed the proportion in tunnels will drop to 50% overall, as only 30% of the section between Hakata and Shin-Yatsushiro will be in tunnels (Nihon Tetsudō Kensetsu Kōdan 2002a).

The use of tunnels may be more than just a matter of dealing with the challenges of the country's topography. For example, the Tōhoku Shinkansen extensions use of record-breaking length tunnels may be to appeal to a Japanese desire for firsts. There are other more practical reasons too. Although tunnel construction is expensive and time-consuming, it is easier now than it was only a few decades ago when the Jōetsu Shinkansen was being built, for example, due to the improved construction methods, known as the New Austrian Tunnelling Method (NATM), having been adopted. Furthermore, there are less problems in obtaining the land than if it were open residential or farming land, for example. Also, once completed, although speeds on all new lines are initially restricted to 260 km/h, when this restriction is lifted, the top speed can be raised much higher than on open-air sections. The tunnels also help to overcome other environmental problems (see Chapter 7). One issue that may need to be addressed by some companies in due course is whether the passenger's boredom, as there is no view, will become a deterrent to shinkansen usage (see Chapter 7), and research is even being conducted into improving customer comfort in tunnels (RTRI interview 2001).

Keeping the difference in nature of the lines in mind, we can now look at the costs of the lines (see Table 5.1). However, it is necessary to remember that many of lines have seen extra stations being constructed following the initial opening of the line (see Appendix 2). On top of this, the length of lines does not include extra sections that were constructed to maintenance facilities, etc. that are included in the initial cost of the line. It is clear that costs have varied a great deal. Although Sogō resigned in part due to the overspend on the Tōkaidō Shinkansen, in retrospect the line appears relatively cheap. However, the figure may be distorted by the large amount of land that had already been acquired during the wartime. Furthermore, as there had been pressure to keep costs down, steps were taken to find savings wherever possible. For example, roofing was limited to cover only part of the platform initially and innovations in structural designs for the elevated sections further reduced costs (Kasai 2003:37). However, over time the culture within JNR, fuelled by JNR's links with the construction industry as retiring officials moved into those companies through *amakudari*, became such that this determination to keep costs down was eroded. This undoubtedly had a negative impact upon the costs of the Tōhoku and Jōetsu Shinkansen, as well as other parts of JNR's operations (Kasai 2003:38; Matsuda 2002:84, 88–9, 108–9). Matsuda (2002:75) claims this was still a problem in the construction of the Hokuriku Shinkansen, which went over budget, compared to the Akita Shinkansen, which did not use JRTT and finished under-budget and that JR East, for example, has managed to reduce its material and supplies costs by between 20% and 25% (Matsuda 2002:90; Yamanouchi 2000:227). This is not surprising given that lights, which cost ¥1,200 in the supermarket, were costing JNR ¥8,000 (Noguchi 1990:36).

It is the Tōhoku Shinkansen which has had the greatest variation in cost per kilometre. While the initial section between Ōmiya and Morioka was expensive, it is comparable with the costs of construction of main shinkansen lines post-JNR, particularly when one takes into account the number of stations and challenges posed by some of the route. However, the Ōmiya–Ueno and Ueno–Tōkyō sections were extremely costly due to the challenging nature of the construction work, the problems in purchasing the necessary land, and the cost of land in that area.

Considering the Jōetsu Shinkansen, one can see that the cost per kilometre was about 50% more than that for the section of the Tōhoku Shinkansen which was being built at the same time. This may seem surprising as less than 10% of the route is in tunnels, which one would expect to push up costs. Yet in reality it is bridges (approx. ¥6 billion/km) that are most expensive, while tunnels and elevated sections each cost about ¥2.5 billion/km and cuttings/embankments cost about ¥2 billion/km (JRTT interview October 2004). However what these figures

Table 5.1 Cost of shinkansen line construction

Line	Year completed	Distance (km)	Cost ¥ billion		Price/km (current prices)
			At completion	Current prices	
Tōkaidō	1964	515.4	380	1,593	3.09
Sanyō	1975	580.7	991	1,735	2.99
Tōhoku (Ōmiya–Morioka)	1982	465.2	2,654	3,143	6.76
Tōhoku (Ōmiya–Ueno)	1985	28.0	650	725	25.89
Tōhoku (Ueno–Tokyo)	1991	3.6	130	132	36.71
Tōhoku (Morioka–Hachinohe)	2002	96.6	474	474	4.91
Jōetsu (Ōmiya–Niigata)	1982	206.6	1,700	2,013	9.74
Hokuriku (Takasaki–Nagano)	1997	117.4	842	842	7.17
Kyūshū (Shin-Yatsushiro–Kagoshima-Chūō)	2004	126.1	640	640	5.07
Yamagata (Fukushima–Yamagata)	1992	87.1	38	38	0.44
Yamagata (Yamagata–Shinjō)	1999	61.5	34	34	0.55
Akita (Morioka–Akita)	1997	127.3	60	60	0.47

Sources: Umehara 2002:72, 108, 109, 314, 315, 316, 330; MLIT interview September 2004; JR Tōkai interview September 2004.

do not take account of, which was significant in the case of the Jōetsu Shinkansen, is the cost of land. Furthermore even though there was not much tunnel work in terms of length of construction, compared to the Tōhoku Shinkansen, the construction work itself was extremely challenging, and thus more costly. Both the Tōhoku and Jōetsu Shinkansen construction costs were affected by changes in environmental laws (see Chapter 7), which impacted sections outside tunnels more than those in tunnels and Yamanouchi (2000:202) suggests that the abandonment of the Narita Shinkansen further pushed up the cost of the Tōhoku and Jōetsu Shinkansen as there was a greater need to ensure these lines adhered to environmental restrictions. The Jōetsu Shinkansen's construction, let alone its on-going running costs, were also impacted by the snow counter-measures (see Chapter 7). However, there is also no doubt that some of the stations are larger than they need have been, so seem 'unnecessarily extravagant' (Kasai 2003:38–9). One of the reasons for the drop in average cost per kilometre on the newer lines is the reduction in size of stations. Although some platforms are longer than that currently used, and removing the barriers at the end of the section currently used would allow for longer trains, the lack of passing tracks at some stations may cause problems in the future if services are significantly increased in number.

Given the size of some of the numbers in Table 5.1, it is perhaps difficult to gain a true appreciation of their meaning. The two most recent shinkansen sections to be completed – the Tōhoku Shinkansen between Morioka and Hachinohe and the Kyūshū Shinkansen – have both been about ¥5 billion/km. In comparison, the limited upgrade to part of the West Coast Mainline in England, after it looked as though it would reach £13 billion, is now estimated to be £7.6 billion (Strategic Rail Authority 2004:6, 7). This is equivalent to ¥1.22 billion/km. However, as this work involved the construction of no new stations and will bring relatively minor improvements in travel times, it is perhaps more analogous to mini-shinkansen construction. In terms of full shinkansen construction, the only British example is the new line to link London with the Channel Tunnel. The length of this construction is 109 km, with 25% in tunnels, and 152 new bridges being required. The estimated total cost is £5.2 billion (Department for Transport 2004), which is equivalent to about ¥9.54 billion/km. Naturally costs vary greatly between Japan and Britain due to various factors, so this comparison may not be fair. Table 5.2 compares the cost of construction of the shinkansen with various other recent projects in Japan. This table would appear to confirm that the cost of building the a shinkansen is not inherently expensive when compared to other major 'public works', and that the high cost that became associated with the shinkansen is largely due to the construction of the Tōhoku and Jōetsu Shinkansen lines which were discussed earlier, but no longer remain an appropriate basis by which to do an analysis.

Having considered the costs of construction of the lines, it is worth mentioning the cost of the trains themselves. The cost per train set naturally varies depending on the length and design. For example, some carriages have motors, whereas other do not. The interior design also varies from carriage to carriage. However, the average price per carriage for most series is around ¥250 million. Only the 500-series (around ¥290 million) and the double-decker designs (e.g. E1-series average cost per carriage was about ¥360 million) being significantly different to this (JR Tōkai, JR West, JR East interviews 2003). The average cost per carriage is about the same as that of carriages for British Inter-City trains (Ford 2003:17). Naturally, the final price the companies (or JNR) pays does not fully take into account the costs involved in R&D, for example. This aspect has been further complicated since the reform of JNR. For rather than almost solely relying upon RTRI and train manufacturers, many of the JR companies are now doing their own R&D. Indeed, JR Tōkai has even opened a new state-of-the-art research facility in Komaki to help in the development of its new shinkansen, with the N700-series being the first to be developed there. The results of the work conducted there are still likely to be further tested at RTRI, especially the large wind-tunnel in Maibara (RTRI interview August 2004). The Komaki facility includes a simulator which gives the passenger the feeling as though they are on-board a shinkansen. In October 2004 I was able to experience what the ride on the N700-series is expected to be like. While undoubtedly more comfortable, as the difference was only really obvious when standing, so that it will undoubtedly be a major benefit for those pushing the trolleys along the shinkansen, one cannot help wondering whether this huge investment can be justified for relatively minor improvements in comfort and only a few minutes in journey time between Tōkyō and Shin-Ōsaka.

Table 5.2 Comparison of major construction works

Construction works	Construction completed	Total cost (in ¥ trillion)	Length (in km)	Cost/km (in ¥ billion/km)
Kansai International Airport (Second Stage)	1994	1.54		
Chūbu International Airport	2005	0.81		
Noto Airport	2003	0.27		
Akashi Kaikyō Bridge	1998	0.50	3.91	127.8
Tōkyō Bay Aqualine (road tunnel and bridge)	1997	1.40	15.1	92.7
Seto Ōhashi	1988	1.13	13.1	86.5
Underground (average)				30.0
Second Meishin Expressway (Kinki Highway Nagoya–Kōbe)	(not decided)	3.80	171	22.2
Second Tōmei Expressway (Yokohama–Nagoya)	(not decided)	5.50	285	19.3
Railway Link to Chūbu International Airport	2005	0.07	4.2	16.9
Akihabara–Tsukuba Railway Line	2005	0.94	58.3	16.1
Seikan Tunnel	1988	0.69	53.9	12.8
Shinkansen (Kyūshū)	*2004*	*0.64*	*126.1*	*5.1*
New Narita Airport Railway Line	(2010)	0.13	51.4	2.5
Mini-shinkansen (Morioka–Akita)	*1997*	*0.06*	*127.3*	*0.5*

Sources: MLIT www.mlit.go.jp/tetudo/shinkansen/shinkansen6_QandA.html; MLIT interview September 2004; http://en.wikipedia.org/wiki/Chubu_International_Airport; http://homepage2.nifty.com/usui-postoffice/sub4-4.htm; http://www.ocab.mlit.go.jp/data/plan/plan.htm; http://www.jhnet.go.jp/aqua-line/About/About.html; http://www.pref.ibaraki.jp/closeup/cl0410_01/; http://www.nui.or.jp/project/toumei.htm; http://www.pref.hokkaido.jp/skikaku/sk-kkkkk/sitetop1/koutuu/trein/3-8seikan.htm; http://www.mifuru.to/frdb/

Yet with access to Haneda Airport by monorail being improved by a few minutes and an ending of the point-to-point flying whereby planes fly to certain beacons could cut several minutes of most flight times (Yasubuchi interview April 2004), the battle between the shinkansen and the airlines is being fought on marginal differences and so such research and improvements to the shinkansen could be key to winning (see Figure 5.2). Although the scope for improvements along the Tōkaidō Shinkansen would appear to be limited due to the nature of the line and areas that it goes through, the research may be a stepping-stone to further improvements in the future, as well as for spin-off products.

Options in expanding the network

Returning to Table 5.1, it is clear that building mini-shinkansen is cheaper than building 'full' shinkansen. Mention has also been given elsewhere to 'super-*tokkyū*'.

Figure 5.2 A 700-series shinkansen passes alongside the monorail that connects to Haneda Airport and the mass of conventional lines in Tōkyō.

Table 5.3 summarizes what the difference is between the four main types of railway and provides a comparison of different types of road, which may be more familiar and easy-to-understand for many readers.

To further illustrate the different ways that a route can be improved, let us use a hypothetical example (see Figure 5.4). There are three major cities (A, B and the capital, C), and three smaller towns (D, E and F). Currently there is a conventional line between A and B (via D, E and F), with a shinkansen between B and C. This means there is a need to change trains at B to get to C on the shinkansen service. The current journey time from A to C is about four hours including the change of trains.

If one were to upgrade the line from A to B to mini-shinkansen, it would get rid of need to change at B and may increase the speed trains can travel on some sections of the track. There would be major disruption during construction as services would be suspended totally (if the line is single track) or partly (if the line is partly or wholly dual track). Tunnel T1 would not be altered as mini-shinkansen are about the same size as conventional trains. Towns D, E and F could expect to see benefits as they would become more accessible to the capital. The total journey time between A and C would be improved by about an hour. The cost of this option would be about ¥125 billion for the construction of the line with a further ¥1.4 – ¥3.6 billion (depending on length, etc.) needed for each new train.

Table 5.3 Comparison of railway line types

Feature	Conventional	Mini-shinkansen	Super-Tokkyū	Shinkansen
Gauge (in mm)	1,067	1,435	1,067	1,435
Typical train width (in mm)	2,946	2,946	2,946	3,380
Level crossings?	Yes	Yes (see Figure 5.3)	No	No
Top speed (km/h)	95	130	160	260–300
Road equivalent	Single carriageway	Dual carriageway but still has roundabouts, traffic lights, etc.	Full dual carriageway without roundabouts, traffic lights, etc.	Motorway

Figure 5.3 (a) An E3-series shinkansen passing a level crossing and station served by wide-gauge-conventional trains along the Yamagata Shinkansen; (b) a 400-series shinkansen at Yamagata station.

Consequently, the mini-shinkansen option is a relatively affordable option for rural prefectures to develop a seamless link to the capital.

With the Super-*Tokkyū* option, a tunnel (T2) would be built together with a new section of track between the junction (J) near town E and city B, meaning that services no longer stop at town F, which is likely to either see its section of line close or more likely operated by a 'third sector' company (see the following

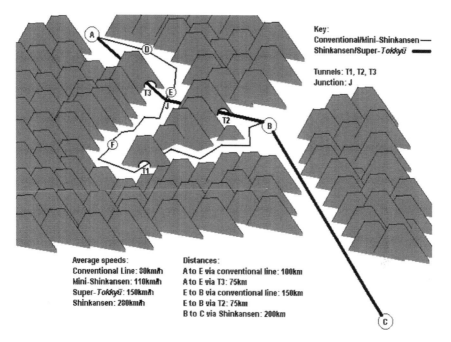

Figure 5.4 Upgrading a line.

sections). Meanwhile, journey times between A (as well as towns D and E) and C would be improved as trains will be using a much straighter stretch of track, with a higher top speed. However, there will still be a need to change trains at city B as the service between A and B would still be using conventional trains. Total journey time between A and C would be improved by 82 minutes. The cost of this option would be about ¥375 billion for the construction of the line. No further money would necessarily be needed for new trains, depending on the limitations of the current rolling stock, since the new track would still be conventional. Following the opening of tunnel T2, tunnel T3 and corresponding section of track could be built once the funds have been secured. This will cost about another ¥375 billion in construction work. As a consequence of this work conventional train services will now miss town D also. A new station near junction J – let us call it Shin-E – would also be needed as the station in E would no longer be served by the trains going from A to B. Finally, when extra funds have been secured, the whole line could be converted to full shinkansen specifications. Like with the conversion from conventional line to mini-shinkansen, there would be some disruption while the conversion work is done, the cost of which is likely to be less than ¥75 billion. New shinkansen trains would also be needed at this stage.

Another option is to extend the shinkansen from B towards A. Again a new station would be required to serve town E. If there is a need to do the work in stages, Shin-E could become the terminal of the shinkansen until the rest of the line is completed. In this case, the main benefit would be for town E although town D and city A would also see a reduction in the overall journey time to city C, despite a change of train now being required at Shin-E. The amount of disruption to complete this construction work would be minimal. Total journey time between A and C would be improved by over an hour and a half. The cost of this option is likely to be about ¥375 billion for the construction of the line, with further funds needed for each new shinkansen. Once extra funds have been raised the final section between Shin-E and A could be completed. Once the construction of the shinkansen is completed some or all of the conventional line would probably be handed over to at least one-third sector company.

If funds were available, the whole shinkansen line could be built at once. There would be little or no disruption to services during construction. When completed some or all of the conventional line would probably be handed over to at least one third sector company. Although towns D and F may be able to access the capital quicker than before, as they are not on the shinkansen route they may suffer from a reduction in business and visitors. Another disadvantage with this option is that revenue only starts once the whole line has been completed, whereas both of the previous options can benefit from additional income at the completion of each stage of construction. However, when all the work is completed, the total journey time between A and C would improve by about 165 minutes to just 1 hour 15 minutes as shinkansen would be used for the whole journey, there would be no need to change trains, and the total distance from A to C would be shorter. The cost of this option is likely to be about ¥750 billion for the construction of the line with further funds needed for each new shinkansen.

There is a final, theoretical option, but one that JR Kyūshū, JR West and JR Shikoku have been exploring. Rather than making significant changes to the line, this concentrates on the train. At the point where a change from conventional line to shinkansen is necessary, a small connecting track is built, the gauge at one end is conventional, at the other it is 'standard'. As the train moves along this section of track, its wheels change gauge so that it can continue to operate on the next section of track. RTRI began development work on the 'Free Gauge Train' and work has continued in the USA and in Japan. Using the earlier scenario, the improvement in journey time between A and C would not be significant as no improvements are being made to the track itself, so the main difference will be the lack of need for a change in trains and the lack of construction costs. In other words, compared to the mini-shinkansen option, it is likely to be about ¥100 billion cheaper, but not as fast. For companies such as JR Shikoku, where competition with roads is particularly severe due to the better development of expressways on the island (see Kishi 2004), and a lesser degree JR Kyūshū, which serve prefectures that are rural and have limited funds for 'public' transport, the Free Gauge Train may be the only way of expanding their 'shinkansen' network within their area beyond the lines included as *Seibi-Shinkansen*.[2]

As can be seen from Table 5.4, each option has its different advantages and disadvantages. Of the main options (i.e. ignoring the experimental Free Gauge

Table 5.4 Upgrading a line

Option	Construction costs (in ¥ billion)	Train costs (in ¥ billion)	Disruption?	Total journey time (A to C)	Time saving (A to C) (in minutes)
0 – Conventional railway	—	—	—	4:00	—
1 – Mini-shinkansen	125	1.4 to 3.6	Yes	2:59	61
2 – Super-*Tokkyū* stage 1	375	None?	None	2:38	82
2 – Super-*Tokkyū* stage 2	375	None?	None	1:53	45
2 – Super-*Tokkyū* stage 3	75	1.4 to 3.6	Yes	1:15	38
2 – Super-*Tokkyū* total	825	1.4 to 3.6	Some	1:15	165
3 – Shinkansen extension stage 1	375	1.4 to 3.6	None	2:24	96
3 – Shinkansen extension stage 2	375	None?	None	1:15	69
3 – *Shinkansen extension total*	750	1.4 to 3.6	*None*	1:15	165
4 – Full shinkansen	750	1.4 to 3.6	*None*	1:15	165
5 – Free gauge train	25	1.4 to 3.6	None	3:50	10

Notes
Times assume 10 minutes needed to change trains where applicable. Train costs vary depending on length of train sets and number of sets required.

Train option), the mini-shinkansen option is the cheapest. Yet in the long run it offers the smallest overall advantage to city A, although other towns along the route, with the exception of E, arguably benefit most from this option. Considering the Yamagata and Akita Shinkansen lines, neither super-*tokkyū* nor full shinkansen was likely, so mini-shinkansen was the only option. That it has been effective can be measured in many ways. JR East has had to put on extra services and trains have been made longer. Although airlines had been increasing their share of the market up to about 1997, once the line was extended beyond Yamagata, it became harder for the airlines to compete and ANA scrapped its air services on the Tōkyō Haneda–Yamagata route in 2002 (ANA interview April 2004). Yamagata has been so pleased with the results of the mini-shinkansen, with passenger numbers to Tōkyō and surrounding prefectures having increased by about 20% since the opening of the line, that consideration is being given for a mini-shinkansen service to serve Sakata and other cities on the Japan Sea coast of Yamagata, either by extending the current line or by building a mini-shinkansen from Niigata (interview at Yamagata Prefectural government 2003). The situation has not been so positive in relation to Akita for the shinkansen. For while shinkansen services have seen increase in services and train length since the first year, its share of the market has dropped from 59.3% in Fiscal 1998 to 57.0% in Fiscal 2002 (JAL interview April 2004). It is important to note that as the mini-shinkansen are not *Seibi-shinkansen*, there is little or no money from the national government, with the Prefecture, JR East and local companies meeting most of the financial burden. In the case of the Yamagata Shinkansen for example, only about 20% of the construction costs came from the national budget, and there was no money provided in the case of the Akita Shinkansen (MLIT interview October 2004; Yamagata Shinkansen Shinjō Enshin Suishin Kaigi 2003:3).

The super-*tokkyū* option offers modest gains in the short term, while the costs are high. Indeed, if done step-by-step as suggested above, the total cost of this option is eventually greater than building the full shinkansen. However, in practice, the super-*tokkyū* option has only been planned where it is not practical to build full shinkansen initially due to that section being isolated from the rest of the network (i.e. if the northern section via tunnel T3 was built before the southern section via tunnel T2 in the previous example). The current section of the Kyūshū Shinkansen was initially planned to be super-*tokkyū* until approval for it to be constructed to shinkansen standards was given. The extension option is the most preferred and has been the most used, for although the expense is high, the rewards, and the potential to further extend are best. Indeed, although the mini-shinkansen and super-*tokkyū* options both have some advantages, it has to be noted that some local residents were not happy with their introduction. The analogy used is that while they ordered eel (shinkansen), a popular yet moderately expensive dish, what they got was either loach (mini-shinkansen), which is similar in shape to eel but smaller, or catfish (super-*tokkyū*), which has a similar feel as eel but a different shape (Yamanouchi 2000:247).

Prior to 1987 all funds for new shinkansen lines came directly from the government and went to JNR, with the construction work being done by JRCC. Originally central government was to bear 35% and local government 15% (Aoki *et al.* 2000:199). However, as it was recognized that it was extremely unlikely that any JR company would undertake further investment in new shinkansen lines by themselves (Yasubuchi interview 2004), a new formula was developed. Following revision to the Nationwide Shinkansen Railway Development Law in 1997, local government funds 33% with the rest from national government. In practice, much of the national government's funding comes from the income from the lease of the existing shinkansen lines at the time of the break-up of JNR (see Figure 5.5). Upon completion the line is leased to the appropriate JR company, who have to cover costs of any extra rolling stock. Although it may appear that the government's share of the costs had been reduced, it needs to be remembered that by 1987 there were no new shinkansen lines being constructed. Therefore, when work started on the Hokuriku Shinkansen, Tōhoku Shinkansen extension and the Kyūshū Shinkansen, funds had to be diverted from the Ministry of Transport's budget from other areas. However, even now, the share of the MLIT's budget that is spent on the shinkansen is less than 1% of its budget on public works (see Table 5.5). Despite this, one MOF official told me that he was concerned that the approval given in December 2004 for the construction of three sections of shinkansen line would be linked by the public to a rise in income tax and would further damage the shinkansen's image.

Mention has been given above to 'Third Sector Companies'. These companies are neither 'public' nor 'private' in the usual sense of the words, but gain subsidies

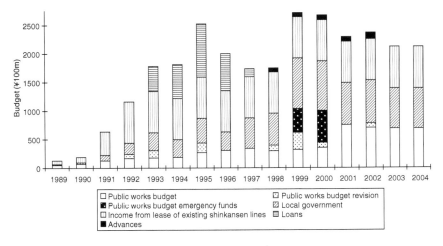

Figure 5.5 Shinkansen construction budget since fiscal 1989.
Source: Based on MLIT (2004d).

Table 5.5 MLIT budget for public works (Fiscal 2004)

Area	Amount ¥ billion (change since fiscal 2003)	Share (change since fiscal 2003) (in %)
1. *Landslides and flood prevention and control*	1,135.9 (−76.0)	14.67 (−0.43)
Flood-control	922.6 (−60.0)	11.92 (−0.33)
Landslide prevention	134.7 (−10.8)	1.74 (−0.07)
Sea defences	78.5 (−5.2)	1.01 (−0.03)
2. *Road construction*	1,802.8 (−276.5)	23.28 (−2.63)
3. *Ports, airports and railway construction*	564.4 (−10.5)	7.29 (+0.13)
Ports	277.1 (−17.4)	3.58 (−0.09)
Airports	164.4 (+10.8)	2.12 (+0.21)
City and main railways	48.8 (−3.6)	0.63 (−0.02)
Shinkansen	68.6 (+/−0)	0.89 (+0.04)
Lighthouses	5.6 (−0.3)	0.07 (+/−0)
4. *Housing and city environment*	1,724.4 (+229.7)	22.27 (+3.64)
5. *Sewerage, etc.*	1,260.7 (−113.6)	16.28 (−0.85)
6. *Agriculture*	834.5 (−44.3)	10.78 (−0.17)
7. *Forestry and fishing*	360.6 (−10.0)	4.66 (+0.04)
8. *Regulatory fees*	59.9 (+20.0)	0.77 (+0.27)
General total	7,743.3 (−281.2)	100.00
Disaster restoration work	72.7 (+/−0)	—
Total public works budget	7,815.9 (−281.2)	—

Source: MLIT (2004b,c); MLIT interview September 2004.

from local government while attempting to operate on private sector principles. I would argue that what could be termed 'the third-sectorization of Japanese railways' is probably one of the most negative aspects of the continuing development of the shinkansen network and on-going maturation of the Japanese railway industry. Not only are local governments having to raise the funds for the construction of new shinkansen lines, which could effectively be seen as a one-off payment, but they also have the ongoing burden of having to support the existing conventional line through the third-sector company. On top of this, those people that have to continue to use the conventional line often find the number of services reduced and the cost of travel significantly increased. Should they transfer to the shinkansen for part of the journey, the fare is even greater. For JR East the expansion of its shinkansen network has allowed it to get rid of loss-making sections of its conventional railway network. It was only due to this being made a possibility that the JR companies even began to contemplate new shinkansen construction. While Smith (2004) is generally positive about third-sector railways, he notes that only five of the 32 companies are making profits. For many the financial challenges are severe, and when visiting Komoro on the Shinano Railway, I noticed signs asking for people to even sponsor railway sleepers!

Red lines or profitable railways?

On the eve of the JR companies birth in 1987, total JNR debt stood at ¥25 trillion, with a further ¥12.2 trillion in debt relating to the construction of the Jōetsu Shinkansen, Seikan Tunnel, Seto Ōhashi bridge, and pension obligations. The number of employees was 277,000 (Aoki *et al.* 2000:189). The various bodies that had been involved in the break-up and privatization of JNR had recognized that it would be impossible for the new companies to take on all of this debt or all of the employees. The excess labour force were found jobs, given retraining or encouraged to take early retirement. None were made unemployed. Indeed, it was only due to the promise that no person would be made unemployed that the reform of JNR managed to proceed so rapidly. As for the debt, it was necessary to come up with a mechanism of sharing the burden. To handle some of the debt, the JNR Settlement Corporation (JNRSC) and Shinkansen Holding Corporation (SHC) were created. The debt was then shared so that JNRSC, SHC and the three Honshū JR companies together with JR Freight had ¥25.6 trillion, ¥5.7 trillion and ¥5.9 trillion respectively. The JR obligations became not just ¥5.9 trillion, but also the SHC's debt and a further ¥2.9 trillion to be paid to JNRSC to take account of the replacement value of the shinkansen lines. Of the remaining ¥22.7 trillion, the government would take on ¥13.8 trillion (the equivalent of ¥353,571 per household), ¥7.7 trillion would be raised from the sale of land, and ¥1.2 trillion would be raised from the sale of JR stock (Aoki *et al.* 2000:187).

In relation to the shinkansen, essentially the debt was transferred to the SHC from whom the JR companies leased the lines. The income from the lease charges, which were to be reviewed every two years taking into account changes in transport volume on each line, would then allow the SHC to pay off the debt and interest over a period of 30 years. However, the lease charges were not done on an equal basis, which allowed for 'profit adjustments' (i.e. cross subsidies) to be made (Kasai 2003:165). In other words, as the Tōkaidō Shinkansen improved its performance relative to the other lines, so JR Tōkai would have to pay an increasing amount in lease charges, despite the fact that other, non-shinkansen, lines were also able to make large profits (e.g. JR East's Yamanote Line in Tōkyō). It was a perverse system. That it was deemed necessary reveals just how bad the JNR debt had become and how any solution, however imperfect, was needed as a means to reform JNR. However, it also meant that the JR companies would inevitably be relying on each other, and many on JR Tōkai. Furthermore, becoming a truly private company was almost impossible due to uncertainty over future charges and ownership. Indeed some even suggest that the existence of the SHC showed that complete privatization of the JR companies was mere *tatemae* and was not really believed possible (Kasai 2003:167). It was a system that had the potential to lead the JR companies ultimately on the same track to failure as JNR had been on.

Another problem with the SHC system was that the JR companies could not allow for any depreciation on the shinkansen ground facilities which they were operating. With the shinkansen being so important for JR Tōkai in particular (see following sections), it meant that total depreciation costs were a mere 7% of total revenue. This ultimately would have meant that to maintain the facilities, funding had to come from income or extra loans (Kasai 2003:168). In other words, either debt would have increased or compromises would have had to been made on maintenance, for example, which could have undermined the safety of the railways. On 1 October 1991 the SHC was abolished with its assets and liabilities being transferred to the JR companies. Again, it was JR Tōkai that shouldered the greatest burden. In 1987, the book price of the shinkansen lines was ¥5.65 trillion. This had become ¥8.54 trillion under the system whereby extra debt was transferred to SHC to be paid off by the shinkansen-operating companies. However, while the valuation of the Tōkaidō Shinkansen was ¥0.47 trillion, JR Tōkai was initially expected to pay ¥5.02 trillion. With the abolition of the SHC this figure became ¥5.09 trillion despite the valuation of the Tōkaidō Shinkansen having only risen to ¥2.96 trillion (see Figure 5.6). Although JR Tōkai may not be happy about taking on this extra debt, that some of Tōkaidō Shinkansen lies within the territories of the other two Honshū JR companies (Shin-Ōsaka–Maibara for JR West and Atami–Tōkyō for JR East) has also been made as an argument as to why JR Tōkai should not complain too much about taking on the debt of others, since effectively the way the break-up of JNR was done, although being simple, does mean that the other two companies are losing some revenue on these sections of track (anonymous interviews 2003).

Of course, the Tōkaidō Shinkansen's burden is greatest due to its potential to make much greater profits (see below), but the current system still does not take account of each JR company's total earning potential. Naturally this is something that does not please JR Tōkai, but Kasai (2003:168) notes that it was impossible to overcome 'the tendency of bureaucrats to stick to something unless it is

106 The bottom line

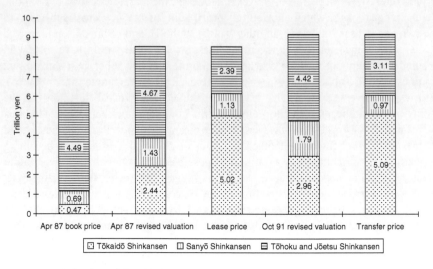

Figure 5.6 Valuation of the shinkansen lines.

Source: Based on Kasai (2003:215).

Note
This only applies to sections of shinkansen lines built prior to 1987.

afflicted with a truly fatal flaw'. Although without the current set-up investment in the Tōkaidō Shinkansen could be further improved, the amount invested each year has already been increased from ¥40 billion p.a. to ¥90 billion p.a. since 1987 as it is no longer having to subsidize small rural lines, for example (Kasai 2003:188). But the Tōkaidō Shinkansen will effectively continue to support the other JR companies in much the same way as it had supported JNR in the past too. For although the cost of construction of the line ended up being nearly double the amount initially forecast (see Chapter 2), the revenue from it had already surpassed this figure by the time the Sanyō Shinkansen opened (Semmens 2000:19). While it is true that JNR first went into the red when the Tōkaidō Shinkansen began operations, to suggest that the shinkansen was the reason for the change in JNR's fortunes would be a fallacy (see Kasai 2003:21). By 1987 shinkansen revenue accounted for about 40% of all of JNR's revenue (Semmens 2000:38), while shinkansen-related debt, even if the Seikan Tunnel is included, accounted for less than 20% of JNR's long-term liabilities (Aoki *et al.* 2000:187).

Yet Kasai points out that the collapse of JNR was looking more and more likely as the work on the Tōhoku and Jōetsu Shinkansen came to an end. Indeed Yamanouchi (2000:201) suggests that if the Tōkaidō Shinkansen 'symbolizes the passionate and energetic JNR', then the Tōhoku and Jōetsu Shinkansen 'embody a stumbling railway struggling with financial and labour problems'. Therefore,

> If we interpret JNR's mission as having been to construct Japan's comprehensive post-war rail network it could be argued that the company collapsed due

to sheer exhaustion, having had to complete its job taking on ever-increasing amounts of debt within the irreconcilable framework of financial self-sufficiency. As this contradiction became ever more apparent, so did the apathy and lack of discipline of the company's management.

(Kasai 2003:39)

There is no doubt that change at JNR was needed. However, despite this being obvious in hindsight, it is unlikely that anything would have happened had it not been due to the luck of various individuals being in the right place at the right time. It needed the intervention of those such as Prime Minister Nakasone, who saw JNR reform as a means not only to deal with JNR's debt but also the railway labour unions, although these two issues are not completely unrelated (see Chapter 6). Other key figures went on to become leaders of the JR companies – for example, Kasai (JR Tōkai), Ide (JR West) and Matsuda (JR East). Without them, the JNR reform may not have been successful and would have been merely an image change 'as happened with NTT or railway reform in the UK' (anonymous interview 2003). Given the success of JNR reform, it is hardly a surprise that it has been suggested as a model for other reforms in Japan. However, without the same sort of individuals involved and without taking into account the difference in the relevant businesses, success in these areas is not guaranteed.

Although the way in which JNR was reformed was not ideal, as Kasai (2003:181) concludes it 'succeeded in correcting the fundamental defects that had plagued JNR. The Chinese saying, "With the root established, the tree will grow" is quite appropriate in summing up the reform of JNR.' JR companies are run on the basis that they should seek to make profits. However, they are also expected to act in the 'public interest'. Essentially this was the same basis upon which JNR was established. That it has apparently worked this time is in part due to the maturation of the railway industry following the completion of much of the necessary modernization work, including the construction of some of the shinkansen lines (Kasai 2003:162).

Given the size of the debt, it is no wonder why the JR companies are so keen for the interest rates to remain at their remarkably low level. In the case of JR Tōkai, for example, interest payments account for 20% of revenue. Despite this, in the 10 years since the abolition of the SHC, JR Tōkai was able to reduce its debt by ¥1.2 trillion to ¥4.3 trillion (Kasai 2003:169). JR Tōkai has also been able to repay its debt relatively quickly in recent years due to a revision in the tax law and the JR Company Law in 2001. This allowed the company to create a ¥500 billion reserve for large-scale repair work to be done on the Tōkaidō Shinkansen. However, as this work is not yet being done, the money can be used to reduce debt repayments (Kasai 2003:169–70). JR Tōkai's performance in reducing its debt appears exceptional. The danger may be for JR Tōkai, that although at this rate it will take about another 30 years to repay all the debt, this level of performance may suggest that, contrary to what JR Tōkai suggests, government funding would not be needed for construction of the Chūō Shinkansen – which is estimated would cost ¥8 trillion to construct (Kasai 2003:197). However, despite JR Tōkai's past record, there are no guarantees that this repayment performance can be maintained, so it is unlikely that the company would be able to find the funds for financing the Chūō Shinkansen privately as such a level of extra debt would concern investors (Yasubuchi interview April 2004).

Table 5.6 shows the different levels of importance of the shinkansen for each of the JR companies. Although beyond the scope of this study, it is worth mentioning that as a result of the reform of JNR, the JR companies are now free to behave more like 'ordinary' railway companies in Japan. There has always been a history for these companies to diversify, with shops, housing and leisure facilities being the typical ventures that are also pursued. Income from the railways is often no more than these ventures, and in the most extreme case, Tōkyū's railway revenue accounts for only 7% of their total income (Semmens 2000:38). In the case of the JR companies, diversification has led to the opening of restaurants (including in Britain), hotels, and a variety of shops and services. Despite continuing to diversify (see Figure 5.7), JR Tōkai remains the most shinkansen-dependent. It is little wonder, then, that the company has been investing in its improvement in recent years, for example, with the introduction and expansion of *Nozomi* services. While this investment cost ¥120 billion, the revenue increased by about ¥15 billion per annum (Semmens 2000:59). Such a return, about 12.5%, is exceptional in the railway transport.

While the Tōkaidō Shinkansen has always been successful, initially the Tōhoku Shinkansen was loss-making. Prior to 1987 it was losing over ¥120 billion per year. Due to changes, which will be discussed in the next section, it is now profitable. Similar changes have also occurred on the Jōetsu Shinkansen which saw its annual losses rising from ¥35 to over ¥100 billion by 1987 (Semmens 2000:35). With shorter trains being operated by JR West and JR East, and flexible designs used by JR East, the 'break even point' for a shinkansen service is now about 30% occupancy (interviews with JR staff). Above this level, the service will be helping to bring in profits for the company that help to subsidize other parts of the company's operations or cover the long terms debts relating to the construction of the shinkansen, for example. However, to concentrate on the financial impact of the shinkansen for the companies, is to overlook some of its wider significance.

The shinkansen and regional economies

The shinkansen's relative importance varies from company to company with the different lines having different markets. The reform of JNR has enabled the companies to better respond to these differences. However, the fact that Japan has also changed during the past 40 years should not be forgotten. The 1960s saw a great migration from rural areas to the metropolitan areas. Tanaka's plan to remodel Japan was in part a response to this. But the shinkansen network was not built and the Kantō plain, in particular, continued to develop. The Tōhoku and Jōetsu Shinkansen lines have become beneficiaries of this. While in 1987 the passengers on them were primarily going long-distance, with increases of 8–13% annually, the number going shorter distances was falling (Semmens 2000:34). Indeed, while the forecast for daily usage on the two lines had been for 90,000 and 70,000

Table 5.6 Importance of the shinkansen for JR companies

Item	JR East			JR Tōkai			JR West			JR Kyūshū		
	¥ billion	%	%	¥ billion	%	%	¥ billion	%	%	¥ billion	%	%
Operating revenue	2,542	100		1,128	100		1,216	100		158	100	
Total railway revenue	1,666	65.5	100	1,064	94.3	100	751	61.8	100	118	74.7	100
Shinkansen revenue	466	18.3	28.0	958	84.9	90.0	308	25.3	41.0	10	6.3	8.5
Conventional line revenue	1,200	47.2	72.0	106	9.4	10.0	442	36.3	58.9	108	68.4	91.5
Non-railway income[a]	876	34.5		64	5.7		465	38.2		40	25.3	
Operating costs and expenses	2,227			801			1,089			157		
Profit on operations	316			327			127			1		
Number of companies in group	98			38			145			36		
Percentage of shares sold by government	100			60			100			0		

Sources: East Japan Railway Company (2004:58, 91); Central Japan Railway Company (2004b:4, 29); West Japan Railway Company (2004b:28, 33); Kyūshū Railway Company (2004); JR Kyūshū interview September 2004.

Notes

All figures are for fiscal 2003 other than JR Kyūshū whish are fiscal 2004, the first full year of operations of the Kyūshū Shinkansen. Some sub-totals do not add up to the total due to rounding.

a Includes revenue from land leasing fees at stations, track usage fees, advertising fees, etc.

110 *The bottom line*

Figure 5.7 A 700-series shinkansen at Nagoya station. In the background is JR Tōkai's impressive Central Towers development that has completely transformed the station and the surrounding area.

respectively, they were only 57,000 and 41,000 (Yamanouchi 2000:201). However, many of the intermediate stations have now become part of the capital's commuter belt. Although distances of up to 100 km would not normally considered to be part be commuting distances, these areas enjoyed an increase in the number of commuters by 35–75% between 1990 and 1995 alone (Ja and Konami 2002). As a consequence, which has undoubtedly helped this process develop further, JR introduced double-decker shinkansen, particularly emphasized by the introduction of the E4-series (see Figure 5.8), which can easily be joined to other shinkansen for use on busier sections at peak times during the day. However, it is perhaps the impact of the Niigata-Chūetsu Earthquake in 2004, when the Jōetsu Shinkansen was suspended between Niigata and Echigo-Yuzawa, that revealed how important the line has become. The two month closure saw the number of people travelling each day from Niigata to Kantō fall from about 11,600 to 3,800. The cost to JR East in lost income from the suspension of the Jōetsu Shinkansen and other conventional lines in the area was ¥13 billion, with a further ¥20 billion needed to repair the line itself (*Japan Times*, 29 December 2004).

Railway investments are huge, not only in financial terms but also in terms of their impact. Kasai (2003:12) argues that their impact on the regional economy is actually likely to be greater than the return on investment in the short term, while

The bottom line 111

Figure 5.8 (a) A double-decker E4-series arriving at the popular commuter city of Utsnomiya; (b) an E4-series linked with a 400-series passing Kita-Yono station in Saitama – note how the windows on the lower deck of the E4-series are almost totally obscured by the sound-protection wall in this area where there was much opposition to the construction of the shinkansen line; (c) the E4-series in profile reveals some of its links to the 700-series design; (d) two E4-series shinkansen joined together – the highest capacity high-speed train service in the world.

over time passenger numbers should increase due to the increase in economic activity in the area, so that in time, perhaps 'decades', 'the utilization rate reaches the break-even point'. In other words, the Tōkaidō Shinkansen, as already suggested, is exceptional, rather than typical of railway investments. The regional, and subsequent national, impact of railways, which allows people to commute to work, to visit clients, etc., is its 'external benefit'. Kasai (2003:19) notes that the ' "external benefit" which society enjoys far exceeds the "internal benefit" enjoyed directly by passengers'. The demand for train usage is largely 'derived demand'. The majority of passengers use trains to get from A to B rather than for the sake of taking a train journey itself. In other words, the real 'role and function' of the shinkansen 'lies in the social activities of the people that use it when they reach their destinations' (Konno 1984:79). It is because of this 'external benefit' that the passenger railway often receives 'large amounts of subsidy and is regulated accordingly' (Burchell 2000:10). Indeed, Cole (interview 2004) suggests that the issues of subsidies and how to reduce them has been the main focus of railway operations and why there is such a small body of literature on railway economics. Perhaps it is due to a lack of a full understanding of all the issues surrounding railway economics that leads to a situation whereby Kasai (2003:73) argues airlines have been able to enter the market due to the way in which fares have been set (see following sections), and that this has harmed the economy and peoples' lives as 'passengers had to pay unreasonably high fares, enabling inefficient transportation modes to enter the market'. Kimata (2001) has estimated that the 'external benefit' of the opening of the Tōkaidō Shinkansen, by calculating the time saved by travelling by shinkansen and multiplying it by the number of passengers and the average salary, was equivalent to ¥5.2 trillion. Clearly, the shinkansen brings huge savings to the national economy, although the losers should not be forgotten if trying to come up with a truly representative figure for the 'external benefits' of the line.

There are various ways to look at the impact that the shinkansen has had on regions. One can look at the economic output of the area while keeping in mind the changes in the national economy at the same time. One can also look at population changes in towns and cities served by the shinkansen, while bearing in mind national changes and that for some areas – particularly if one was to consider the popular of the ward where Shinagawa is located – the daytime population may be more significant than the residential population. One can also look at changes in visitor numbers where stations serve areas used by tourists. However, even doing this may be over-limited. For not only does one need to remember that cities near those served by the shinkansen may also see changes due to the shinkansen, due to the straw effect, whereby 'some of the urban functions may be pulled out to adjacent big cities because of the convenient shinkansen service' (Ja and Konami 2002), but also we need to be aware that 'the Shinkansen's contribution to culture is its area of greatest influence' (Konno 1984:80). Some of its cultural impact has already been suggested in this study and further examples will be considered in the following chapters. But let us concentrate on the issues of population change, impact on business, and its impact on tourism here.

The effect of the shinkansen upon the populations of cities and towns along the Jōetsu Shinkansen are shown in Table 5.7. Although there are a few anomalies, the general trend is for stations which have the best connections to Tōkyō, as services also serve Tōkyō rather than terminating at Ōmiya, have seen population growth over the past two decades. Although the population of Honjō has fallen in the run up to the opening of the station, one has to bare in mind that the station is some distance from the centre of the city and that it is probable that the area around the station will be developed over the coming years. The only real surprise is the continuing fall of the population of Yuzawa town, which, considering its size, remains one of the best served towns by the shinkansen in the whole of Japan, as can be seen from Appendix 2. The severity of the weather of the area and the drop in the cost of land and living in the Kantō region are likely to be two major reasons for this, although a falling population and its impact on

Table 5.7 Population of cities and towns along the Jōetsu Shinkansen

City/town	Station(s)	Population						Change between 1980 and 2003 (in %)
		1980	1985	1990	1995	2000	2003	
Saitama[a]	Ōmiya	354,000	373,000	403,800	433,800	456,700	474,321	34.0
Kumagaya	Kumagaya	136,800	143,500	152,100	156,400	156,200	155,626	13.8
Honjō	Honjō-Waseda[b]	53,500	59,100	60,800	61,500	62,000	59,082	10.4
Takasaki	Takasaki	221,400	231,800	236,500	238,100	239,900	242,359	9.5
Tsukiyono town	Jōmō-Kōgen	*	10,800	11,000	11,300	11,500	11,368	5.3[c]
Yuzawa town	Echigo-Yuzawa and Gāla-Yuzawa[d]	9,500	9,500	10,000	9,600	9,100	8,968	−5.6
Minami-Uonuma[e]	Urasa	14,600	15,400	15,500	15,900	15,600	*	6.8[e]
Nagaoka	Nagaoka	180,300	183,800	185,900	190,500	193,400	191,212	6.1
Tsubame	Tsubame-Sanjō	44,200	43,900	43,900	43,600	43,400	43,989	−0.5
Sanjō		85,300	86,300	85,800	85,700	84,400	85,510	0.2
Niigata	Niigata	457,800	475,600	486,100	494,800	501,400	515,192	12.5

Source: Based on table 1-3-13 in Hiraishi 2002:101; http://www.gazetteer.de/fr/fr_jp.htm (accessed 1 October 2003); *Nihon Tōkei Nenkan*; http://www.kamimoku.com/tsukiyono.html (accessed 1 October 2003); Email from Japanese Embassy in London; Email from Niigata Prefectural government; Populations for March 2003 and was taken from http://www.towninf.co.jp/

Notes
* Refers to missing data.
a Ōmiya city merged with Urawa and Yono to become Saitama city in 2001 – the area that was Ōmiya city is now made up of Ōmiya, Kita, Minuma, and Nishi wards of Saitama city (population 1,038,100 in 2003).
b Station opened 13 March 2004.
c Data comparison between 1985 and 2003.
d This station is only open during the Winter and early Spring.
e Urasa served Yamato Town until it merged with Muika Town to become Minami-Uonuma city on 1 November 2004, population change based on 1980–2000.

local services, including schooling, for example, is itself likely to perpetuate the cycle.

Clearly the shinkansen, although being linked with regional investment is not always successful. Given its location it is perhaps surprising that Maibara's population has remained so low (see Appendix 2). Annaka, served by Annaka-Haruna station on the Hokuriku Shinkansen, has become a focus of JR East's investments. With the increasing distances people are prepared to commute, this small city on the side of a mountain with views across the Gunma plain is clearly in a desirable location. With the opening of the shinkansen station, research parks, sightseeing facilities and shopping centres have all been built (see Ja and Konami 2002). JR East even inaugurated a tree-planting festival in the area in a move to not only help prevent subsidence but also to encourage people to visit the area, which would hopefully entice some to move there having seen what was on offer (JR East interview 2003). This activity may have been successful as it has also become the location of the first housing land to be placed in front of a shinkansen station according to an advertisement I saw for the land during a visit in October 2003. This points to an important difference between Japan and many other countries in that people actually want to, or are prepared to, live near stations, rather than have to commute some distance to them. On top of this the relatively small size of Japanese homes (that is the building and garden, which tends to be particularly small) together with the incredibly close proximity, although terraced and semi-detached housing do not exist, in which they are built to neighbouring houses, means that the number of people, let alone businesses, within a given radius from a station in Japan tends to be much greater than in countries such as Britain, for example. Yet despite the trend for Japanese to be prepared to commute longer distances, the local government for Izumizaki had to offer financial deals of up to 60% of the commuting costs to Tōkyō for three years to encourage people to its village, just three stops from Shin-Shirakawa (*Mainichi Shimbun* 12 September 2004). Although the scheme is relatively popular, it would appear that it is still too early to suggest that there will be a significant decentralization of the Japanese population.

There is no doubt that the shinkansen has a huge impact upon regional economies. On top of this, particularly now that regional funding is needed to support new construction, its development also reflects changes in regional economies. That the shinkansen can impact the regional economy is perhaps best demonstrated by the fact that 400 companies, mostly from outside Kyūshū, opened up branch offices in Fukuoka prior to the Sanyō Shinkansen's extension opening there in 1975 (Semmens 2000:46). What such figures do not fully reveal is the extent to which these branch offices were new investments, and how many were transferred from other cities. Although there was likely to have been a predominance of the former, cases of the latter almost certainly existed too.

Considering the important role that the shinkansen can play in regional economic development, it is no wonder that local prefectural governments have been keen to promote the construction and extension of shinkansen lines. For example, in the case of the Hokkaidō Shinkansen, the expected cost, if built to Sapporo, is about ¥1.53 trillion and ¥420 billion for the initial section to Shin-Hakodate. The line would cut journey times between Sapporo and Tōkyō from the current 9:46 hours to 4:23, and the time between Hakodate and Tōkyō from 6:40 hours to 3:40. Consequently, the number of people visiting Hakodate and Sapporo by train is expected to rise from about 1.3 million and 3.2 million a year to 3.5 million and 13.6 million per year respectively. Significantly, the shinkansen will be expected to take about 54% of the share of the Tōkyō–Sapporo market, whereas currently 95% travel by air. However, as the line will initially only go as far as Hakodate, which is expected to see the rail share of the market grow from 6% to just 14%, there may be a danger that the line's impact will be much less and this could endanger the prospects of the line being further extended. However, even with the line only going to Hakodate, an extra ¥36 billion is expected to be added to the Japanese economy every year, of which ¥12 billion will be for Hokkaidō, ¥17 billion for Tōhoku, ¥4.2 billion for Kantō and ¥2.7 billion for other regions. These figures demonstrate the importance of the shinkansen both to inter-regional, non-Tōkyō, trade and to the Japanese economy as a whole. This could be an important development for Hokkaidō given its relative state of underdevelopment, but the full benefit of the shinkansen for Hokkaidō is only likely to be felt if the line is extended to Sapporo, which is expected to add ¥272 billion to the Japanese economy annually, and ¥147 billion, ¥39 billion, ¥33 billion, and ¥54 billion to the other regions respectively (all data provided by JR Hokkaidō and Hokkaidō Prefectural Government during interviews in January 2004). Given that the expected costs for construction, the potential returns look quite favourable.

At the other end of the country, strong economic arguments are also made for the Nagasaki Shinkansen, estimated to cost ¥410 billion. While currently on average 9,655 people a day travel Hakata–Nagasaki, the shinkansen would raise this to an estimated 15,400 – which is higher than the 7,500 to 8,000 estimated for the Hachinohe–Shin-Aomori section of Tōhoku Shinkansen or the 3,000 to 3,500 estimated for Nagano–Jōetsu of Hokuriku Shinkansen. After 10 years of opening, ¥1.19 trillion is expected to be added to the economy of the region, and an extra 10,000 people are likely to have moved to the region (*Nagasaki Shinkansen Kensetsu Kiseikai*). However, with regional governments expected to meet some of the cost of shinkansen line construction, Saga prefecture has generally been opposed to having to fund some of the Nagasaki Shinkansen as it already has a good conventional line link with Hakata and feels as though there would be little benefit from having a shinkansen link to either Hakata or Nagasaki.

An increasingly important economic activity is tourism. In fact, tourism has a long history in Japan, where pilgrimage was one of the original reasons for travel (Sudō interview 2004; Buruma 2001:209; see also Hendry 1999:36). With the shinkansen's ability to link major urban areas with tourist spots around the country, so its role in promoting tourism is significant. Of course, there are also tourist destinations either in big cities, or, in the case of Tōkyō, a visit to the city itself is probably of greater tourist significance than the relatively bland sites it has to offer. Indeed, although JR Tōkai's main market is the businessman (see following paragraph), many adverts have concentrated on the tourist appeal of Tōkyō (often using the imagery of Tōkyō Tower) and the temples, shrines and gardens of Kyōto (JR Tōkai interview 2001).

With a shinkansen network serving smaller, more rural towns and cities, JR East is perhaps more dependent than JR West and JR Tōkai on the tourist market. Although not offering many of the famous sites that are found in Western Japan, JR East's territory contains many sites of natural beauty – such as lakes, mountains and *onsen*. These sites remain popular throughout the year due to the popularity of winter sports in Japan. So great is this demand for winter sports that JR East has its own dedicated ski resort built into a shinkansen station at Gāla-Yuzawa. On top of this, in 2004, JR East's marketing campaign was 'JR Ski – We want snow', a concept that would be alien to many railway operating companies.

While some tourist-related demand is annual, there can also be one-off demand due to events. The Tōkaidō Shinkansen and Hokuriku Shinkansen were both opened to coincide with hosting of the Olympics. The extension of the Tōhoku Shinkansen also occurred in time for the Asian Winter Games being held in Hachinohe. As mentioned in Chapter 2, 'EXPO 70' had a significant impact upon the Tōkaidō Shinkansen. Although, unlikely to rise to that level, the expected extra demand due to EXPO 2005 in Aichi meant that JR Tōkai ordered extra 700-series shinkansen to be built so that they would be able to put on an extra *Nozomi* services. The JR companies also benefited from Japan's co-hosting of the World Cup in 2002, although it did lead to some difficulties (see following paragraph and Chapter 6).

It is important not to get carried away with the apparent positive affect that the shinkansen has on regional economies. A local government, local businesses and local people raising funds so that a shinkansen line can be built to their region are likely to be disappointed if they believe that it will automatically lead to untold wealth. When a line is built, there are potential losers too. Only a limited number of stations can be built, and the route may not always follow the conventional line (which tends to go via smaller towns). An example of a city that has been badly affected by the shinkansen is Komoro. This city was along the conventional route between Tōkyō and Nagano and was a popular destination for tourists in early Spring in particular, who flocked to see the cherry blossom at Kaikoen park on the site of the remains of Komoro Castle. Having had a peak of around 1.9 million visitors per year prior to the opening of the Hokuriku Shinkansen, the figure fell to 1.4 million visitors within four years. While occupancy rate of shops in the city centre stood at 85% in 1992, by 2003 this had fallen to just 46%. The number of passengers handled by the station in Fiscal 2003 was only 65% that for the year in which the Hokuriku Shinkansen opened (interviews at Komoro city government, October 2004). The population of the city has also fallen 2% from 45,711 in 1995 to 44,705 by 2003. Although the city is just a short distance from Karuizawa, and many visitors have always got there by car rather than train, not being on a main line to Tōkyō has helped to erase its name from the public consciousness. There is a real danger, despite occasional campaigns by the city and JR East and the struggling Shinano third-sector railway company, that the city itself will cease to exist in a few decades' time, just as many stops on the original Tōkaidō route all but disappeared when locals, not realizing the importance of the railways, did not allow construction of a railway there in the Meiji period (Ericson 1996:60–1). The city of Saku, served by Sakudaira, on the other hand, has benefited from finding itself on a main route to Tōkyō, and the population has swelled by 8% from 62,003 in 1992, five years before the Hokuriku Shinkansen opened, to 67,025 in 2003. Across Nagano Prefecture, the average increase in population of cities and towns on the shinkansen route was 6% between 1987 and 2001, while it was less than 2% for those not on the shinkansen route (the national average during this period was just over 4% increase) (Kōsoku Tetsudō Kenkyūkai 2003:48). Some of the problems being experienced by Komoro may have been averted, while Saku could have still benefited from its shinkansen station, had the name 'Komoro' been added to the name 'Sakudaira' as some had apparently suggested (Yamanouchi 2002:45).

Even in towns and cities where there is a shinkansen station, the influence may not be wholly positive. In Karuizawa, for example, many of the shops and other businesses in the central area of the town around Naka-Karuizawa station, just one short stop from Karuizawa station itself, have seen a significant decline in business since the opening of the Hokuriku Shinkansen, and many have subsequently closed (Komoro city government interview October 2004). It would appear that in some resorts railway passengers are lazy, choosing to use the train to get to their destination and then going from that station to major sights, without exploring the whole town or city to which they have travelled. Takaoka is likely to become another example of this. Already the area around the main station seems deserted with many shops boarded up. When it becomes connected to the Hokuriku Shinkansen, and a new station is built away from the current station and city centre, but closer to one of the city's main attractions, it is likely that the central area will be further emptied. Hakodate too, with its shinkansen station due to be built several kilometres from the city, is likely to experience this hollowing out in the future. Meanwhile Karuizawa reveals another potential problem the shinkansen can present. By having no level crossings, the number of access points to traverse the line tends to become greatly limited when compared to conventional lines. Effectively the shinkansen can create physical barriers within cities that did not previously exist.

Chapter 2 also referred to there being losers when the Tōkaidō Shinkansen opened. Indeed, the shinkansen has a negative impact upon hotels – for as day trips become possible, the need for accommodation all but disappears. Even in Komoro, an increase in visitors in 2003 to 1.8 million was largely due to an increase in day-trip visitors taking advantage of new *onsen* facilities catering to such visitors. While there are undoubtedly financial savings for businesses that see their employees managing to complete day trips, the impact upon hotels, as well as potentially upon the health of those workers needing to spend time travelling long distances, should not be overlooked.

Although the shinkansen, like other transport developments before it, have made some places more accessible in a shorter time, one has also to consider the journey itself for this is part of the experience and adventure of a trip. I used to travel from Seto, near Nagoya, to Kinosaki in northern Hyōgo Prefecture about twice a month for a year. While about half of this journey, in terms of distance, was done by shinkansen, the rest was on, at that time, a relatively slow 'express' train on the San'in Line, much of which was single-track. Although the journey time could be significantly reduced if the San'in Shinkansen were ever built, and the *onsen* resort of Kinosaki would probably gain much extra business, I cannot help but feel that something would be lost with such a 'development'. The journey that I took was like a journey back in time, so that by the time I reached the *ryokan*-filled streets of Kinosaki it was as if I had arrived in the late Tokugawa Period or early Meiji Period. Perhaps this is the way that it should be. But who has the right to deprive such towns of the 'benefits' of modernity and potential to make huge financial gains? Perhaps if we can make things easier and increase opportunities, we should. Yet I cannot help but think that there is something wrong about it. Similarly, although I have been to the top of Mt Fuji twice and if I could find the time would do it again, doing so by massive fast-moving escalators, if they were available, though convenient would somehow feel wrong.

The shinkansen and the making of modern Japan

As different lines have different markets, their response to changes in the national economy also varies. While it is natural to think this is particularly the case for the Tōkaidō Shinkansen as about 70% of its users are businessmen (JR Tōkai interview October 2004), and indeed JR Tōkai often show the close link between GDP growth and shinkansen usage, one has to remember that during times of limited economic growth, as occurred during the 1990s in Japan, the amount of money that people spend on tourism tends to fall, which also impacts the usage of lines such as the Jōetsu and Tōhoku Shinkansen lines. However, despite the suggested link between the economy and the shinkansen, during the recession of the mid-1970s, even as passenger numbers fell, the shinkansen's profitability rose. This was largely due to a rise in fares, showing that the demand for the shinkansen is relatively inelastic. It was not until 1979 that *Kodama* services were reduced from 16 to 12 carriages as a measure to cut expenses (Semmens 2000:24).

Although the correlation between Japan's economic performance and usage of the Tōkaidō Shinkansen is inconclusive, JR Tōkai President Kasai certainly believes that the Tōkaidō Shinkansen is crucial to the national economy, for he says that the company is 'responsible for the operation of Japan's most vital transportation artery that is at the heart of the nation's social and economic activity' (Kasai 2003:182). Given this, it is no wonder why JR Tōkai see the Chūō Shinkansen as also being an important investment for Japan. Based on the argument that the government should take responsibility for some of the 'external benefits' of the line, as with other large-scale projects such as airports that are of 'strategic importance for the nation', JR Tōkai argue that funding for the line's construction should come from the government. Indeed, the argument made by

Kasai (2003:197) is that as the line would be in competition with the Tōkaidō Shinkansen, which could continue to make large profits forever without this competition, it is not in JR Tōkai's interest to take on the debt of constructing the line – a view that would be supported by investors (Yasubuchi interview 2004). Yet, it has been JR Tōkai more than any other body that has been pressing for the Chūō Shinkansen through ongoing research. Given the current state of the nation's finances, I think it is unlikely that the government would be willing to cover all of the estimated costs of the line. However, when one bears in mind that JR Tōkai has had to take on an additional ¥2.13 trillion in relation to the debt of other shinkansen lines, and given that JR West and JR East are also profitable companies, some of the cost of construction could effectively be released to JR Tōkai by redistributing this money, which would appear to be appropriate. This would still leave a ¥5.87 trillion shortfall. Coincidentally, this shortfall is almost exactly the same cost as the shinkansen lines that were constructed during the days of JNR and about double that constructed in the JR days up to 2004 – much of which was funded by sources other than national government. By the time the other *Seibi Shinkansen* have been completed in about 2015, it is imaginable that the Chūō Shinkansen may become a channel for MLIT shinkansen-construction funds rather than that part of the budget being cut back to zero, although there is a question mark over how much funding would be made available from prefectures and cities along the route, especially Nagano.

The idea of the Chūō Shinkansen and the use of linear technology has a long history and was even supported by Tanaka (1973:122) in his plan to remodel Japan. Part of JR Tōkai's support for the plan and why it believes funding should come from the government is that, like the Tōkaidō Shinkansen 40 years ago, it is likely to further encourage economic activity and innovation as new technologies are developed and exploited (Kasai 2003:201). As the route is likely to be based on the Nakasendō, it is likely to help economic activities in areas that have not fully benefited from the development of the railways and shinkansen in the past. By making Ōsaka and Tōkyō only one hour apart, it is likely to further help the business operations between these two cities. This assumes of course that the fare is not so high that people, or perhaps more significantly their companies, do not consider using it. This must be a concern given the nature of the market along the Tōkaidō route at present (see following paragraph).

In recent years JR Tōkai has been stepping up its campaign for the Chūō Shinkansen by allowing the public to visit and travel on the test route (see Figure 5.9). Since 1998, about 80,000 people have taken a ride on the demonstrations (Central Japan Railway Company 2004b:25). At the same time, work has continued on speeding the vehicles up. Although tests have now reached a record of 581 km/h, the need to respond to energy efficiency and infrastructural costs and limitations means that speeds are likely to be limited to 500 km/h if the Chūō Shinkansen were to become operational (Okumura 2001). The capacity of the line would be about 10,000 people per hour each way (Okumura 2001), while the Tōkaidō Shinkansen now has a capacity of nearly 20,000. However, given the increases in the past few decades, it is uncertain whether the Tōkaidō Shinkansen

120 *The bottom line*

Figure 5.9 (a) A linear shinkansen on display during one of the public open days, which continue to attract people, although not in the numbers perhaps hoped for. (b) The route of the test track disappears into one of the many tunnels (this one, unlike some, has a noise-reducing hood) at the point where a station may be constructed in the future for easy transfer to the private Fuji-Kyūkō line, which crosses beneath the line, that would provide a link to the Chūō conventional line at Ōtsuki and to the tourist destinations of Kawaguchi-ko and Mt Fuji. Naturally the Fuji Kyūkō company is keen to seen the construction of the Chūō Shinkansen as it is likely to significantly boost their business as this area would become a suburb of Tōkyō (Kawashima interview 2004). Visible on the right is the enormous electricity transformer station – the variation in sound pitch, which is audible from some distance, reflects changes in the speed of the train.

will be able to continue to meet demand – especially if one takes into account the number of the potential passengers that currently travel by plane. As well as addressing future capacity problems, the Chūō Shinkansen would also be less prone to earthquake damage.

It should be noted that the Chūō Shinkansen would also have other benefits for JR Tōkai. Looking at the possible route, it is noticeable that much of it runs at first in JR East territory – from Tōkyō via Kōfu to Shiojiri. This would allow JR Tōkai to compete directly with JR East on a route which currently has many *tokkyū* travelling between Tōkyō and Matsumoto, just north of Shiojiri. From Shiojiri to Nagoya, the route would be in JR Tōkai territory – but close to JR Tōkai's Chūō Line, much of which is loss-making and so would probably be handed to a third-sector company. Finally, the route passes into JR West territory as it continues via Nara into Ōsaka. Given the historical importance of Nara and how it is a popular tourist destination, this would allow it to not only compete with JR West, but also one of its other rivals in the region, Kintetsu. The Chūō Shinkansen is more than a national project and a means to show the world Japanese technology, this is a means for JR Tōkai to directly compete with its rivals.

The debate over the Chūō Shinkansen raises two important questions. First, is it really that important for the shinkansen to become faster? Second, what is the real competition for the shinkansen? These two questions are of course linked. If the shinkansen is competing with airlines, then extra speed, assuming customers are basing their travel choice on time taken, would allow the shinkansen to increase its market share. While much mention is made of speed when discussing the shinkansen, perception of speed on shinkansen is hard. I referred to it in Chapter 1, but this was as much due to looking at the map and being aware of how fast I must be travelling on that basis as a real feel of speed from the G-forces in the train or by looking at the scenery passing by. Only when passing objects close to the line does one get any real feeling of speed. Even when in the cab with the driver of a 500-series travelling at 300 km/h (see Figure 5.10), there is not a great sensation of speed. Indeed, it reminds me of when I visited the cockpit of

Figure 5.10 (a) A 500-series shinkansen crossing a river in Okayama prefecture; (b) the speedometer of a 500-series shinkansen doing 300 km/h; (c) the cab of the 500-series shinkansen; (d) a 500-series shinkansen passing Odawara station.

122   *The bottom line*

an ANA Boeing 747 and was only aware that we were moving at all when the occasional cloud went by or another plane passed in the opposite direction. Ironically, rather than the 700-series or 500-series, it was in the cab of the old 'Dr Yellow' (see Chapter 6), which was based on the original 0-series, when travelling at night that I most felt speed, despite travelling some 60–90 km/h slower than in the other trains.

Speed, together with the danger of derailment, was obviously a concern for the original designers of the shinkansen. However, since then, much research has continued into improving the speeds of the shinkansen. As a consequence of the 'dark valley', no significant progress was made for many years. However, spurred on by the reform of JNR and the shock of France's TGV taking the shinkansen's mantle of being the world's fastest train (Yamanouchi 2002:80) which had been 'an ongoing battle' since 1964, significant developments have occurred in recent years. Yet, when new lines begin operations, due to the *Seibi-Shinkansen* law that dates back to when maximum operating speeds were 210 km/h, the maximum speed is set at 260 km/h. Only following a successful trial period can speeds be increased to more 'normal' speeds for current shinkansen of 275 km/h or above.

Regarding international comparisons, there are various ways in which these can be done. While top speed is probably what many would expect comparisons to be based upon, the more significant figure is the average speed between two stations, whether that be the starting point and terminal or two stations along the route. This basis is used for comparison as it avoids the problems caused by particularly slow areas of track (e.g. Tōkyō–Ōmiya) and distortions by only considering the top speed, which may not be maintained for much of a journey. Table 5.8 shows how the shinkansen is still considered to be the fastest service in

Table 5.8 The fastest passenger trains in the world (2003)

Position	Country	Train	From	To	Distance (km)	Average speed (km/h)
1	Japan	500-series	Kokura	Hiroshima	192.0	261.8
2	France	TGV	Valence TGV	Avignon TGV	129.7	259.4
3	(International)	TGV Thalys Soleil	Brussels Midi	Valence TGV	831.3	242.1
4	Germany	ICE	Frankfurt Flughafen	Sieburg/ Bonn	143.3	232.4
5	Spain	AVE	Madrid Atocha	Sevila	470.5	209.1
6	Sweden	X2000	Alvesta	Hassleholm	98.0	178.2
7	United Kingdom	IC225	York	Darlington	71.0	177.5
8	Italy	ETR500	Roma Termini	Firenze SMN	261.0	166.6
9	United States	Acela	Wilmington	Baltimore	110.1	165.1
10	Finland	Pendolino	Salo	Karjaa	53.1	151.7

Source: http://www.railwaygazette.com/2003/2003speedsurvey.asp

Figure 5.11 (a) A 300-series passing the RTRI large-scale low-noise wind tunnel facility in Maibara, outside of which three experimental trains are housed; (b) 300X; (c) STAR 21; (d) WIN350.

the world in terms of average speed between two stations on a route. Speed, while being the 'primary and most obvious indicator of technological level' is just one part of the picture (Yamanouchi 2000:170, 179). After all, as Yamanouchi (2000:211) questions, 'Who is to say whether the technology behind a Toyota or a car like a Porsche or Ferrari is more advanced?' One has to remember that the trains have very different capacities, for example. While many shinkansen have capacities in excess of 1,200, the TGV seats less than 400, and the Tōkaidō Shinkansen carries seven times the number of passengers daily than the Paris-Lyon TGV (Yamanouchi 2000:4). So comparing Japan with many other countries is like comparing oranges and apples.

On top of operational trains, there are many experimental trains (see Figure 5.11) that have gone even faster than the speeds mentioned in Table 5.8. Naturally, it is the linear shinkansen that leads the world in this respect. It has only been through the development of experimental trains that further improvements to the shinkansen in service today has been possible, although standard train-sets have also done various speed tests when there have been no fare-paying passengers on board (see Table 5.9).

While it is speed that is usually discussed, acceleration can be nearly as significant. Although technically capable of more, shinkansen on many lines are limited

124 The bottom line

Table 5.9 Japan's fastest experimental trains

Speed (km/h)	Train	Date	Location
256	Set B test train	30 March 1963	Odawara test track (Tōkaidō Shinkansen)
286	951 test train	24 February 1972	Sanyō Shinkansen
319	961 test train	7 December 1979	Oyama test track (Tōhoku Shinkansen)
325.7	300-series shinkansen	28 February 1991	Tōkaidō Shinkansen
345.0	400-series shinkansen	19 September 1991	Jōetsu Shinkansen
350.4	WIN350	8 August 1992	Sanyō Shinkansen
425.0	STAR21	21 December 1993	Jōetsu Shinkansen
443.0	300X	26 July 1996	Tōkaidō Shinkansen
581.0	MLX01 linear Shinkansen	2 December 2003	Yamanashi test track

Source: Kōsoku Tetsudō Kenkyūkai (2003:226); JR Tōkai press release December 2003.

to an acceleration of about 1.6 km/h/s. In other words, it takes just under three minutes for a shinkansen to get from 0 to its maximum speed. Braking, unless under emergency conditions, takes longer. Stopping at a station, therefore, when one includes the even the shortest waiting time of 50 seconds, adds at least eight minutes to the journey time. If waiting for another shinkansen to pass, as often happens for *Kodama* services on the Tōkaidō and Sanyō Shinkansen, for example, the waiting time is often much longer. By the time such a stopping service has reached its top speed, the other train will already be nearly 15 km down the track. Meanwhile, of course, as the stopping service has been accelerating, there will be another shinkansen closing on it from behind. This was a particular problem on the congested Tōkaidō Shinkansen when *Kodama* services were operated by trains with lower top speeds than the *Hikari* and *Nozomi*. Indeed JR Tōkai spent ¥800 billion (Kasai 2003:188–9) to speed up the withdrawal of its 100-series shinkansen. However, on the JR West and JR East lines, where there are also many shinkansen with very different performances, it is less of a problem due to the lower usage. Only between Ōmiya and Tōkyō would it be an issue, if it were not for the top speed being well below even the slowest shinkansen.

Let us now consider the shinkansen's competitors. Although car ownership is about 86.4% in Japan (Asahi Shimbun 2004:181), the cost of using the expressways, most of which have tolls, is high. Top speeds are also relatively low compared to their counterparts in Europe, for example, and when coupled with the problems of traffic jams, tend not to offer a practical alternative. As a consequence, the car tends to be used mostly for shorter distances. As the distance increases, so trains gain an advantage. However, the issue is not merely distance. Cost and time are also important. It is for this reason that the shinkansen has to compete in some areas with other trains. For example, JR Tōkai effectively competes with itself, as well as with the private Meitetsu company, along the corridor between Toyohashi and Nagoya as relatively fast and cheap conventional services compete with the shinkansen. JR Tōkai's shinkansen further has competition from Kintentsu in the Nagoya–Ōsaka market. The nature of privatization of JNR also means that JR West's shinkansen competes with JR Kyūshū's conventional trains between Hakata and Kokura, JR Tōkai's shinkansen competes with JR West's, as well as with other private companies, conventional trains between Kyōto and Ōsaka, and with JR East's conventional trains between Tōkyō and Atami. Should the Hokuriku Shinkansen ever be completed as between Kanazawa and Shin-Ōsaka, it will be interesting to see whether the JR East and JR West services along that route will be allowed to directly compete with JR Tōkai's shinkansen by offering reduced prices to encourage passengers to take the longer route.

However, while there are a few areas where the shinkansen may be competing with conventional services, generally the shinkansen's competition is for longer journeys – whether that be in terms of distance or time. Although in Kyūshū the expressways have made the bus competitive, the main competitor for longer distances for the shinkansen is air travel. In considering 'longer distances', the shinkansen is said to have an advantage over planes for journeys of about three hours. This can be seen by looking at the comparative share of the market between Tōkyō and various cities along the Tōkaidō and Sanyō Shinkansen. While the shinkansen has 82% of the Tōkyō–Ōsaka market, which takes 2 hours 30 minutes, it is 50% for Tōkyō–Hiroshima, which takes 3 hours 51 minutes (Central Japan Railway Company 2003d:17). Why is this the critical distance? The answer comes from needing to take account of the *whole* journey, not just the time between the shinkansen stations or airports. For while the shinkansen takes 3 hours 16 minutes from Tōkyō station to Okayama station, and the plane takes 1 hour 10 minutes from Haneda to Okayama, most passengers will be travelling to and from other points to get to and from these stations and airports. Indeed, it is likely that at least one other form of transport, whether it be local train or car, for example, will be needed at least once during the process. On top of this, greater times tend to be needed for checking-in before flights in comparison to using the shinkansen. As most passengers, at least businessmen, are likely to be going to city centres, and with airports tending to be located away from the city centre, so the shinkansen begins to gain the advantage, despite having a lower top speed than the plane. So in the case of Tōkyō–Okayama, as the total journey time by plane inclusive of check-in time and transfers is about 3 hours, the shinkansen can take 73% of the market, despite taking about an extra 2 hours to cover the 676 km distance (Central Japan Railway Company 2003d:17).

Access to shinkansen stations is as much an issue as access to airports. While Tanaka (1973:121) envisioned Niigata and Toyama becoming as accessible as the suburbs of Tōkyō, or for Matsue to 'be like a suburb of Kōchi, Okayama, Osaka', this overlooked the fact that most people in those three cities do not live next to the main station. Access to this can often take in excess of 20 or 30 minutes, even in a relatively small city. In Tōkyō, even Shinjuku station, in one of Tōkyō's key business districts, is nearly 15 minutes from Tōkyō station, and one still has to access that station as well as allow for time to change at Tōkyō, all of which is likely to take more than 10 minutes. While this is an issue whether the longer journey is done by shinkansen or plane, it is clear that the journey time to and from the shinkansen station or airport at either end of the journey can become the most significant area in time taken. Although not necessarily a means to improve overall journey times, it is only recently that 'park-and-ride' has begun to be promoted in Japan as a means to make the journey to the shinkansen station more convenient.

Considering the apparent advantage that the shinkansen has over planes in terms of time, it is a surprise that planes fly on some routes at all. In part there are historical reasons for this situation. Due to the high fares that the government made JNR set in the past, it meant that it was possible for airline companies to enter the market and charge lower fares – a situation which added to JNR's worsening finances and led to further fare increases as attempts were made to try to boost revenue (Kasai 2003:73). The problem for the shinkansen is that it cannot compete on price. One also has to remember that JR Tōkai is having to repay debt on lines other than the Tōkaidō Shinkansen. If this were not the case, it is likely that JR Tōkai would be in a position to reduce fares on the route by about 20%. Meanwhile planes have managed to increase their share of the Tōkyō–Ōsaka market in the past ten years, despite the advances made on the shinkansen in the period.

So while the shinkansen cannot compete with planes on price, there is a need for the companies to find other means by which to develop its market share. Speed is one. Convenience is another. As already mentioned, airports are not conveniently located in Japan, whereas most shinkansen stations are within easy access of city centres. On top of this, the shinkansen tends to have a significantly greater number of departures per day, which allows people to be more flexible in their scheduling. For example, while there are around 76 flights from Tōkyō to Ōsaka every day, there are nearly over 100 shinkansen departures. Airlines find it hard to compete in this respect. Although JAL (interview April 2004) claims that it is quicker to travel by plane between these two cities, both JAL and ANA have had to be smart in the way they market their product. For example, the use of 'air miles' and dealing with companies directly to try to encourage them to 'force' their employees to travel by air rather than shinkansen for certain journeys as it is cheaper has helped (interviews at JAL and ANA in April 2004). Similarly employees at some public institutions have to travel certain distances by air, unless they can come up with a valid way to break the journey and so have it considered as two shorter journeys, even if their area of research is railways (Mizutani, Nakano and Shōji interview April 2004). JR companies have so far tended to focus on general media campaigns and still have not introduced an equivalent to 'air miles', which some feel may actually lose them revenue since so many shinkansen passengers *are* prepared to pay the full price (Yasubuchi interview 2004).

Another area where the shinkansen can compete with planes is comfort. Planes tend to have the advantage when it comes to being able to take luggage, particularly in the case of the Tōkaidō Shinkansen where there is no significant space for cases.

However, as courier services are relatively inexpensive, with a typical 20 kg suitcase costing less than ¥1,500 for a next day delivery between Tōkyō and Ōsaka, most Japanese tend to use this option rather than carry it with them, so the plane's relative advantage in this respect is unlikely to be a deciding factor. However, the shinkansen has the potential to offer a more comfortable journey in terms of its facilities. Many of these features reflect changes in Japanese society and lifestyles and so will be looked at in Chapter 7. Similarly, although unlikely to influence most people's choice of transport mode, another area where the shinkansen compares favourably with planes is in terms of its environmental impact (see Chapter 7).

Returning to the issue of speed, it has already been mentioned how the number of stations along many lines has been increased. Although the reasons for such stations seem obvious, when one looks at the time savings and the reality of what such stations mean, it would appear that these stations are not as convenient as one might assume, and that perhaps they have more symbolic value than anything else. To not have a shinkansen station is perhaps a much greater negative, than having a shinkansen station is a positive. For example, let us consider Kakegawa station on the Tōkaidō Shinkansen. Kakegawa is served only by *Kodama* shinkansen, which have to wait for faster *Hikari* and *Nozomi* services to pass, in a process known as *taihi*.[3] This process occurs on other shinkansen lines too, as well as on many conventional lines.[4] If going to Tōkyō, for example, the passengers may need to change trains at Shizuoka, requiring a wait on the platform, for the next faster service.[5] What is more, it may be cheaper to take a conventional train for part of the journey, while not taking significantly longer and saving a wait at a shinkansen station.[6] It would also appear that sometimes it makes more sense to travel in the opposite direction initially to enable a faster service to be caught to the destination,[7] although I have not heard of people doing this.

It has to be remembered that not everyone is going to Tōkyō or even the major *Nozomi* or *Hayate* stops. While the relative lack of congestion on the Sanyō Shinkansen means that there are still many *Hikari* and *Kodama* services, the October 2003 timetable change to a predominantly *Nozomi* service on the Tōkaidō Shinkansen has clearly come at a price for those who need to use at least one non-*Nozomi*-stop station during their journey. Clearly when taking a train in Japan, whether it be a conventional one or a shinkansen, it is not always the case that the first train will get you to your destination quickest, and at smaller stations there is a need to study the timetable more carefully as, even on the shinkansen, there may be a need to catch specific trains. Consequently one can understand why having access to timetables, and being able to understand its use, is so important and why there is software for mobile phones, PDAs, etc. designed to help the travellers in Japan and that timetable-wielding business people running to catch trains is such a common sight in Japan. Those not used to these timetabling issues can get caught out. For example, a group of football fans were left stranded at Shizuoka station during the World Cup in 2002, leading to another shinkansen having to make an unscheduled stop to pick them up (*Mainichi Shimbun*, 12 June 2002).

The issue of fares has been touched upon on a number of occasions already. Fares in Japan are, on the whole, easy to understand. If we consider only JR fares, the system is the same for all six passenger companies. The basic fare is calculated on the basis of the shortest route between the two stations. Prices are decided for groups of distances, for example, 1–3 km, 4–6 km, and so on. As the distance increases, so does the range between the lower and upper figure. The average cost also decreases. In the case of the three Honshū companies for main line routes, if the journey is up to 3 km, the cost is ¥140, whereas for a 2,000 km journey the costs is ¥19,320 – the average cost per kilometre is ¥46.67 and ¥9.67 respectively. A ticket based on the distance charge alone does not allow the passenger to use faster services, for which a supplementary charge has to be paid. The amount is based on the type of service being used and the distance being travelled. Initially these special fares were introduced to help mend the government's finances following the expensive victory in the Russo-Japanese War of 1904–1905 (Nishio interview 2001). However, as the system worked so well, without a significant impact upon usage, it has remained ever since. Although there have been changes recently, the system has tended to mean that passengers using the shinkansen, for example, would actually have two tickets for the journey – one for the basic fare, and one to allow them to use the shinkansen.

There are no significant discounts for return journeys and reserved seats have to be paid for. There is a further supplement for using the 'green carriage' (equivalent to first class). However, prices hardly vary depending on the time of day or year travelled.[8] Timetables carry details of how fares are calculated, and generally it is understandable for the layman. This is not the case in Britain, for example, where fares often appear to have been calculated by multiplying the scores of three darts thrown randomly at a dart board. With different fares for different trains at different times of the day and on different days, let alone the problem of staff not always giving the same information, it is hard to have faith in the British system and it makes it hard to make a price comparison with Japan. However, if we take a journey such as London–Cardiff, the cost of a ticket can vary from around £24 for a return journey if the seats are booked at least a week in advance, to £110 for a return ticket that allows travel on any train ('open return'), a one-way 'open' ticket is £55. The published service time is 2 hours 5 minutes for the 232 km journey. This is about the same as Tōkyō–Nagano (which includes the relatively slow Tōkyō–Ōmiya section). This journey will cost ¥7,870 (about £40.36) for an unreserved seat one way, and takes between 1 hour and 25 minutes and 1 hour 56 minutes depending on the number of stations the *Asama* service stops at.

One point that needs to be noted in the calculation of shinkansen fares is that the distance between stations, as marked in timetables, is not the real distance. In order to help JNR finances initially as well as to try to prevent people from choosing to use conventional lines for part of the journey due to the reduced cost, the fare between shinkansen stations is based on the distance as if doing the same journey by conventional trains. The difference between this 'operating distance' and the 'real distance' can be quite significant, with the former usually being longer (see Appendix 2). In the case of Tōkyō–Shin-Ōsaka, for example, the difference is 37.2 km. If the charge were based on the 'real distance' it would save the passenger ¥530 each time they travelled. Ironically, it is the 'real distance' which is marked down by the track, as can be seen from the platform of stations, but passengers either do not worry about the difference or have merely come to accept the system.

In conclusion, although the construction of shinkansen lines is undoubtedly an expensive undertaking, and some lines were probably more expensive than they need have been, these costs were not the reason for JNR's demise and all lines are now all able to operate on a profitable basis. Furthermore, this chapter has shown why it is important for the research and development in improving the speed, in particular average speed between stations, of the shinkansen is so significant in terms of the benefits to the economy as well as to the finances and pride of the JR companies.

6 The need for training

The five principles of being a professional driver:

1 A professional is a person who takes pride in their job.
2 A professional is a person who takes on their job earnestly.
3 A professional is a person who does their job responsibly.
4 A professional is a person who always makes 'the best conditions'.
5 A professional is a person who gives their all in order to advance their abilities.

(Sign in JR West Shin-Ōsaka
Shinkansen Personnel Office)

Chapter 5 has considered the degree to which the shinkansen as a whole makes economic sense. Yet it is too simplistic to merely consider this issue when evaluating the shinkansen. To gain a full appreciation of how the shinkansen operates and how it reflects different aspects of Japanese society, it is necessary to consider other aspects too. This chapter will look at some of these key areas and how they further enable the JR companies to improve the performance of the shinkansen, and their own business performance, as they continue to compete with other forms of transport and try to encourage more people to travel.

When one thinks of the shinkansen, it is natural to picture a train. If one allows the image to develop, it may include the line, a station, passing scenery and the like. From there, it may also be the case the image will begin to include people, particularly passengers. I wonder how many people would picture the train crew or the JR personnel at the stations. While it is certainly the case that many visitors to Japan that I have spoken to have commented on the politeness of the train crew, often mentioning the bowing as they enter and exit a carriage, it is nonetheless the case that when asked about outstanding features of the shinkansen, the list tends to be more technically based. This technical side, referred to here as the 'hardware', is the train itself, the track, the stations and so on. It is the hardware where much of the money that was discussed in Chapter 5 is spent. It is the hardware that travels at speeds of up to 300 km/h. It is the hardware that almost always arrives on time. It was the hardware, in particular, that was the focus of my initial fieldtrips for this research. However, during one of these early trips, one employee at JR Tōkai, who is an engineer and is responsible for developing and improving the hardware, stressed that rather than the hardware, it was the 'software' that was key to the shinkansen's success (JR Tōkai interview January 2001). The software are the employees, the ones who maintain the trains, drive the trains, interact with the passengers both on the trains and at the stations. According to Kasai (2003:182), 'the human factor, the loyalty, skills and morale' of employees is an 'essential prerequisite' for running a company, which 'combined with the judgment of management and improvement of facilities, can work together to establish a permanent advantage over competitors and further enhance our raison d'être.'

Having been told of how important the software is in the operations of the shinkansen, I began to look into more detail about the way in which the employees are trained. As I did this, and as I spent more time at each JR company, so I became more and more aware of the apparent differences in approaches between each company. Not only that, there appeared to be completely different 'cultures' at each company. Yet, with these different approaches and cultures, the shinkansen operated by each company performs to the same high level. These are some of the areas that will be covered in this chapter.

Safety

Safety, more than any other, is the area of operations that I have found that all of the JR companies appear to emphasize. An example of this is can be seen in the words of Shima Hideo in a publication commemorating the thirtieth anniversary of the opening of the shinkansen, 'I hope that you'll always keep in mind that you have the lives of many passengers in your hands. This is not a light responsibility. Never forget it' (Semmens 2000:111). This sentiment was also held by Sogō. When I asked one of his sons about whether his father was disappointed not to have been at the opening ceremony in 1964, he said that his father's comments were 'I am happy as long as the shinkansen runs safely. That's all I wanted' (email correspondence with Sogō Shinsaku in 2004). The reason for this focus on safety may not seem unnatural, but it appears to go deeper than that. JNR's poor record, particularly up to the 1960s, and concerns amongst the public that the reform of JNR would lead to cost-cutting in the pursuit of profits and this would increase accidents (Matsuda 2002:95), appears to have engrained a deep, but healthy, concentration on the issue.

So how safe is the shinkansen? Perhaps the most graphic answer to this question is the statement that appears every year in the JR Tōkai data book and annual report; 'zero passenger fatalities and injuries due to train accidents such as derailment or collision' during passenger operations since operations began in 1964 (Central Japan Railway Company 2003c:13). The record is the same for each of the other shinkansen-operating companies. Furthermore, as people travelling do not have to use cars, perhaps the most dangerous way to travel, the shinkansen reduces the number of deaths on Japanese roads by about 1,800 annually and the number of serious injuries by 10,000 (*The Economist*, 21 February 1998).

132 *The need for training*

How is this level of safety achieved? Based on the comments that I have heard from staff at each of the JR companies, the answer would appear to be through the high quality training of their personnel, who are backed up through investment in appropriate hardware. The most symbolic of these investments are 'East-i', and the better-known, 'Dr Yellow'. Although this is how the trains are commonly referred to, their proper names are *denki kidō sōgō shikensha*. These trains are capable of performing numerous tests on the track, overhead power cables, communications, etc. while the train travels at normal shinkansen operating speeds. During the course of this research I have had the privilege to travel on both the old Dr Yellow, officially designated as T2, which was based on a 0-series shinkansen, and also its replacement, officially designated as T4, which is based on a 700-series shinkansen.[1] Being in T2 was a bit like a step back in time in terms of its technology, with instruments using pens to pick up vibrations from any minor defects in the line, for example, which were then drawn onto a long scrolling piece of paper, which is later analysed by staff back at JR Tōkai. One carriage was also stripped bare, contained items for maintenance, and was not fully pressurized.[2] T4 (see Figure 6.1), on the other hand, has all the latest technology on board. With the use of lasers and computers, it is able to make

Figure 6.1 (a) The T4 Dr Yellow at Kokura station – a video camera for recording the track is visible in the area beneath the headlights; (b) viewing points in the roof of the train allow first-hand observations of the connection between the pantograph and overhead power cables – a live video feed is also provided to a screen in one of the carriages; (c) staff monitoring information on board Dr Yellow.

much more accurate measurements. The data is downloaded onto computers at the end of the journey, although some analysis is also done on board. The seven carriage T4 and its equipment cost ¥5 billion (JR Tōkai interview 2004), about three times the cost per carriage as a standard 700-series shinkansen. Where necessary, maintenance workers can be easily dispatched to the appropriate section of track, confident in the knowledge that they are being sent to exactly the right point. The T2 Dr Yellow often operated in the evenings during its latter years, as its top speed was slower than the new series of shinkansen that had been introduced in recent years and so it needed more headway between it and any following passenger service. Neither East-i, nor the new T4 Dr Yellow, on the other hand, has this problem and so can operate at any time of the day, although one JR Tōkai employee said that using it during the day may lead to some passengers thinking that Dr Yellow was largely symbolic rather than making any real difference. The new Dr Yellow is also painted in a brighter shade of yellow (a colour chosen due to its association with maintenance vehicles)[3] than its predecessor, making it stand out more. Despite posters at stations pointing out what its role is when it was introduced, based on my observations at Kokura station when it happened to pass, it would seem that other than little boys, their parents and rail fans, most people are unaware of what this yellow train is.

The shinkansen system was built with safety in mind. For example, there are no services between midnight and six in the morning, except in very special circumstances – such as during the 2002 World Cup when some shinkansen returned to Tōkyō past midnight or when there have been disruptions due to a typhoon. This 6-hour period is reserved so that maintenance work can be carried out on a daily basis. It has the added bonus of meaning the shinkansen do not cause any noise pollution in urban areas during normal sleeping hours. Furthermore, as pointed out in Table 5.3, there are no level crossings and even on the mini-shinkansen their number was reduced and others redesigned during the conversion process.[4] In 2003, level crossings accidents accounted for 50% of the 833 railway accidents across Japan on JR and private lines (MLIT interview October 2004). One problem has been their relatively poor design (often just a coloured pole rather than a gate), while the lack of patience and irresponsibility of people crossing is a key factor. These are not problems for the shinkansen.

Another piece of hardware that backs up the shinkansen are the Centralized Train Control (CTC) centres, based in Tōkyō and Ōsaka (Tōkaidō and Sanyō Shinkansen), Tōkyō (JR East Shinkansen), and Fukuoka (Kyūshū Shinkansen). Although the appearance varies between the four centres, in essence they all bare a greater resemblance to space control centres, such as in Houston, than what one would expect to see at a railway operations centre. Through use of COMTRAC (Computer-Aided Traffic Control System), introduced in 1972 when the Sanyō Shinkansen opened, trains' progress can be constantly monitored with screens or boards showing where each train is. Another screen, for example, shows how behind schedule all shinkansen are along the line. This system helps to ensure that there are no collisions, as CTC can clearly see where all trains are, what speed they are going at, and where they are going to. Furthermore, the Leaky Coaxial

134 The need for training

Cable (LCX) by the track that allows this system to work also allows CTC to communicate directly with any shinkansen at any time by picking up a telephone. This was demonstrated to me on one visit as the CTC phoned a driver to warn him that a driver on a train in front had called to say that there was poor visibility in one area due to a line-side fire. As the electricity is the shinkansen's blood, so this communication cable is the shinkansen's nervous system.

Mention has previously been given to ATC. ATC essentially divides the track into blocks. The greater the number of blocks there are between two trains, the higher the speed allowed of the second train. If a train is above this speed, the brakes come on automatically and stay on until the speed drops to below the limit.[5] Many drivers try to keep the speed below the maximum as the ATC taking over can lead to a jolt and reduce the ride comfort for the passengers. This is quite a feat, when one considers that effectively it means remembering exactly where the speed limits change for over 500 km of track in some cases. On some services the train is not even expected to be going at the maximum possible speed (see Yamanouchi 2000:235). This is what I refer to as 'comfort driving'. ATC ensures that no signals can be passed at danger (SPAD), which has been an issue in Britain recently and has been a source of some disquiet amongst the general public, although I suspect many have themselves jumped a red light or gone faster than the legal limit while driving their car. This perhaps reveals a tendency for people to wish for a particular characteristic in general terms, but a tendency not to comply with this characteristic oneself. It is for this reason why, when it comes to the shinkansen, not only highly trained professional drivers are needed, who will obey rules, but also a mechanism such as ATC is needed in case there is a problem with the 'software' for any reason.

The ATC system itself is evolving. A new digital ATC has been introduced on the Tōhoku Shinkansen north of Morioka and on the Kyūshū Shinkansen. The technology allows braking in a more uniform and comfortable manner by calculating

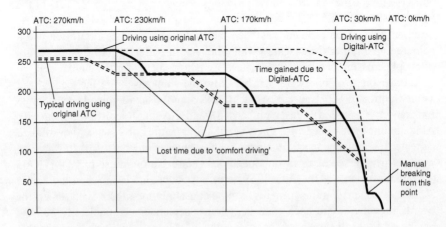

Figure 6.2 ATC and digital-ATC.

Source: Adapted from Central Japan Railway Company (2003d:11)

the safe distance to the preceding train or next station-stop rather than using the system of blocks. The result is not only an improvement in passenger comfort, although I suspect the assumption made by many is that this is due to improvements in the construction of the train, but also that trains can travel closer together and stay at higher speeds for longer. While the digital-ATC has been developed by JR Tōkai so that journey times along the Tōkaidō Shinkansen be further improved, the company was happy for it to be used on these other lines first so that it could be tested before introduction on that line (interviews with JR Kyūshu, JR East and JR Tōkai). The difference between the two systems can be seen in Figure 6.2. Even compared to driving on the limit, the digital-ATC provides a significant time saving. When one adds the three areas of 'lost time' due to 'comfort driving', although some services even after digital-ATC is introduced are unlikely to be expected to go at the maximum speed all of the time, the potential saving is even greater.

It should be noted that although ATC is an automated system, the shinkansen is not automatic. Without a driver, the train would go nowhere. Many modern conventional line trains in Japan,[6] and in other countries,[7] are more automated in this respect than the shinkansen. ATC merely applies the brakes as it approaches a point where it needs to stop, whether it be a station stop or due to a stopped train on the line. In the case of stopping at a station, the driver still has final control. As the train passes down the platform, and the speed drops below 30 km/h, the driver has to press a button in the shinkansen so as to prevent it coming to a halt at the wrong place as the automatic stopping point is before the end of the platform. The driver then uses the brakes to ensure that the train stops at exactly the correct point. Should the driver miss that spot, the train will reach a further '0 km/h' point beyond the end of the platform. That the system works was demonstrated in February 2003 when a shinkansen stopped 100 m from its proper place at Okayama station. When the reason was investigated, it was found that the driver was unconscious, and probably had been so for about eight minutes, during which time the shinkansen would have travelled some 26 km. However, thanks to ATC, the passengers of this train, and others, were never in any danger. That this incident is so rare and that the quality of the shinkansen service is so high is perhaps best demonstrated by the level of interest this story caused, not only in Japan, but also worldwide. However, these articles were not without their mistakes.[8]

Some people I have spoken to about the incident, have suggested that there was danger as the driver was unable to react to something that would lead to an emergency stop. This is true. However, even if awake and alert, a driver would need a straight track and remarkable eyesight to do anything about an obstacle on a track. At top speed, a driver cannot see the limit of a shinkansen's braking distance. Indeed, so useless can it be looking out of the front of a shinkansen, that I have noticed drivers only applying the windscreen wipers on rainy days when they are making their final approach into a platform. Furthermore, apparently lights were only put on the front of 0-series shinkansen as it was felt that passengers would not wish to travel on a train at night if they did not have them, rather than them serving any practical purpose (Nishio interview 2001). That this is the case was demonstrated to me when I travelled in the cab of a shinkansen at night and could only see any significant distance out of the front when I turned on the infrared feature on my video-recorder. That a driver cannot see far enough is one of the reasons why there are no level crossings and such strong measures are taken to prevent animals or people getting on the line.

There are three safety issues that need to be considered. First, there are environmental safety issues. These problems, which in Japan are significant, are addressed in Chapter 7. Next, is the maintenance of the trains themselves. Maintenance is done on a regular basis (see Table 6.1). Although there appears to have been some extension of when maintenance occurs, one has to bear in mind that there have also been improvements in design that mean that checks are not needed on such a regular basis.[9] Having visited both JR Tōkai's main maintenance depot at Hamamatsu and JR East's at Rifu near Sendai, the one thing that has impressed me more than anything else is how neat and tidy the depots are considering the nature of the work done there. Such an environment, which is apparently different to how it was during the JNR days and was improved, in the case of Hamamatsu, thanks to someone from Toyota being brought in to make recommendations (JR Tōkai interview January 2001), is clearly something that not only leads to greater efficiency but also helps maintain an excellent safety level. On top of this, trains are cleaned after every journey. In asking the JR companies what they would do if a shinkansen were delayed in its arrival at a terminal and cleaning it would mean that its departure on its next journey became delayed, the response has been that in most instances the train would be cleaned. The reasons for this are that cleanliness is what the passengers expect and deserve, and not providing it is likely to make a greater impression than if the train is delayed, especially as there may be an opportunity to make up lost time. Furthermore, shinkansen trains are significant investments and not maintaining them properly will shorten their working life and so increase the costs of the company.

The final safety issue that needs to be considered is unwanted interference with the line or trains. As already mentioned, there are no level crossings on the shinkansen lines. However, in areas where access would be relatively easy (i.e. other than tunnels and elevated sections), the lines have high barriers or fences by them. So high are these barriers that rail enthusiasts know of the spots where their position is such that it is still possible to get a good view and photograph of the shinkansen. These barriers are there not only to prevent accidental incursion into the area, but also deliberate. The issue of suicide will be addressed in Chapter 7, but terrorism is also a concern, especially since Japan experienced a terrorist attack on the Tōkyō Underground in 1995 at the hands of the Aum Shinrikyō cult. One of my fieldtrips to Japan was shortly after '9/11', and one JR official was quick to point out that one of the advantages of trains over planes is that they could not be used in this way. This may be the case. Yet, with over 1,200 passengers on board, the shinkansen itself, especially given its symbolic status and its economic importance, could potentially be a target. Indeed, this has been the theme for many of the books and the film that were mentioned in Chapter 3. President Kasai of JR Tōkai has suggested that anyone trespassing on the Tōkaidō Shinkansen would be arrested within 15 minutes. My experience in taking

Table 6.1 Shinkansen maintenance

Type of inspection	1964	1971	1984	1997	2002
Light inspection	Within 48 hours	Within 48 hours	Within 48 hours	Within 48 hours	Within 48 hours
Regular inspection (general condition and equipment)	Every 30 days or within 20,000 km	Every 30 days or within 30,000 km	Every 30 days or within 30,000 km	Every 30 days or within 30,000 km	Every 30 days or within 30,000 km
Bogie inspection	Every 12 months or within 240,000 km	Every 12 months or within 300,000 km	Every 12 months or within 300,000 km	Every 1 year or within 450,000 km	Every 18 months[a] or within 600,000 km[b]
Complete inspection	Every 24 months or within 750,000 km	Every 30 months or within 900,000 km	Every 36 months or within 900,000 km	Every 3 years or within 900,000 km	Every 3 years[c] or within 1,200,000 km[d]
Special inspection	As necessary	As necessary	As necessary	As necessary	As necessary
ATC inspection	Within 48 hours	Within 48 hours	Within 48 hours	Within 48 hours	Within 48 hours
Complete ATC inspection	Every 60 days or within 40,000 km	Every 60 days or within 60,000 km	Every 60 days or within 40,000 km	Every 60 days or within 40,000 km	Every 60 days or within 40,000 km

Sources: Kōsoku Tetsudō Kenkyūkai 2003:175; JR Tōkai interview January 2001; JR East interview August 2003.

Notes

a New sets are first inspected after 30 months.
b 450,000 km for 0-series shinkansen.
c New sets are first inspected after 4 years.
d 900,000 km for 0-series shinkansen.

photographs near the line in Shizuoka (mentioned in Chapter 1) may suggest that this is correct. That drivers are able to communicate with CTC so easily, and with a train passing any particular point every couple of minutes, should also mean that the chances of getting a response there quickly are high. Although symbolically the Tōkaidō Shinkansen would probably be the most prone to attack, all the companies are having to take extra steps to be careful. For example, JR Kyūshū has now followed the lead of airline companies by not allowing any visitors in the cab with the driver of its shinkansen, while JR East closed all the dustbins on its shinkansen, with those on platforms replaced with see-through fronts to make inspections easier as well as make it easier for people to see which bin their rubbish should go in (see Chapter 7). However, it would be naïve to suggest that a determined terrorist could not find a means to hijack or attack a shinkansen train, line or station. The supports of the elevated sections of the shinkansen lines, which are often relatively easy to access, are particularly vulnerable. But the companies appear to be trying to keep things in perspective. In my opinion this approach is the sensible course. To start checking the baggage of all passengers before they board, for example, would likely cause greater anxiety amongst passengers and is still unlikely to be fool-proof against the determined terrorist.

Punctuality and preciseness

Perhaps the most well-known aspect of the shinkansen is its punctuality, it has probably become *the* example of Japanese efficiency. Even Queen Elizabeth II was reported to have said that 'I hear the shinkansen is more accurate than a watch' when she travelled on a shinkansen in 1975.[10] Littlewood (1996:51) comments that 'In part, it is Japanese efficiency that provides a focus for western suspicions; we've always been uneasy about people who make the trains run on time', while Marzuki (2002) suggests 'The perfect timing of the arrival and departure of the Shinkansen is a classic example of how the Japanese strive for perfection and punctuality.' Furthermore, Konno (1984:74) argues that the Tōkaidō Shinkansen's success was due to the elimination of waiting time, 'for the Japanese are an impatient nation', which Konno puts down to its agricultural past and the nature of Japanese seasons. Although one will see people waiting for trains in Japan, often taking the time to use their mobile phone or practice their golf swing with an umbrella, more often than not the passengers will arrive only a short time before the shinkansen enters the platform. The announcement of a late running shinkansen is met with a mixed reaction of shock and derision.

So how is this punctuality achieved? First, it has to be noted that most shinkansen services operate with spare capacity. In other words, if a train is late, it is likely that it could go faster than normal, while still being within the ATC limits, and make up time. In Britain, as I discovered when travelling in the cab and talking to one driver, the system is 'once there is a green light, go as fast as possible until arriving at a station or a red light'. If the train becomes delayed, it will not make up time and it is likely to have an impact on other services. Based on my observations in the cab of shinkansen, even many *Nozomi* services, for example, particularly on the Tōkaidō Shinkansen, will not travel at their maximum speed all of the time. So when a service is late, particularly those that have greater spare capacity built in (e.g. a *Hikari* service takes about 30 minutes longer than a *Nozomi* service between Tōkyō and Shin-Ōsaka despite sometimes making a similar numbers of stops), there is potential to make up time. I had the opportunity to experience this for myself in March 2000 when my shinkansen left Tōkyō 17 minutes late. By Shin-Ōsaka the train was still 9 minutes behind schedule – for which the JR Tōkai conductor apologized profusely. When the shinkansen arrived at Okayama terminal, the JR West conductor continued the apologies although by now the train was only one minute late, having covered the final stage in about the same time as a *Nozomi* service despite stopping at two extra intermediate stops. While Chapter 5 mentioned that the battle with airlines was being fought over margins of just a few minutes difference in time taken, clearly the JR companies have identified the predictability of an on-time service as being the shinkansen's greater weapon. It is because there is spare capacity that the definition of a 'late train' is one that is more than one minute behind schedule, whereas it is 14 minutes in France, 15 minutes in Italy and 10 minutes in Britain (Mito 2002:7).

While Japan's traditional agricultural customs has been suggested as one reason for the punctuality of railways, another suggestion is due to geography. Mito (2002:52) claims that as travel was traditionally done on foot, towns tended to be located at fairly regular intervals of 16 km. This in itself does not guarantee a punctual railway. However, having stations so regularly spaced helped when the need to have the railways better run due to an increase in demand while the number of coaches remained limited during the Taishō period (Mito 2002:58). In other words, rather than sensitivity to the seasons, it was industrialization that demanded punctuality that was the driving force, so that now not arriving on time is not a matter of 'bad luck' but is an 'incident' (Mito 2002:115).

Another important factor is that drivers are taught to respect the clock. Upon entering the company all employees receive a pocket-clock. Even in the most modern shinkansen there is a space carved into the dashboard where this clock sits (see Figure 6.4b). Drivers check that their clock is accurate with the national clock when they collect their work orders shortly before going to board the shinkansen (see Figure 6.4a), and the time is constantly checked throughout the journey. Indeed, on the Tōkaidō shinkansen, there is a readout on one of the monitors that shows how many seconds ahead or behind schedule the service is. Whenever I have travelled in the cab, it has spent most of its time hovering near '0', with drivers expected to stay within 15 seconds of schedule. Such a timing mechanism is not present in all shinkansen. For example, when I asked the driver of an E4-series about it during one journey in the cab, he confirmed that he did not have this device and so had to use his own judgement for much of the journey in relation to his progress compared to the timetable. Should the service deviate too much from the scheduled time, the driver makes necessary gradual adjustments so that the service is back on time some minutes later, rather than making sudden adjustments that may adversely affect passenger comfort. As a shinkansen passes a station, whether it stops there or not, the driver checks what time the train was supposed to have passed that station, what time it did pass it, and what time it is due to pass or reach the next station. This constant attention to the time keeps the trains on schedule. It is worth pausing for a moment to compare the situation I have just described to the one often encountered in Britain. In many intercity trains, there is only a clock in the cab if the driver is wearing their own wristwatch. Although I am sure experience teaches them how long a journey is likely to take, there is less interest in the time. A clock, after all, is something traditionally given to British train drivers upon their retirement!

While it is often said that you can set your watch by the shinkansen timetable, this is not strictly accurate. For the time in timetables tends to only show the hour and minute of the departure. The actual time of train departures, which are used by the crew, are detailed to the second. In other words, a 9:54 departure may actually be scheduled to depart at 9:54:30. Yet, as most watches lose or gain a few seconds each week, assuming they even show the seconds, it is usually safe to use the time according to the shinkansen. However, for the railways to work effectively and efficiently, it needs not only high-quality staff, but also the co-operation of the passengers that are using the trains (see Mito 2002:101). This is something that does not always exist in other countries. For example, in Britain, people do not move along the platform, especially at cramped underground stations, but crowd near the entrance/exit, which naturally means that the carriages near those points become crowded while others remain relatively empty. In Japan, people are usually lined up along the platforms – particularly in the case of the shinkansen, where many passengers will have reserved seats – and remain in an appropriate queue as indicated by painted lines on the platform depending whether they are taking the first or second train to leave that platform (see Figure 6.3).[11] Furthermore, while in Britain, people will try to board the train before letting other passengers alight, in Japan passengers are aware that the process is less stressful and quicker if alighting passengers are allowed to get off first.[12] Perhaps these cultural differences explain why when a shinkansen is about to make a stop at a station, in Japanese only the station name is mentioned, but the English announcement says that 'a *brief* stop' will be made at the station.

In fiscal year 2003–2004, the average delay on the Tōkaidō Shinkansen fell to 0.1 minutes, that is, 6 seconds, and JR Tōkai has set itself the target of getting the average to 0 (JR Tōkai interview October 2004). Naturally, when nearly 300 trains are operating per day, with most being within 1 minute their scheduled arrival (delays are calculated to the minute, not the second, for this basis), even if one train were delayed by 5 hours, the average delay for the day would still only be 1 minute. The record on other lines is equally impressive – especially considering that with fewer services, the possibility of the average being pulled up by a few late-running services is higher. For example, during its first five months in operation, the Kyūshū Shinkansen had an average delay of only 0.4 minutes – although it was higher for those leaving Shin-Yatsushiro due to the occasional late arrival of the connecting conventional line service (JR Kyūshū interview

The need for training 141

Figure 6.3 (a) An E2-1000-series shinkansen at Tōkyō Station. Triangular markings along the platform edge show where the doors of certain shinkansen, depending on the series, etc., will be. Two dark lines (to the right of the column) show passengers where to queue if taking the first train, to its left are two more lines for those passengers queuing for the following train. (b) An E2-series shinkansen passing an E4-series shinkansen at Takasaki station.

October 2004). The Tōhoku and Jōetsu Shinkansen also tend to have average delays of under 40 seconds (Yamanouchi 2000:151).

While undoubtedly the ability to have the shinkansen arrive on time virtually all of the time is impressive, this should still not necessarily, in my opinion, be seen as an example of Japanese efficiency. Rather than being an example of efficiency, I would suggest that it is an example of effectiveness. Indeed, I would argue that Japan is rarely efficient, although recent changes to the economy may be forcing it in that direction. Japan's output has been, at times, remarkable. Japanese people get the job done. But observations at companies in various sectors and markets make it hard for me to accept the idea that Japanese workers or companies, particularly the larger ones, are particularly efficient. I would like to stress that this is by no means a criticism. Indeed, I think the importance placed upon efficiency, as is found in many 'Western' countries, is often misplaced and has been, and continues to be, one of the weaknesses that blights their companies and economics. In the case of the shinkansen, I suspect part of the problem is one

142 *The need for training*

of semantics as 'efficiency' is often used when 'effectiveness' or another word altogether would be more appropriate.

This 'effectiveness' could be described also as competence or professionalism. While, the ability to have the trains arrive and depart at their advertised time is the best known example of it, there are other areas of the shinkansen operations that reveal that this is a trait that continues throughout their work and is not restricted to clock-watching. For example, the shinkansen will always stop in the correct place on the platform (see Figure 6.4e). This is something that impresses many visitors to Japan. It is also something that is important for the passengers wishing to catch the train who are lined up in boxes that are in line with their allotted carriage. This perfect positioning is achieved by the driver who lines up a mark in the window of the cab with a board at the end of the platform. Having been in the cab on a number of occasions, it still never ceases to amaze me how the driver can stop a train, around 400 m in length and weighing around 630 tons in the case of a 700-series shinkansen, without either a significant jolting-stop or

Figure 6.4 (a) Having checked their clocks aginast the 'standard' in the foreground, the crew confirm their orders for their next journey. (b) A driver confirms that he has acknowledged the change in ATC limit on an E2-series shinkanses. (c) A new JR Tōkai recruit practices on a shinkansen simulator. (d) The conductor checks along the platform that there are no problems. (e) Marks on the platform show where the train should stop – the dot in between the numbers relating to the series of shinkansen should be in the middle of the cab door, as in this case.

by merely drifting to a slow stop (which would likely lead to a delay since the time available to stop at a station is only 50 seconds) with such accuracy. While many think that computerization may lead to a lowering of demands on the 'software', with digital ATC drivers will be effectively having to manually stop shinkansen from around 75 km/h rather than the current 30 km/h! Once the train has stopped, the conductor in the rear carriage also checks that the mark appropriate to their series of shinkansen is position in line with the middle of their door before opening the doors to all of the carriages.[13]

Training system – professionalism

Let us now look at how the high quality of performance of the shinkansen is achieved and what it reveals about Japanese society more generally. Although methods vary between the JR companies, a theme which will be returned to in another section, the essential principle is the same; employees start from the bottom and work their way up after having gained experience in key areas of the operations. This is quite typical for a Japanese organization. It is for this reason why many Japanese will identify themselves with their company name rather than their profession, for over a period of years, regardless of what they may have originally studied for or what their aspirations may be, they are likely to be rotated from department to department. Employees are a resource of the company and they have little or no say in where they are transferred to, even if it means being relocated to another part of Japan.

In terms of the JR companies there is an obvious division between the railway operations and the other operations. These two sides of the companies remain largely separate with no cross-over in staff except for managerial and senior positions, where staff will have had experience in some aspect of the railway operations of the company. The only other exception to this appears to be at JR Tōkai, where most employees are expected to get a shinkansen driving license as to achieve this requires hard work and dedication to observing certain rules and expectations of the company, while also experiencing working on the company's most important area of business (JR Tōkai interviews and Nakamura 2004). Over 70% of applicants pass the shinkansen driving licence test (*The Japan Times*, 5 January 2003). In fiscal 2003, 201 people gained a shinkansen license, taking the total to 4,764 (MLIT interview October 2004).

Considering the railway operations, the pattern of training is essentially the same in each company. Upon entering the company employees will undergo an initial induction. Following this, time is spent working at a station, gaining experience in dealing with the public as well as with the arrival and departure of trains. Following this, they proceed on to train crew operations, including time being spent as a ticket inspector and conductor. Having completed this, they are then able to proceed to become a driver. In some companies time will be spent as a driver of a conventional train before being able to become a shinkansen driver. The length of each stage varies between companies. JR Tōkai, for example, has a relatively small conventional line network, and staff are required in greater

144   *The need for training*

numbers in shinkansen operations. As a consequence the process to becoming a shinkansen driver can be relatively quick. JR East, on the other hand, is primarily a conventional line company and so the rate, and opportunity, to become a shinkansen driver is slower and harder – consequently the average age of JR East shinkansen drivers is higher.

This information may give the impression that the shinkansen driver's position is the pinnacle of the railway operating side of the JR companies. In many respects it is, although even JR Tōkai (interview October 2004) claim that there are no differences from the company's point of view, the view of drivers that I have spoken to would suggest that this is the case. This partly reflects the status of the shinkansen itself as the shinkansen is associated with a level of performance, service and safety to be aspired to throughout the company. However, it is also one of the better paid positions as drivers receive a base salary, which is then topped up depending upon the distance they drive. Naturally shinkansen drivers cover greater distances than conventional train drivers.

Visiting a JR training centre is like visiting a school or a university. Indeed, JR East are hoping that their new state-of-the-art training centre in Shirakawa, which cost some ¥26 billion, will become the site of the International Railway University (Matsuda 2002:132, 147). The training centres include dormitories as well as teaching rooms, self-study rooms, and re-creations of various parts of railway operations which can be practiced on. With the development in IT, it has also been possible to develop driving simulators which accurately recreate the conditions in the driving cab, allowing the employee to gain experience in driving a train and be taught and tested on how to cope with various problems that they could potentially face once they have got their license (see Figure 6.4c). On one visit I was allowed to try JR East's new simulator, which gives the sense of being in a real shinkansen cab as the whole simulator, together with other props, moves.

The training centres provide an opportunity to learn about what is expected in all aspects of the job. The time spent there helps in the development of the company culture. The way in which they should dress is emphasized, particularly at JR Tōkai, with signs in corridors and lifts reminding employees of their expected form of dress, hair cut, jewellery etc. Such posters reminds me of those I saw when teaching at schools in Seto, Aichi Prefecture. One driver I spoke to about them said that she believed the rules are a good thing 'especially as it is a nice uniform' and was better than there being too much variation (JR Tōkai interview April 2004). It is hard to imagine such rules in Britain, as some may see them as an infringement upon the individual's rights and individuality. However, in Japan, while there are no shortage of individuals, they learn from a young age that sometimes they have to 'to put their own needs second to those of a wider group' (Hendry 2003:57). Buruma (2001:221) notes that 'consensus may often be a public façade, but that façade counts for a great deal in Japanese life', with 'etiquette' rather than 'morality' or 'decency' preventing violent confrontation in Japan, creating a system 'based almost entirely on known human relationships; without a group to relate it to, it tends to break down rather quickly'. However, we should not get too carried away with painting Japanese organizations as harmonious. Without the rebellious nature of people such as Kasai, Matsuda and Ide within JNR, it is unlikely that the reform of JNR would have ever happened.

Yet identity within the group and the identity of the group is clearly important. The dress code is one way that this identity is developed and maintained. The result within JR is that personnel, like employees at many other Japanese companies, always look smart and presentable. Drivers, like others performing public functions in Japan, wear white gloves (see Figure 6.4b), which Buruma (2001:9) sees as reflecting the 'Japanese feeling for purity' and Hendry (1999:87–8) points out symbolically separates the employee from the public. Consequently train crews, particularly drivers, look 'more like airline pilots than railway men' (BBC TV 2001). Such examples, as Littlewood (1996:55) comments, further help to reinforce stereotypes based on differences as 'we're left with the same old pattern of antithesis: England – coarse, rude, messy, anarchic; Japan – refined, polite, neat, disciplined. At no stage do we relinquish our obsession with difference and catch sight of the points of contact.' However, differences, although sometimes minor, can be representative of more deeply engrained differences in attitudes and ways. So Goldthorpe (1993:xv) suggests Japan may have developed a 'new form of industrial society...combining "modern" economic and technical rationality with "traditional values" and norms that diverge from those followed in the west – and that western theorists of industrialism have wished to see as functionally imperative'.

Although JR companies are now associated with these apparently typical Japanese attributes of being 'refined, neat, polite, and disciplined', the image of JNR was quite different. Indeed, so bad was the culture at JNR, that when JR East was created, for example, its first Managing Director ordered that staff 'for better or worse, do the exact opposite to what you did during the JNR days' (Matsuda 2002:3). Yet despite this, Matsuda (2002:63) points to the '*Poppo-ya Damashii* (Railwayman's Spirit)' as being the reason why the change from JNR to JR overnight on 30 March and 1 April 1987 was achieved without any trains stopping or any accidents. Clearly there were some aspects of the culture at JNR that were not too bad, but there was a cancer that was eating at the heart of the organization. However, even when its image was at its worst, the shinkansen remained the area of JNR's operations that was the most refined, and the area where JNR functioned efficiently (Kasai 2003:16), although it may have not been to the standard of today. Although a piece of fiction, one of the more shocking scenes in the film *Shinkansen Daibakuha* is when a driver drinks a cup of tea while driving. Such a sight today is unimaginable in a Japanese train. Although perhaps safe to do, the message which such actions send across to the public are not considered appropriate.

The environment of the cab of the shinkansen is worth mentioning as it further reveals how the training is done and how staff operate. While the cabs of conventional trains, particularly in countries such as Britain, are filled with the noise of occasional buzzers or alarms that must be responded to, including the so-called 'deadman's handle', otherwise the train will come to a stop. The shinkansen has no need for such a system due to ATC. The only sounds that will be usually heard in the cab of a shinkansen are the bell to notify the driver of a change to the ATC limit, the driver's own voice and any announcements that are being made over the public address system. Although usually alone, the driver will continue to make confirmations that changes in ATC status has been acknowledged. This is done by voice and by hand gestures (see Figure 6.4b). Such actions are not atypical in Japan. Not only does this extra action demonstrate that the appropriate information has been acknowledged, it also helps to keep the person alert. In Japan it is common to learn the correct way to do things with the body, not just the brain (Buruma 2001:156). The result is that there are appropriate forms and postures, known as *kata*, for different actions and jobs. Although in traditional arts this is largely done by 'mimicking one's masters' (Buruma 2001:70), in business it is taught at the training centres. While the shinkansen driver is generally out of sight of passengers, if you stand near the door, the voice acknowledgements can often be heard, suggesting that the actions are done regardless of whether an observer is present or not. Similar actions by other members of the shinkansen crew, station staff, as well as drivers and crew on conventional lines, are more easy to spot. This can lead to the conclusion that 'life in Japan seem highly theatrical to the outsider' (Buruma 2001:70).

One problem that all the JR companies appear to be facing is that culture is changing. Upon one visit to JR East's training centre, I was told that the greatest challenge was how to respond to the 'reset button culture' (JR East interview 2003). These days, when something goes wrong with our computers, Playstations, etc., the immediate reaction of what could be termed the 'reset generation' is to press 'reset' or its equivalent. On a train, there is no reset button or undo feature. Employees have to be taught that they are responsible for their actions, and that the safety of the passengers is in their hands. The issue of safety is further emphasized by the existence of a museum of train crashes at that centre, which reminds the employees of what happens if they get it wrong. It is a graphic, but I suspect, an effective system and reminds me of the huge reports that pilots have to read following the conclusion of investigations into major air disasters. Clearly what the JR companies need, therefore, is employees who can be trained effectively. The result, like at many other major Japanese companies, is to employ a mixture of graduates from university and secondary schools. It is not common practice for employees to come from other companies. For employees to come from another JR or other railway company is almost unimaginable. Japanese companies want 'promising but relatively inexperienced students' or 'raw materials (*sōzai*)' rather than those who have 'previous job-related experience' who would have 'moulding experience' and so requiring the company to 'break him of acquired work habits' through retraining so that they can be adjusted to the company's 'colour' (Ishida 1993:150).

Not a perfect system

The impression given thus far may suggest that the shinkansen is some utopian railway. While undoubtedly highly impressive, it is not perfect. There have been mistakes and there have been accidents. Indeed, as early as the second day of operations there were problems with the ticketing system and problems surrounding the Tōkaidō Shinkansen led to media stories for many months thereafter (Yamanouchi 2000:126, 128). Although Yamanouchi (2000:138) claims it took around ten years for all of the 'bugs' to be eliminated, there have been problems in later years with passengers being injured by breaking windows, houses near lines being struck by ballast (Nishimoto 2003:64) or even parts from trains themselves. However, what has concerned me in investigating some of these is that while companies seem to be well prepared for *expected* incidents, they are not always so good at responding to the *unexpected*. This is a problem that appears to be inherent within Japan. Other problems suggest that, despite the high level of training and professionalism generally, mistakes do happen. Let me give some examples of problems that have occurred.

While living in Japan, I once attempted to travel from Nagoya to Kyōto by shinkansen but was prevented from doing do as the first shinkansen of the day had collided with a maintenance vehicle. The result was that no shinkansen had left Nagoya towards Shin-Ōsaka by mid-afternoon. The reason for this was that the system means that trains have already been allocated service codes and schedules. As a result, a shuttle service between Nagoya and Shin-Ōsaka using the shinkansen sets available could not be established, losing JR Tōkai revenue in the process. While this sort of problem may be uncommon, perhaps demonstrated by the fact that one senior JR Tōkai official could remember this incident when I discussed it with him some eight years later, it does suggest a certain rigidity and inflexibility.

One of the major problems that the shinkansen has to contend with is forces of nature (see Chapter 7) and Chapter 4 has already pointed to how some construction was designed for this purpose. Yet, the JR companies appear to have problems responding sometimes when the shinkansen is affected by nature or some related problem. For example, one acquaintance was caught on a shinkansen that became stranded for about 18 hours on 11 September 2000 due to a particularly severe typhoon. Despite the fact that the toilets were overflowing and there was no food left on board, nothing was apparently done to try to make the conditions more pleasant for those trapped, or to attempt to evacuate the passengers. Following an incident in July 2000 when a crow hit a pantograph, which left passengers without water and air-conditioning for two hours, JR East responded by stepping up the training the drivers receive so that they are better prepared for dealing with unexpected incidents (*The Japan Times*, 5 January 2003). However, this response, like the lack of response in the previous case, reveal an apparent lack of desire or ability to call upon the emergency services or to deal with the problem. This is not a problem unique to the railway industry. Even the government has been relatively ineffective in response to disasters. For example, the JAL flight 123 crash on Mount Osutaka in 1985, the Great Hanshin Earthquake in 1995 and the Tōkaimura nuclear accident in 1999. In both the JAL and Great Hanshin Earthquake cases, foreign help was rejected although both times this could, and probably would have, saved extra lives (BBC TV 1999b). It is beyond

148 *The need for training*

the scope of this study to try to develop a theory for this apparent deficiency, but it is nonetheless necessary to note its apparent existence despite the excellent overall safety record, quality of training and effectiveness of many companies and organizations in Japan.

The problems associated to level crossings in Japan, and why there are none on the main shinkansen lines has previously been mentioned. That this needs to be the way was demonstrated by an accident that occurred on the Yamagata Shinkansen in 2001, when an old man apparently became confused at a level crossing during heavy snow and began to drive up the railway line rather than continue along the road. This lead to a collision with a shinkansen and his death. Although the shinkansen does not travel at normal shinkansen speeds along the line, it is still fast and so the existence of level crossings, particularly in an area that suffers from heavy snowfall and a relatively aged population,[14] was always likely to raise the potential for an accident. Clearly more needs to be done to prevent further accidents.

Since 1995, JR Tōkai has to be careful in the way it defines 'zero passenger fatalities' following an incident at Mishima station when a boy running to catch a shinkansen got his arm trapped in the closing doors while he was still on the platform. Although the boy was seen by staff on the train and the platform, both directly and by using the monitors, it was assumed that he was running alongside the train waving to a friend or relative. Sadly, rather being sure and safe, the train continued on its way and the boy was killed when he collided with the barrier at the end of the platform. JR Tōkai had to pay ¥60 million compensation (Umehara 2002:258). The accident happened despite the fact that announcements are made to remind passengers not to run for trains and despite further announcements, followed by ear-piercing warning buzzers, to stand behind a line on the platform which keeps them a safe distance from the trains as they arrive and depart. It also occurred despite the staff being trained to be prepared for such incidents. As a result of this accident, and one at Shin-Ōkubo station on the Yamanote Line in 2001, barriers are now being installed so that it is easier to segregate those on the platform from those who are actually trying to board trains. At some JR East shinkansen stations this means having moving barriers and/or tapes (some of which are automatic), due to the great variation in train designs and lengths. It should be further noted that while there have been no passenger fatalities on the Tōkaidō Shinkansen, there have been deaths of maintenance workers – with five being killed when hit by the first *Kodama* of the day only a few weeks after the opening of the line (Yamanouchi 2000:128).

Two other minor incidents involving the shinkansen have also made the news in recent years. That they made the news is probably indicative of the importance of the shinkansen in Japanese society and the standards that people expect. Consequently both incidents probably became stories due to concerns about other aspects of Japanese life that the media had. In the first, it was reported how a driver was disciplined after his shinkansen arrived at Tōkyō station over a minute late, having only made the relatively short journey from Oi depot. The story suggested that the reason for the tardiness was that the driver had left the cab to go back to another carriage where he had left his cap when he had washed his face there. Due to ATC there was no danger of an accident, but as the driver was not at the controls, the train slowed on a gradient to a point whereby it became delayed such that the time could not be made up. The implication of the story was that rules are sometimes too strict, with the driver apparently fearful of what would have happened had he arrived in Tōkyō not wearing his cap. Although this is the story that ran in the media, when asked, a JR Tōkai employee suggested that the real reason why the driver had left his cab and gone to the toilets was due to a sudden attack of ill-health.[15]

The other story I would suggest was about link between the individuals involved and more significantly the continuing rise in use of mobile phones, their various functions, and how these are becoming a distraction rather than about the quality of shinkansen staff. In this particular incident, a driver was disciplined in 2003 after he was found to have been taking photographs of the views from his cab on his mobile phone and sending them to 'his girlfriend'. JR Tōkai were alerted after the woman's *husband* found the pictures (*Mainichi Shimbun*, 18 November 2003). Although there was no danger posed by his actions, it does show how the training systems at companies appear to be facing increasing challenges from the 'reset generation' who like to play with their various electronic gadgets, especially mobile phones.

Different cultures and competition between JRs

Although discussions on the reform of JNR tend to focus upon the finances, mention has already been given to the bad culture that was felt to have existed at JNR, and there are many who believe that this was no small part due to the labour unions, particularly Kokurō and Dōrō (Kasai 2003:172). So reform was a means to deal with them, whether they were a bad thing or not (Mutō interview April 2004). This seems plausible given that the person who took responsibility for JNR reform was Nakasone, who was also a supporter of privatization more generally, and who also embarked on education reform as a means to deal with the militant Nikkyōso (Japan Teachers' Union) (Hood 2001:84). While the alterations at JR companies has led to people speaking of the 'Miracle of privatization' as the 'the attitude of JNR workers seemed to change overnight', Kasai (2003:172) argues that rather than a significant change, the process allowed the workers 'to express themselves more freely'. This would certainly confirm the comments made by Matsuda earlier. It also confirms the suggestion of Bradshaw and Lawton Smith (2000:8–9) that while deregulation and competition can stimulate improvements, when union power is weakened efficiency can rise dramatically.

Various labour unions now exist within the JR group. Kokurō still exists, but has been joined by JR Rengō (Japan Railway Trade Unions Confederation), which is a part of Rengō (Japanese Trade Union Confederation), and JR Sōren (Japan Confederation of Railway Workers' Union). Various other smaller unions also exist. While JR Rengō's 'guiding principles are those of the democratic labour movement within the private sector', JR Sōren is more 'hard line' like

150 The need for training

Dōrō (Kasai 2003:174). The differences in union membership at the JR companies are shown in Figure 6.5. Using Kasai's views as a base, it could be said that JR Hokkaidō, JR East and JR Freight are closer to JNR in relation to union membership and those unions' positions. Consequently, these companies' cultures, since union membership both reflects and helps shape attitudes of employees, may be similar to JNR's. This is something that Kawashima (interview October 2004) agrees with. Regarding the shinkansen, the companies' cultural differences can be seen in the different ways in which the shinkansen is operated. For example, JR East, while being a 100% private company now and having introduced the innovative mini-shinkansen which are joined with other shinkansen while on the Tōhoku Shinkansen, is more traditional and like JNR. This can be seen by its reluctance introduce shorter but more frequent shinkansen services on less-used areas of its network as JR West has done, and having its operations divided on a regional basis rather than on a shinkansen/conventional line basis as is done at JR Tōkai.

Yet at all the JR companies, the attitudes and the culture since the JNR days have clearly changed. Noguchi (1990:184) suggests that the phrases that encapsulate this change is that prior to 1987 the public felt the JNR workers' attitude was one of '*notte yaru*' ('I'll let you ride'), whereas what was desired and is expressed now is '*notte itadaku*' ('Please ride with me'). As an example of this, it is now usually possible to see into the drivers cab of many conventional trains if you stand in the front carriage, while during the JNR days it was normal for the driver to pull down the curtain, as still happens in the evening or areas where there are many tunnels due to reflections, so that they could not be seen. But more significantly, strikes are now almost unimaginable upon the JR lines. In 1975

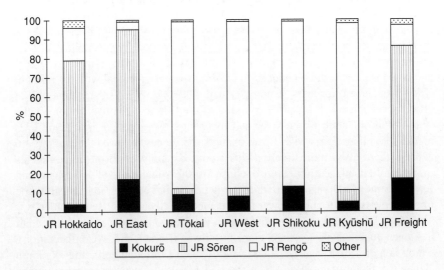

Figure 6.5 Union membership in JR companies (January 2004).

Source: Japanese Trade Union Federation – interview April 2004.

The need for training 151

there was a strike, and for eight days the whole JNR network stood still. The assumption made by the unions was that this would paralyse the economy and lead to their demands being met. In reality, it further harmed JNR's image and finances as people and companies sought alternatives to using JNR (Kasai 2003:48).

While Kasai suggests that changes were due to giving employees the opportunity to express themselves, even during the JNR days there were opportunities for employees to do this. One form of expression that exists in most Japanese companies is *kaizen* or *gemba kaizen* (see Imai 1997). This process gives employees the chance to make suggestions about the way in which improvements may be made to the operations of the organization. In its simplest terms it is similar to a customer-feedback system, which Japanese companies also use. However, with *kaizen* the advantage is that it encourages workers to feel a part of their organization and to want to be involved with its development. The most notable example of *kaizen* in relation to the shinkansen is probably Gāla-Yuzawa station and resort (discussed elsewhere), which was put forward in the final days of JNR by seven section men in the area (Oikawa and Morokawa 1996:67). Such is the importance of *kaizen* that when visiting some areas of the shinkansen-operating companies, for example the Hamamatsu works of JR Tōkai, the number of suggestions made and implemented were included within the introductory documentation.

In terms of developing a connection to the company, clearly the training centres play an important role. At the same time they begin to instil the company's culture by training the employees in the company's ways and rules. However, one advantage that the JR companies have in comparison to many other service sector companies, such as hotels (see Ogbonna and Harris 2002), is that they have a large '*ura*', or 'reverse side', which is out of sight of the customers. As a consequence, many of the signs that I saw at training centres were also visible, together with other slogans (such as the one at the start of this chapter), in parts of stations and employee facilities that customers would not normally see. This continual exposure to company messages is likely to increase their chances of adoption.

Not only are the cultures of the JR companies different, which may be surprising given that they were all part of JNR fairly recently and that have to comply with various national laws, as well as rules and regulations monitored by MLIT, but there is also a real rivalry between them. As mentioned elsewhere, the locations where the JR companies are in direct competition with each other are limited, so it is unclear what the motivations for this rivalry is, other than pride and an overwhelming sense of identification with one's own company. I do not want to overemphasize the rivalry, as clearly the JR companies do co-operate and even tend to refer to each other as '*Tōkai-san*', '*Kyūshū-san*', etc. where '*san*' is the suffix normally added to people's names to mean Mr, Mrs, Miss, Ms, etc. However, the difference in culture and rivalry is worthy of mention as it may be significant when considering whether the shinkansen can be imported by other countries (see Chapter 8).

It is difficult to express how the cultures of the four shinkansen-operating JR companies' cultures vary. Culture by its very nature is something that is difficult to gauge and express (see Chapter 1). Let me first attempt to express the differences by turning to Japanese history. In relation to three of Japan's most significant figures, Ōda Nobunaga, Toyotomi Hideyoshi and Tokugawa Ieyasu, it is said that their characters, and those of most people, can be summarized in how they would deal with a nightingale that does not sing. Ōda's policy would be to kill it, Toyotomi's to ignore it, and Tokugawa's to be patient and wait for it to sing. In relation to the shinkansen-operating companies, I would suggest that JR Tōkai would be Ōda, JR West would be Toyotomi and JR East would be Tokugawa. As for JR Kyūshū, their emphasis on good design and being Japanese-like, as demonstrated by the design of many of their trains and the use of traditional material on the inside of the 800-series, would suggest that they would replace the nightingale with a crane or a swallow, appreciate its beauty and write a *haiku* about it.

Having travelled extensively throughout Japan, I have begun to wonder whether the cultures of the JR companies reflect their regions as much as their training programmes. This is certainly a possibility since one of the positive aspects of the reform of JNR was to establish a system whereby the JR companies are more responsive to their customer base, which is largely intra-regional rather than inter-regional. Employees also tend to come from the respective regions. JR Tōkai, for example, has many of the elements of conservatism that I experienced when teaching in Aichi, the heart of the JR Tōkai region. Similarly, Nippon Sharyō, one of the shinkansen manufacturers who appear to have a particularly close relationship to JR Tōkai, are also based Aichi and appear to have a more conservative approach to manufacturing than Hitachi, for example (see Chapter 3). Japan is not a homogeneous society, although it is often treated as such by Japanologists and Japanese alike.[16] There are significant differences, which date back to times before Japan became unified as one country and a standard language was established. Yet today dialects still exist and are regularly used in day to day conversations. So different are they from standard Japanese, that visitors from outside some regions, whether Japanese or foreigners that have learnt standard Japanese have little chance of understanding what is said. While such differences can be found through the archipelago, it is the Kantō – Kansai differences that are most celebrated, encapsulated in the rivalry between Tōkyō and Ōsaka, in as diverse ways as the language used, rivalry between Yomiuri Giants and Hanshin Tigers baseball teams, and even the usage of escalators at stations.[17] I do not want to suggest that regional cultures are the only influence on JR cultures – clearly there are other significant factors, such as the culture created by the differences in customer base (i.e. business travellers, tourists, short-distance travellers etc.), but it does strike me that the regional culture is one that is highly significant.

Figure 6.6 shows the cultures of the shinkansen-operating companies and their relationship with some other cultures. While each of the company cultures is presented as a single entity, one has to bare in mind that each almost certainly has various sub-cultures within it. If seen from a distance and a rather simplistic level, differences may not seem apparent. This 'unified' culture could perhaps be

The need for training 153

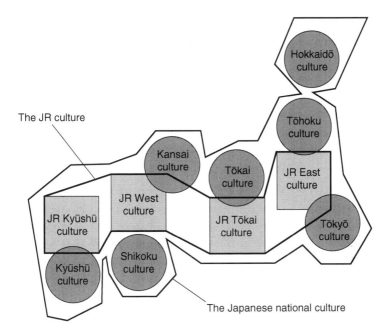

Figure 6.6 Cultures of the shinkansen-operating JR companies.

described as a 'JR culture' or it may even be a 'railway company culture' if it appears similar to that of other railway-operating companies. Examination of other companies and organizations would, I suggest, lead to similar relationships being developed – for example, 'a car company culture' which is made up of quite distinct Toyota, Honda, Nissan, Daihatsu and Mitsubishi cultures, and 'a beer company culture' which is made up of quite distinct Asahi, Suntory, Sapporo and Kirin cultures. As suggested earlier, looking at the cultures of the companies may also draw one to see similarities between companies from certain regions – that is, the existence of regional cultures. The degree to which the cultures of the JR companies overlap with these regional cultures is debatable and may vary from company to company. For many companies (the case of JR East is shown in Figure 6.6), there is the existence of at least two regional cultures that it is probably linked to. But if these regional cultures were not looked at in detail, or if one concentrated on the similarities rather than the differences, what one is left with is what could perhaps be described as the 'Japanese national culture'. In this way it is possible to see how when asked about what 'Japanese culture' is, people may give somewhat different answers although there is likely to be significant overlap, which tend to be the foundations of stereotypes and clichés.

Although JR Tōkai is the *Central Japan* Railway Company, it is JR West that has the most borders with other JR companies. This need to deal with so many other companies, let alone arguably having the greatest balancing act of the four

154 *The need for training*

shinkansen-operating companies in terms of whether its railway focus should be on commuter trains, rural trains or the shinkansen (Yasubuchi interview 2004), appears to lead to its culture being stretched in many different directions. This may partly explain why JR West has reminded me of teaching second year students at lower secondary school in Japan. These students are no longer free and easy-going first year students, which if continuing the analogy could be paralleled to JR Kyūshū, nor have they become exam-orientated third year students, cf. JR Tōkai or even JR East. The need to keep good relations with five JR companies appears to impact upon JR West's culture. However, having a border with another JR company does not necessarily mean there are good relations. I have found the relationship between JR Tōkai and JR East to be very competitive. This rivalry is despite the close relationship between many of the senior staff at the two companies and their work which was critical in bringing about the reform of JNR. Indeed, Kasai (2003:175) himself notes that in the reform process the JR companies were 'treated as the offspring of JNR who would amicably and flexibly use the assets after the break-up, as if they were brothers and sisters'. However, like brothers and sisters, there is squabbling and a desire to have the most attention. Initially I thought that much of this was jovial banter, such as when one JR Tōkai employee I was with laughed when walking through Tōkyō station and it was announced that there were delays on the Yamagata Shinkansen. Yet, when JR Tōkai opened their shinkansen station at Shinagawa, JR East put a large sign next to the shinkansen entrance and several hundred metres from JR East's conventional line entrance, proclaiming that the station is 'JR East Shinagawa Station'. Later, JR East opened a ticket office at the entrance, taking business away from JR Tōkai's nearby ticket office and ensuring that they receive a share of the income from the sale of shinkansen tickets at the station. Indeed, Shinagawa appears to have been one of the contributing factors to the rivalry between the two companies. The plan for a shinkansen station at Shinagawa dated back to 1975, but when JR Tōkai resurrected the plan, JR East voiced opposition to it and only after much negotiating over costs, etc. could an agreement be reached (Umehara 2002:342–3). While the dispute was going on, JR East was investigating the possibility of a connection between the Tōhoku Shinkansen and Tōkaidō Shinkansen. This was something that had been proposed during the JNR days (Kawashima 2004:96–9) and although JR East was apparently keen for this link to be built as it would increase the capacity of its part of Tōkyō station by being able to send trains to use the Oi depot rather than having to have trains cleaned at Tōkyō or reverse direction and take them to Ueno station or the Tabata depot, JR Tōkai was less keen. In the end, due to technical issues (the difference in electrical currents in particular), the high usage of the Tōkaidō Shinkansen in the Tōkyō area, the cost and the relationship between the two companies at the time, the plan never materialized (Nishimoto 2003:56; Umehara 2002:244–6).

While many of the cultural differences between the JR companies are out of sight of the public and may not appear particularly obvious, some differences are more visible. For example, if you stand at the end of platform 14 or 23 at Tōkyō station it is possible to see both JR Tōkai and JR East shinkansen platforms at the same time. If you observe the behaviour of the train crew and the actions of the station staff, it is often possible to see the difference between JR Tōkai's relatively regimented approach and JR East's more chaotic approach. I do not want to suggest that one way is better or worse than the other. Both appear, based on both company's excellent safety record, effective. However, they are clearly different and reflect the differing nature of their companies. These differences are also reflected in the designs of trains – only three (one of which is a JR West train) different designs are visible at JR Tōkai's platforms, but six, with further differences in formations, at JR East's. Just as owners are said to look like their dogs and vice-versa, perhaps JR companies reflect the look of their shinkansen and vice-versa. Variety is also reflected in the design of *meishi* (business cards) which are the window to someone's identity at a Japanese organization. JR Tōkai designs tend to be very uniform in design. JR East *meishi* tend to vary in colour a great deal, with an emphasis placed on them being made of recycled material. The greatest diversity, with the most emphasis placed on exciting pictures and design, are those of JR Kyūshū. Clearly the *meishi* are not only a window to the individual's identity, but also reflect the organization's identity too.[18]

While there are differences between JR East and JR Tōkai, for example, there are areas of co-operation too. For example, one will hold the last shinkansen of the day for passengers of the other company if their shinkansen has been delayed for any reason and there are passengers wanting to transfer. Also JR Tōkai provides all the station staff at Shin-Yokohama, even for the conventional lines, despite the station being in JR East territory (interview with former station master of Shin-Yokohama, April 2004). It is easy to become too focused upon differences and lose sight that much of the time there is similarity. Yet at the same time to concentrate solely on the similarities would to be to lose sight of the real diversity that does exist, and to merely treat the companies as a homogeneous entity. The JR companies are different. They are rivals much of the time, but they are not enemies, and this 'rivalry itself is not a negative force' (Yamanouchi 2000:207). They do co-operate with each other too. However, the differences do mean that we cannot always speak of 'JR companies', let alone 'Japanese railways', as a single entity.

Marketing the shinkansen

Although there is co-operation, particularly through the JR Group activities whereby joint marketing campaigns take it in turns in promoting travel to one of the JR company's territories, given the differences in cultures and the different markets they serve, it is hardly a surprise that each company has subtle differences in the way it markets the shinkansen. As already mentioned, there is diversity in design of the shinkansen itself. Its external appearance may be significant as in Japan the 'wrapping' can be extremely important, 'a veritable "cultural template" or... "cultural design" ' (Hendry 1999:172). The wrapping of a gift can be just as critical as the contents. That the shinkansen's wrapping is significant is reflected not only in how many comment on how clean the trains tend to be on both the inside and outside, but also that Sanrio uses the slogan '*itsumo pika pika*' ('always sparkling') in relation to one of its characters.

The original shinkansen was ivory with a blue stripe along it. Although this may not seem particularly eye-catching compared to some of the colours that can be seen on shinkansen and conventional trains now, compared to the rather dull brown and cream colours, although a source of some nostalgia now, that all JNR express trains used to be, the shinkansen's colours must have seemed quite refreshing. When the shinkansen network was expanded north of Tōkyō, the new shinkansen were given a green stripe on the ivory base rather than a blue stripe, so as to reflect the colour of Spring and the hope this brings for the people living in the 'snow country'. However, this choice of colour was apparently a source of disappointment to some people that this shinkansen served, as they had hoped to have the same colour as used for the Tōkaidō and Sanyō Shinkansen (Umehara 2002:123). It should be noted that the perception of what is 'blue' and 'green' is different for Japanese people than it is for others.[19] The word '*aoi*' means 'blue'. However, this word is also used to refer to the colour used to mean 'go' on traffic lights, as well as the colour of grass, suggesting a light 'green' would have been appropriate for the Tōhoku and Jōetsu Shinkansen. However, it was '*Midori*' ('green'), a darker shade of green that was used.

Since the original colours were introduced to the shinkansen, changes have been made to this 'wrapping'. Some of these have been subtle, such as changing the design of the stripe or making the background colour a little different, while others have been more pronounced. When the 400-series mini-shinkansen was introduced, it became the first shinkansen not to use ivory/white as its base colour, instead using silver. However, it still kept the green stripe, with green having also been chosen as JR East's corporate colour. The 500-series became the first shinkansen on the Tōkaidō and Sanyō Shinkansen not to use ivory/white as its base, again being predominantly silver. It kept the familiar blue stripe, though a little paler than in the past to make it closer to JR West's corporate colour, together with a larger grey stripe above it. But JR West has also made changes to its other shinkansen that remain solely on the Sanyō Shinkansen. Interestingly, none of these use JR West's corporate colour,[20] but tend to use a combination of grey and green or yellow. The difference is quite striking. For although the *Hikari Rail Star* is a 700-series, most agree that it looks more striking than the standard 700-series, although the only difference is the colour it is wrapped in.

This process of changing original colours, which often includes other modifications to the interior décor and technical upgrades, is referred to as a 'renewal'. It has now taken place on 0-, 100- (see Figure 6.7), 200- (see Figure 6.9), and E1-series shinkansen. The advantage that this 'renewal' gives is that it makes trains more recognizable for passengers. In the centre of the platform it is not always easy to recognize the series of the shinkansen. By using colours to represent certain services, although the difference between an E2-series *Asama* service and E2-series *Hayate* service is only subtle, further helps to ensure that passengers catch the correct train. However, these alterations can also improve business. This

The need for training 157

Figure 6.7 A 100-series-renewal shinkansen – the familiar blue stripe of old has been replaced by grey and green stripes, with further improvements made to the seating as JR West tries to encourage people to use its regular *Kodama* services.

idea is encapsulated in the Chinese saying 'Although Peach and Plum Blossoms Do Not Speak, a Path Will Naturally Appear Underneath Them', meaning that 'if you produce something of beauty, a small road is formed naturally by people who are attracted to it' (Kasai 2003:183). This fits with JR Kyūshū's unofficial motto of 'good design is good business'. For this reason, companies pay a lot of attention to the appearance of their trains. While I have heard employees at one JR company joke that JR Kyūshū can never hope to make a profit so they have to concentrate on winning awards mentioned in Chapter 3, JR Kyūshū may have the last laugh as it looked to make its first profit in Fiscal 2004.

A recent development in the 'wrapping' of the shinkansen in the JR companies' marketing mechanism has been the use of adverts. JR West, for example, placed large badges of the Hanshin Tigers and Daiei Fukuoka Hawks on the front of its *Hikari Rail Star* during the baseball championship which the two teams, representing areas at either end of the Sanyō Shinkansen, had reached in 2003. JR East, in celebrating the twentieth anniversary of Tōkyō Disneyland, and in trying to encourage people to use its trains to go there, posted pictures of the Disneyland castle and characters on some of its older, non-'renewed' 200-series shinkansen (see Figure 6.8). Although one may think that it would be hard to see such adverts when a train is moving at over 200 km/h, one has to remember that they are visible

158 *The need for training*

Figure 6.8 (a) A 200-series shinkansen during the campaign promoting Tōkyō Disneyland; (b) a 200-2000-series shinkansen, which was based on the 100-series design rather than the 0-series design as used for the 200-series.

when the train is at a station or moving relatively slowly along the urban stretch south of Ōmiya, particularly between Ueno and Tōkyo. JR Tōkai, who tend to be rather conservative, and have resisted altering the colours of their shinkansen, to the extent that using their corporate colour, orange, rather than blue stripes on their shinkansen would never be considered (JR Tōkai interview October 2001), celebrated the opening of Shinagawa station and the change to a predominantly *Nozomi* service by placing advertisements on many of their shinkansen. This campaign was given a name, the 'Ambitious Japan!' campaign (see Figure 6.10). The campaign included posters, advertisements in all media and even the release of a single by popular boy-band Tokio. The lyrics of the song included the words '*nozomi*' and '*hikari*' in their normal meanings of 'hope' and 'light' but no mention of 'shinkansen' or '*densha*' (train). Breaking with JR Tōkai's conservatism, many shinkansen also had the slogan written on the side of carriages 1 and 16, while related logos appeared on the sides of many other carriages. This must have been particularly galling for those in Shizuoka since shinkansen promoting the campaign stopped there, despite the fact that *Nozomi* continue to pass by (see Chapter 4). The slogan 'Ambitious Japan' is derived from the well-known words of William Smith Clark, an American who taught at Sapporo Agricultural College (now Hokkaidō University) when he said to his students upon his return to the

Figure 6.9 A 200-series-renewal shinkansen passing typical Tōkyō landscape – a commuter train, an expressway with its high noise-reducing walls, interesting as well as bland architecture, and a golf-driving-range with its enormous nets.

States, 'Boys, be ambitious' (Matsuda 2002:153; Kawashima interview October 2004; JR Tōkai interview April 2004). Although the opening of Shinagawa station would not appear at first to deserve a link to Clark's invitation to be pioneering, it should be noted that the area around the station has seen much redevelopment in recent years and is now home to many of Japan's leading companies.[21] Furthermore, the change, which was the origins for the proposal, allows JR Tōkai to run an extra four trains per hour from Tōkyō to Ōsaka, although this has not yet happened.[22]

It is interesting to reflect upon JR Kyūshū's campaign that was launched at the opening of the Kyūshū Shinkansen. The main slogan was '*Minami-kara*', literally 'From the South', as, despite the fact that officially the shinkansen distances are measured from Hakata and Shin-Yatsushiro, due to its larger size and importance, it was being suggested that the line was coming from Kagoshima in the south. Riding upon the train, one is soon aware of how JR Kyūshū has to respond to a different market than the other JR companies, for announcements, which are accompanied by a somewhat more modern and funky jingle than on other trains, are given in Korean and Chinese as well as Japanese and English, which are standard on all shinkansen.[23] Another announcement also appears to reflect JR

160 *The need for training*

Figure 6.10 (a) A 700-series shinkansen with the Ambitious Japan! livery passes a school in Tōkyō; (b) commuters using the new Shinagawa station; (c) the Ambitious Japan! Logo on a 300-series shinkansen; (d) checking the platform at Shinagawa station shortly before a train arrives.

Kyūshū's connections with JR West, who trained the new shinkansen drivers. For just as JR West drivers used to announce when the 500-series had reached 300 km/h, which is still done on the digital display, so the driver of the Kyūshū Shinkansen announces to the passengers when they have reached their top speed, just as a pilot tells passengers of speed, altitude, weather conditions etc.

Yet, I would say that JR Tōkai is perhaps the company that is best at marketing, particularly in the area of PR. For although the safety record and performance record, in terms of lack of delays and cancellations, is exemplary at each JR company, it is the record of the Tōkaidō Shinkansen that is most well known and is featured in the media. This is doubtless in part due to the nature of the line, its symbolic nature as well as being the busiest high-speed railway line in the world. This can mean that it gets the brunt of the media attention when services on it and other lines, for example, are affected by bad weather. However, the role of JR Tōkai cannot be overlooked in producing press releases that the media want to print. A prime example of this is that the media, including foreign media, gave the impression that JR Tōkai was the first to have female shinkansen drivers, while in fact it was JR West (see Chapter 7). Yet in other respects, one has to be critical of what JR Tōkai has done in its marketing. The introduction of the *Nozomi* service is similar to the process whereby a product may be kept at the same price while it is made smaller, then some time later a 'new' version of the product is introduced, at about the same size as the original product used to be, but at a higher price. This practice is often seen with confectionary. Although the *Nozomi* service is faster than *Hikari* services have ever been, in terms of where the *Nozomi* stops, it is what the *Hikari* used to be. The *Hikari* has been devalued, and only with the change to the timetable in 2003 has the price difference between them been reduced.

In conclusion, this chapter has shown that while the shinkansen was originally introduced during the days of JNR and was one of the only consistently good parts of JNR operations, since the creation of the JR companies, although there have been some relatively minor problems, the high standards have not only been maintained, but improved upon. This has been achieved by a keen focus on the human element – the software – of the shinkansen, while at the same time supporting this with continued improvements in the hardware. All of these developments and training have been done by using subtly different techniques that have both created and reflected differences in the cultures of the JR companies.

7 Mirror of Japan

The previous chapters have looked at the shinkansen and what this tells us about the JR companies, Japan and Japanese society. In this chapter, I will change the focus by considering various significant areas of life in Japan and considering how the shinkansen responds to or reflects these. However, while the focus has altered, it will be seen how in many areas the relationship between society and the shinkansen, as with those covered in previous chapters, is not one-way and that the shinkansen may have a role to play in promoting further changes in society or even maintaining the status-quo and traditions of old.

Coping with forces of nature

Japan is a country of exquisite natural beauty. But it presents huge challenges to the shinkansen. While the shinkansen has to cope with Japan's demanding topography, it is the more occasional forces of nature which are unleashed upon Japan in a particularly ferocious manner that offer the biggest challenge to the Japanese people and to the shinkansen. Nature has to be controlled as best as possible. Although the Japanese have a love of nature, with seasonal tourism being an obvious symbol of this, Buruma (2001:65) suggests it is not 'simply a matter of love' but is 'tinged with a deep fear of the unpredictable forces it can unleash' and this leads to an 'abhorrence' to 'nature in the raw'. I think this view is perhaps overstated, though it is true that much of what is 'natural' in Japan, as Buruma suggests, even some of that which is 'worshipped', such as beautiful gardens, are not natural, but are man-made.

Japan is a wet country. While there are not many rainy days, when it does rain, it can do so with great force. So concentrated is the rainfall during the early summer, that it is referred to as the 'rainy season'. Figure 7.1 shows the average precipitation for four cities around Japan. For comparison, data for London is also provided. Heavy rainfall is a problem for trains as it reduces adhesion with the track, increasing braking distances. Should the rainfall become so great that it leads to flooding, there is a risk of derailment. On the shinkansen lines rain is a particular problem on elevated sections as the rain cannot be absorbed in the same way as through ballast into the ground. The shinkansen's response to this problem is to have sensors alongside the track that measure the rainfall. When the amount that has fallen within a one-hour period or over 24 hours reaches certain points, the speed limit of the shinkansen is reduced, until it reaches a point where services are suspended.

Rainfall can bring two further concerns. The first of these is landslides, which have caused a number of accidents over the years on conventional lines. This is a particular problem in Japan with the tendency for lines to be washed away about four times that in Europe or the United States (Noguchi and Fujii 2000:52). Landslides are not a problem for the shinkansen, however, as cuttings are reinforced, and usually covered, with concrete (see Figure 7.6a). Although this is not particularly pleasing on the eye at first, the design of these concrete banks, using a grid, does allow for accumulation of little soil and vegetation over time so that, in due course, people will be unaware that there is concrete under the surface. The other problem that can come at the same time as rain is, however, a concern for the shinkansen. With some trains around 400 m in length and 4 m in height, the shinkansen has a large profile, which makes it vulnerable to wind. As with rain, there is not much that can actually be done to combat this problem, although the walls installed to reduce noise pollution help to a degree. The shinkansen's main response is to have sensors along the track and the speed limit to be gradually reduced, to zero if necessary, should the wind speed get too high.

In the Summer each year Japan is buffeted by typhoons, with wind speeds in excess of 120 km/h and heavy rainfall. During such severe weather, the Japanese turn to the TV for information. When appropriate, 'yellow' cautions and 'red' warnings are given. Yellow cautions merely mean that people need to take extra care, while red warnings mean that travel is not recommended, and schools, for

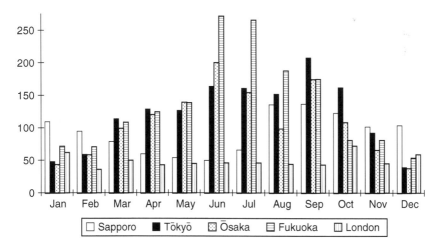

Figure 7.1 Average precipitation (mm) in selected cities.

Source: Nihon Tōkei Nenkan (2004:23). Data for London taken from www.worldclimate.com and is average for Heathrow Airport between 1981 and 1990.

Note
Japanese data are averages for 1971 to 2000.

example, will remain closed. TV coverage of typhoons tends to follow a certain pattern; a map will show the current position and strength of the typhoon and its possible and likely trajectory; images will be shown of the coast and cities being pounded, often with some hapless reporter trying to narrate what is occurring as they are soaked and blown about; and details are given of the typhoon's impact upon transport services. With this begins a PR battle for the shinkansen with its arch the rival, the airplane. Planes have the advantage that they can fly around typhoons, but if the airport is being hit, they have nowhere to land, and, as consequence, will not be allowed to take off if the destination is in the strike zone of the typhoon. While the shinkansen can do nothing about its route, it is less vulnerable to the wind than planes, and so sometimes services continue even when the planes have been grounded. However, there are times when the shinkansen does become stranded due to typhoons (see Chapter 6), so it is by no means invincible.

While rain and wind are occasional problems throughout the year, snow is purely seasonal. However, the challenge that it brings is very different as merely reducing speeds is not an effective option. While rain will drain away and wind will eventually die down, snow continues to accumulate. It also has the associated problems of ice and low temperatures. It is easy to forget just how destructive and disruptive snow is. Just as the shark has gained an undeservedly bad image thanks to popular works of fiction and the big screen, so authors, musicians and the big screen have much to answer for in terms of snow's overly positive image. That snow can be a source of beauty,[1] amusement and money-making ventures has undoubtedly fuelled this image. It is partly for these latter reasons, together with the need to maintain a transport link between major cities, that the shinkansen not only has to combat the snow problem, but there is also a desire on the part of the JR companies to do so. In no other country can I imagine a railway company, especially a high-speed railway company, using 'We want snow' as a marketing slogan, as JR East did in Winter 2003/2004.

Anyone who has seen BBC World's weather will have seen the statistic that 10 mm of snow is equivalent to 1 mm of rain. However, while there may be an appreciation that snow does build up, perhaps most are not aware just how much snow can fall. Table 7.1 shows the problems that Japan and the shinkansen has to contend with every year. So great is the snowfall in some areas that houses have effectively an extra entrance on the first floor as the one on the ground floor is inaccessible during Winter. Every morning people will help to shift the snow from the road in front of their house, and will climb on the house to push the fresh snow off the roof before the weight becomes unbearable. Deaths from such activities are not uncommon. Snow is a real problem. It is one that can particularly affect the railways. In 1981, the Hokuriku Line was closed for ten consecutive days, while airports and roads were only closed for 24 hours. The Tōkaidō Shinkansen also suffered due to 'moderate snowfall' around Sekigahara, with delays, cancellations and damage to trains (Konno 1984:78; Kōsoku Tetsudō Kenkyūkai 2003:176, 7; Yamanouchi 2000:131–2). As a consequence the shinkansen gained an image of being 'weak' in snow. Consequently, the Tōhoku, Jōetsu and Hokuriku Shinkansen, which pass through areas with far more significant

Table 7.1 Snowfall on the shinkansen

Line	Location	Average year		Record for 1993–2003	
		Total snowfall (in cm)	Highest snowfall in one day (in cm)	Total snowfall (in cm)	Highest snowfall in one day (in cm)
Tōkaidō Shinkansen	Ōmi Nagaoka[a]	45	35	81	44
Tōhoku Shinkansen	Kōriyama	20	18	29	28
	Fukushima	26	26	51	51
	Sendai	16	16	28	28
	Furukawa	18	15	35	30
	Ichinoseki	20	19	30	29
	Kitakami	51	34	73	54
	Morioka	33	25	59	45
	Iwate-Numakunai	43	26	78	43
	Ninohe	37	25	67	47
	Hachinohe	29	22	62	39
Jōetsu Shinkansen	Jōmō-Kōgen	32	28	63	40
	Echigo-Yuzawa	271	91	415	122
	Urasa	227	79	375	112
	Nagaoka	141	55	261	89
	Niigata	40	28	88	56
Hokuriku Shinkansen	Nagano	30	—	50	—

Source: Kōsoku Tetsudō Kenkyūkai (2003:177).
Note
a A conventional line station between Sekigahara and Maibara, near the Tōkaidō Shinkansen.

snowfall, need to have countermeasures to deal with the snow problem. Countermeasures common to all these lines are the use of tunnels and snow sheds. The trains themselves are also fitted with special heating devices that work to melt the accumulation of snow that can build up under the train. The 200-series shinkansen, the first shinkansen used on the Tōhoku and Jōetsu Shinkansen, had extra bodywork under the carriage compared to the 0-series to further help prevent the build up of snow and ice. Points on the line are also warmed up to prevent freezing, with hot air fans used to blow snow and ice off the overhead power cables (Kōsoku Tetsudō Kenkyūkai 2003:176–87; Umehara 2002:66–9).

Line-specific countermeasures also exist. This is not merely due to the fact that the levels of snowfall are different, but also due to the differing nature of the snow. Although 'the wrong type of snow' reason for problems on British railways became the source of some ridicule, there *are* differences and the ways to deal with them also need to be different. The snow on the Tōhoku Shinkansen and Tōkaidō Shinkansen usually comes from the Pacific and tends to be relatively dry and light (Kōsoku Tetsudō Kenkyūkai 2003:179). Although this type of snow can build up, and in some areas sprinklers are used, in most areas it is possible for the shinkansen to continue operating with much of the snow on the track being blown out of the way by the draught caused by the train itself. However, the snow that falls on the Jōetsu Shinkansen comes from the Japan Sea and tends to be heavier and more moist in nature. To deal with this snow, sprinklers have to be used. Sprinklers do exist in some areas on other lines too, with cameras installed under the platform at certain stations so that station masters and CTC can see whether there is any build up of snow on shinkansen. In particularly bad cases, trains can be held at the station while snow and ice are removed. However, on the Jōetsu Shinkansen, these sprinklers are of significantly greater need and strength. For example, around Echigo-Yuzawa station, sprinklers come on automatically when either there is a certain amount of snowfall or if the temperature falls below a certain level. When the sprinklers come on, it releases 0.71 for each $1 m^2$ of track each minute. This is equivalent to 42 mm of rain falling in an hour (Kōsoku Tetsudō Kenkyūkai 2003:183), about eight times the power of those used on the Tōkaidō Shinksansen. However, it is not as great as the equivalent of 73 mm/hour which the sprinklers on a 3 km section of track around Kitakami on the Tōhoku Shinkansen produce. However, as this area does not suffer from the same levels or type of snow as on the Jōestu Shinkansen, they only need come on for a relatively short time. That such strong sprinklers are needed is in part due the use of concrete slabs under the track rather than ballast, which helps to reduce accumulation of snowfall and melted snowfall. I visited Echigo-Yuzawa in January 2004 and observed how the sprinklers work, despite it being a comparatively mild winter. Although the statistics given are impressive, nothing can prepare you for the sight or incredible noise of these sprinklers in action. After only a few minutes the area resembled the railway line in *Sen to Chihiro no Kami Kakushi* (*Spirited Away* (2003)) where the train ploughs through the water (Figure 7.2).

The system is highly effective, as reflected by the slogan '*yuki date makenai zo*' ('when it comes to snow, I do not lose') used for one of the Sanrio shinkansen characters. In its first year only one Jōestu Shinkansen was cancelled and hardly any delayed. However, the sprinklers are not cheap – neither the initial outlay nor the continued running costs. The cost per winter of the system is about ¥1.3 billion (Yamanouchi 2000:133). Part of the problem is that the water has to be warmed, to about 10° C (although in one area it is raised to 45.2° C), before being expelled from the sprinklers. Without access to plentiful hot spring water, which is used to keep some roads clear of snow and ice in *onsen* resorts like Kinosaki, a significant amount of warming needs to be done. In the case of Echigo-Yuzawa, water is taken from the Dai-Shimizu tunnel (which JR East also bottles and sells) and also from what has just been pumped onto the line, so helping with its disposal. However, this latter water is colder than the fresh water and also contains rubbish and dirt, so requires extra heating and cleaning. While snow presents a challenge for the shinkansen, as it is also an important source of revenue so it will remain a challenge the shinkansen will battle to overcome.

While all of the given natural phenomenon are, to some degree, predictable, the shinkansen's greatest challenge from nature is not, other than the knowledge that they will happen sometime, somewhere. Japan sits upon one of the most tectonically active zones of the world. Consequently 10% of earthquakes each year occur

Figure 7.2 (a) Sprinklers in action at Echigo-Yuzawa – note how the main line appears to disappear into a river in the distance; (b) An E1-series shinkansen at Gāla-Yuzawa station, where gondolas take passengers direct from the station to the ski slopes; (c) An E1-series shinkansen at Gāla-Yuzawa station with the impressive mountains of the region in the background.

in Japan (Hadfield 1995:20). The reason for this is that four plates (Eurasian, North American (or Okhotsk), Philippine Sea and Pacific) converge there creating several faults, as well as numerous sub-faults. For those that have not experienced the full enormity of an earthquake, it is hard to describe its impact or to explain the sense of helplessness it creates. London's Science Museum has tried in a display that recreates the scene in a convenience store in Kōbe that became one of the many much-used images of the devastation of the Great Hanshin Earthquake that struck on 17 January 1995, killing nearly 6,500 people and destroying over 500,000 buildings (Asahi Shimbun 2004:218).[2] However, that display is largely counterproductive in its attempt to educate about earthquakes as it reduces the earthquake to a trivial form of entertainment by completely understating the strength of the quake. Furthermore, due to the regular pattern on which the display repeats itself, it takes away one of the most terrifying aspects of an earthquake – its unpredictability.

There are three factors relating to an earthquake that need to be taken into account when considering the impact upon a society. First are the intrinsic factors, such as the magnitude, type and location. Second are the geological factors, including the type of soil and the distance from the event. Finally, there are the societal factors which relates to the quality of construction and how prepared a society is. Although the quake itself can cause damage through its direct impact, it is usually the secondary impact, through the seismic shaking that causes the majority of the damage. This can lead to problems such as liquefaction, when soil effectively begins to behave as a liquid leading to buildings and other objects sinking where they stand, fires and tsunami. In terms of measuring the impact of an earthquake, it is common to speak in terms of the Richter Scale. However, this system is primarily only concerned with the intrinsic nature of the quake. In other words, two 7.0 magnitude quakes can have totally different impacts depending on the geological and societal factors. To help take this into account, Japan usually uses its own system, the Japan Meteorological Agency Scale (JMS), similar to the Modified Mercalli Intensity Scale used in the United States, which measures the intensity of the quake. While the Richter Scale is a logarithmic scale with each number meaning a factor of 30 increase in magnitude, that is, a 7.0 quake is 30 times stronger than a 6.0 quake, the JMS ranges between 0 and 7 depending on the intensity at any given point.[3] In other words, for a single earthquake, different JMS figures will be produced depending on the intensity of the quake at different locations.

Given that earthquakes are expected in Japan, one would assume that the shinkansen is prepared to cope with it. To some degree it is. Sensors across the country pick-up the primary (P) wave given off by an earthquake, which travels at around 6 km/s. When this is detected by the UrEDAS system (Urgent Earthquake Detection and Alarm System), a signal is sent to all trains to halt. This process takes about 3 seconds. Depending on the location of the quake, this will allow trains to slow before the main secondary (S) wave, that travels at about 3.5 km/s and causes the damage, strikes. However, it is unlikely that there will ever be more than 20 seconds warning, whereas it can take around 2 minutes to stop a shinkansen from top speed. If a major quake strikes the shinkansen network, there will be trains still moving relatively quickly. It is necessary, therefore, to build the infrastructure to cope with the severe shaking that can occur. The way in which the different types of construction react to quakes, and consequently are built varies. For example, a train passing through a tunnel during a quake will not be as affected as one travelling along a track that is on an embankment. A significant issue for the shinkansen is that large sections are concrete elevations. These are supposed to withstand the effects of all but the most severe shaking. However, the Great Hanshin Earthquake damaged large sections of the Sanyō Shinkansen. Although, having seen pictures of the damage, one can marvel at the speed – a mere six months – at which re-construction was completed and services resumed, much of the damage should not have occurred. These problems, due to poor construction, will be discussed in the following lines. Yet, it is also the case that the impact of the quake was much greater than had been expected for any quake, let alone for that area of Japan. As a consequence, RTRI has revised its design principles to take account of smaller inland faults and for construction to be done so that structures may sustain damage, but that they should not collapse (Noguchi and Fujii 2000:59). The devastation caused by the Great Hanshin Earthquake led to a review of all the supporting piers of elevated sections of shinkansen lines. In

2003, some 4,390 were still found to be in need of reinforcement (*Japan Today*, 2 June 2003). Worryingly, some 3,600 of these were on the Tōkaidō Shinkansen which runs along the area that is likely to be struck at some stage by a major earthquake. Indeed, the Tōkaidō Shinkansen's possible vulnerability may explain why JR Tōkai was so keen to set aside funds for its repair, let alone why they hope to build the more earthquake-resistant Chūō Shinkansen. Only 90 were found to be inadequate on the Tōhoku Shinkansen,[4] which was hit by a large quake in May that year, leading to cracks developing in 23 piers. On that occasion services were suspended overnight, and although passengers were provided with food and drink, no alternative sleeping accommodation or transport was provided (*Kyōdo News*, 27 May 2003).

On 23 October 2004, a Jōetsu Shinkansen was derailed – the first ever derailment of a passenger-carrying shinkansen – following an earthquake, the Niigata-Chūetsu Earthquake, near Ojiya. The derailment happened on an elevated section and there was significant damage to the track in the area. As the train was only 17.7 km from the epicentre, the driver had already begun to apply the brakes before the UrEDAS system could react as the S Wave struck just 2.1 seconds after the P Wave. The train came to a halt, with some carriages at an angle of 30–40 degrees, and the passengers, none of whom were injured, were eventually escorted on foot to the nearest station. Research about why this train derailed will continue for sometime. As with the Great Hanshin Earthquake there was a degree of luck. That quake occurred before the shinkansen services had begun operating that day. In the case of the Niigata-Chūetsu Earthquake, the timing was such that no other train was travelling in the other direction, thus a collision was avoided and the design of the Jōetsu Shinkansen, which has a large trench in places between the up and down tracks for catching melted snow and water, helped to prevent the train from falling over. While some (e.g. Shimizu 2004) have suggested that this has ended the 'myth' of shinkansen safety, I would suggest that it has done the opposite. A train derailed during a major earthquake while it was doing over 200 km/h and no-one was seriously injured, let alone killed. How can this be seen as anything other than an endorsement of the safety of the shinkansen? There were some suggestions in the media (e.g. Asahi Super Morning TV News 25 October 2004; *Mainichi Shimbun*, 25 October 2004) that the derailment would have been worse had the train been a more modern, light design, such as those used on the Tōkaidō Shinkansen, rather than an older, heavier 200-series, the accident may have been worse. This seems unlikely given the low-centre of gravity of most shinkansen, though there must be a concern about the vulnerability of the double-decker E1-series and E4-series shinkansen.

The opening of the Kyūshū Shinkansen provided a further natural challenge to the shinkansen. Sakurajima is an active volcano in Kagoshima bay which regularly deposits ash on the city and surrounding area, as I discovered to my expense on my first trip there in 1992. To deal with the potential damage and problems that this ash could cause the shinkansen, the Sendai depot has a special machine designed to suck out any ash from the machinery under the shinkansen carriages. Clearly forces of nature pose a significant challenge to the Japanese and to the shinkansen. As understanding about these forces continues to improve, so it is possible to improve the way in which the products of human society can respond to them. The shinkansen has benefited from these improvements over the first 40 years of its operations. Yet, it is clear that work still needs to continue to ensure that improved standards are adhered to, while at the same time we have to acknowledge that in a country that is hit so much by such forces of nature it will never be possible to say that any system or construction is completely safe.

Environmental friend or foe?

When considering Japan's environmental record, the summary suggests a mixed, inconsistent pattern. Environmental politics did not become a major issue in Japan until the 1970s (Groth 1996:220), despite the fact that various problems had been coming to light since the 1950s. The most notorious case was in Minamata in 1953 when a local company allowed deadly methyl mercury waste to be poured into the local water supply, but it was not until 1973 that compensation was ordered to be paid. In Niigata, *itai-itai byō* (literally 'ouch-ouch disease') caused by cadmium poisoning came to light during the 1960s. Many other cities, notably Kitakyūshū, Kawasaki and Yokkaichi, became noted for their clouds of pollution and dirty seas. Although increasing public concern, fuelled by media campaigns, raised awareness and began to change industrial policy, it was arguably the 1973 oil shock that was the stimulus for industry to clean up its act. While cities have become noticeably cleaner and Kitakyūshū has been keen to promote its success story to other cities that are facing environmental challenges, many visitors to Japan still comment on the smell of pollution in the cities. These smells tend to be due to the number of small-scale factories that are intermingled with residential areas, and so bringing people in closer contact to these unfamiliar smells than may be the case in many other industrial nations. However, the perception of Japan smelling polluted may also be due to a relatively poor understanding of what certain smells are as I have come across many visitors to Japan commenting on a smell which was not pollution but the odour given off by cars, particularly taxis, that use gas rather than petrol.

Despite many apparent improvements, there remains much to be concerned about. While the image of the shinkansen passing Mt Fuji may be one of the best-known images in relation to Japan and the shinkansen, when on the shinkansen it is rarely possible to see Mt Fuji. The reason for this is summed up by Palin as he travelled on the shinkansen:

> Pass the austere, perfectly formed icon of Mount Fuji, reduced to an indistinct blur by clouds of pollution. The Japanese may wear face masks to avoid spreading germs when they have a cold, but still seem happy to allow industrial chimneys to belch away.
>
> (Palin 1997:62)

While many of the chimneys around Fuji are those of paper mills, and some of the 'smoke' may be steam, the smell is unpleasant. But, Palin's comment does touch upon an interest dichotomy. For the Japanese do try to avoid passing their own pollution on to others, yet not only have many companies continued to be able to pollute Japan, some have even passed on their pollution to other countries by moving their factories there. Perhaps the reason to this lies in the difference between personal, individual responsibility and the face-less responsibility of the organization, company or state. For in the Japanese organization, 'real power' is 'diffused as much as possible so that nobody has to take complete responsibility for anything, and thus risk losing face' (Buruma 2001:151). The individual is and always has been concerned with pollution. Indeed Buruma (2001:5) suggests that 'pollution is the Japanese version of original sin', which is why the *kami* (deities) that created Japan according to Shintō 'enjoyed sex with impunity' but were 'terrified of pollution, especially the pollution of death'. However, Buruma also points out that

> Like earthquakes and other natural calamities common in the Japanese isles, jealousy, pollution and death simply happen. They will always be with us. But they do not occur because of a sinful act. The concept of sin was, and still is, alien to Japanese thought. The Japanese gods (*kami*) are like most people, neither wholly good, nor completely bad. There is no Satan in Japan.
> (Buruma 2001:6)

Given the apparent dichotomy between individual and non-individual behaviour in relation to pollution, and whether it is something that can be avoided or is merely to be reacted to and coped with as best as possible, it is perhaps no wonder that Japan's environmental record today does appear so mixed. For while Japan is one of the world leaders in the recycling of materials (Planet Ark Environmental Foundation 2004:12), its international environmental image is plagued by the debate over whaling, which has been used like a political football by both those in favour of it and those opposed to it, while many of the facts surrounding the issue have apparently been ignored in favour of the emotive-subjectivity of the issue. Since the 1990s, and particularly in relation to the Kyōto Protocol, which as the host country of the conference has given Japan a further impetus to be a world leader in environmental issues, it is greenhouse gases, seen as a cause of global warming, that have become a particular concern (Nagatomo *et al.* 1997). In 2001, Japan discharged the equivalent of 1,299.4 million tons of carbon dioxide (CO_2) (Asahi Shimbun 2004:190), while on a per capita basis, it was 9.4t per person in 2000. This compares favourably to the figures for the United States (19.8), Australia (18.0), Canada (14.2), Russia (9.9), Germany (9.6) and the United Kingdom (9.5), though not as good as France (6.2) (Asahi Shimbun 2004:190). If one considers where Japan's CO_2 is being discharged from, while it is the industrial sector that is the largest (37.2%), the transport sector accounts for 22.0% (Asahi Shimbun 2004:190). Similarly the transport sector accounts for nearly a quarter (24.8% in 2001) of Japan's energy consumption, while Japan's per capita consumption is 4.13t of petroleum equivalent in comparison to the United State's 8.35t, France's 4.25t, Germany's 4.13t and United Kingdom's 3.89t (Asahi Shimbun 2004:147).

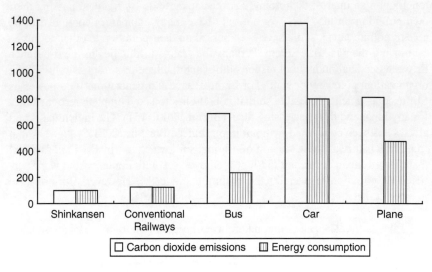

Figure 7.3 The shinkansen's environmental record.

Source: Based on data from Central Japan Railway Company (2004b:16).

Notes
Carbon dioxide emissions are based on amount emitted to transport one person 1 km. The shinkansen figure of 13.7 g-CO_2/passenger-km has been given a base figure of 100. Energy consumption is calculated as the number of Kilo Joules needed to transport one person 1 km. The shinkansen figure of 349 kJ/passenger-km has been given a base figure of 100.

If we look at the shinkansen's record with respect to CO_2 emissions and energy consumption, it is clear to see that its record is good (see Figure 7.3). Indeed most recent documentation from JRTT and the JR companies, particularly JR Tōkai who are so dependent upon the shinkansen, tends to emphasize this impressive record of the shinkansen's. However, as discussed in Chapter 5, as fares do not fully reflect environmental impact, the airline industry has managed to increase its share of the market on the Tōkaidō/Sanyō Shinkansen. If 700-series shinkansen were to replace the flights between Tōkyō, Ōsaka, Okayama and Hiroshima, it would reduce CO_2 emissions by 200,000 t/year, which is the equivalent of one month's CO_2 emissions from all domestic flights in Japan, whilst also increasing capacity at Haneda Airport for longer-distance domestic and international planes (Kasai 2003:171). It should be remembered that while planes have a relatively poor environmental record, much of the pollution around airports is caused by people travelling there by car or bus. Even in Japan, railways only account for 43% of the share of passengers travelling to airports (Hirota 2004:9), and, with the abandonment of construction of the Narita Shinkansen, there is no airport that is directly served by the shinkansen. The divisions referred to in Chapter 4 clearly have an influence in this respect (Figure 7.3).

While CO_2 emissions and energy consumption are the most standard ways of considering environmental impact, in recent years attention has been turning to Life

Cycle Assessment (LCA) which takes account of a product's impact throughout its life cycle (Nagatomo *et al*. 1997). However, when one considers the LCA of the shinkansen, as it is made up of a large number of parts and many consumables are needed in its maintenance, 'it is difficult to take account of all the processes and all component parts in the scope of evaluation' (Nagatomo *et al*. 1997). The life cycle of items relating to the shinkansen vary. While maintenance is done on a regular basis with necessary work being done as appropriate (see Chapter 6), the underlying assumption is that the trains are expected to last 20–40 years, while the track structures should last 50–100 years (Kirimura *et al*. 1997). Furthermore, new models of shinkansen are being constructed in a way that will allow for more of the trains to be recycled and for there to be less toxic or hazardous waste. However, even older models of shinkansen can be recycled to a relatively high degree. For example, 91% of the 200-series is recyclable (JR East interview August 2003). Increasing the amount that can be recycled is particularly important in Japan where disposal sites are limited (Kirimura *et al*. 1997). That shinkansen lines are also being increasingly constructed with concrete sleepers which have benefits in terms of length of usage and low cost, though are harder to dispose off than most wooden sleepers, which are not always from sustainable sources (Kirimura *et al*. 1997).

Another significant environmental issue in Japan is noise pollution. Despite the quietness associated with Zen, Japan is a noisy country. There almost appears to be a fear of silence. It seems that wherever you go there are announcements and background music, let alone the clatter caused by modern living. The shinkansen is no different in this respect. Shinkansen stations, for example, tend to be filled with the noise of announcements and music. However, while certain noises appear to be accepted, there is obviously a limit to what is tolerated. It is partly for this reason that airports have been constructed on man-made islands or in locations far away from the cities that they supposedly serve, while military air bases, many of which are located in densely populated areas, have continually had to face pressure from citizens trying to restrict night flights in particular. Regarding the shinkansen, rather than the station, it has been the train that has been the focus of debate. Noise is created by the shinkansen in four areas; the pantograph, the train body, the underbody and the track (RTRI and EJRCF (eds) 2001:121). The noise created around the pantograph is due to both the contact between the pantograph itself and the overhead wires, as well as the air friction upon the pantograph and its shields as the train moves at speed. With the train body, the shape of the front of the train is significant. Indeed, this tends to be the area that is emphasized in publicity material about attempts to reduce noise pollution. However, as some shinkansen are about 400 m long, the degree to which the side is aerodynamic and free from any protrusions or indentations is highly significant (RTRI interview August 2003). This closely relates also to the noise from the shinkansen's underbody, such as that created by the running of the wheels on the track. It is noticeable how the underbody of a 700-series shinkansen is almost completely 'filled in', so that most the equipment and bogies are hidden from sight, and so helping to prevent both the creation and the escape of noise pollution. Noise pollution in relation to the track is particularly concerned with the concrete elevated sections of the shinkansen lines and the noise that is created under these.

In 1975 the Environment Agency issued new standards relating to shinkansen noise. In Category I areas (primarily residential areas) the standard was set at 70 decibels, while the standard for Category II areas (non-residential areas) was set at 75 decibels. It is the responsibility of prefectural governors to determine which areas are considered residential or non-residential (RTRI and EJRCF (eds) 2001:119). A grace period was allowed, with the length dependent on the current level of noise and whether the line had already been built or was under construction. However, results of surveys done in 1991 and 1994 by the Environment Agency reveal that in some priority areas, standards were still not being met although the grace period has long passed for all lines (see Table 7.2). The challenge to keep noise pollution down is being made harder by the increase in operating speeds that have been seen during the 1990s. In preparation for further advances, some areas of the Hokuriku Shinkansen, for example, have 3 m high sound-barriers, which currently keep noise levels below 70 decibels (RTRI and EJRCF (eds) 2001:121).

The noise pollution created by the shinkansen as it enters and exits tunnels has been a great concern. Due to the speed at which the shinkansen moves, compression waves are created and move through the tunnel at the speed of sound (RTRI and EJRCF (eds) 2001:125). As these waves leave the tunnel it makes a loud bang. Many tunnels are now having special hoods installed at the entrance which reduces the pressure gradient of the compression waves (see Figure 7.4b), or have shelters with slits in them to allow the compression waves to escape (RTRI and EJRCF (eds) 2001:128). Standing next to tunnels along the Sanyō Shinkansen (noted for its tunnels and great variation in rolling-stock) it became possible after about an hour to predict what series of shinkansen had entered the other end of the tunnel based on the sound that came out as well as the strength of the gust of wind. While the 500-series shinkansen was travelling at 300 km/h, it was often the 0-series travelling some 90 km/h slower that made the most noise and created the greatest air displacement. Indeed, when it entered service, the 0-series' noise level was 90 decibels (RTRI and EJRCF (eds) 2001:124). Although, as was

Table 7.2 Shinkansen's noise pollution in priority areas

	Number of sites (measured)	1991 average noise level (% sites below 75 dB(A))	1994 average noise level (% sites below 75 dB(A))
Tōkaidō Shinkansen	120	73.4 dB(A) (78%)	72.1 dB(A) (98%)
Sanyō Shinkansen	62	73.1 dB(A) (84%)	72.3 dB(A) (98%)
Tōhoku Shinkansen	26	75.8 dB(A) (50%)	72.9 dB(A) (100%)
Jōetsu Shinkansen	15	74.1 dB(A) (67%)	71.0 dB(A) (100%)
Total	223	73.7 dB(A) (76%)	72.2 dB(A) (98%)

Source: RTRI and EJRCF (eds) (2001:120).

Mirror of Japan   175

Figure 7.4 (a) A 700-series shinkansen exits a standard tunnel design. (b) A 300-series shinkansen exits a tunnel with a noise-reducing-hood installed. Given the location of the newly constructed house, one can assume the hood has had some positive impact. (c) Noise is clearly not an issue for all in Japan – this house is surrounded by the taxi-ways at Narita Airport.

discussed in Chapter 3, its shape appears 'natural', it is in fact better suited as an aeroplane design, and a skirt, much like the one used on Formula 1 cars, was added to help combat the aerodynamic challenges created by the main body (NHK 2001; Haraguchi 2003a:136–7). A further issue in why newer generations of shinkansen are less noisy is that they have generally become less heavy (see Figure 7.5 and Appendix 3) due to the use of lighter materials, such as aluminium for the train bodies and plastics on the inside, and improvements in construction techniques.

Reducing noise pollution and improving the aerodynamics of the shinkansen continues to be a key area, especially as there is a desire to raise speeds of shinkansen even higher. That some of the work has been successful can be seen by the continued construction of housing near shinkansen lines (see Figure 7.4b). Yet the construction of new housing poses a potential problem to the JR companies as it is likely that such construction will lead to those areas becoming Category I rather Category II, which will mean that noise pollution in that area will have to be reduced even further. To put such a burden on the JR companies

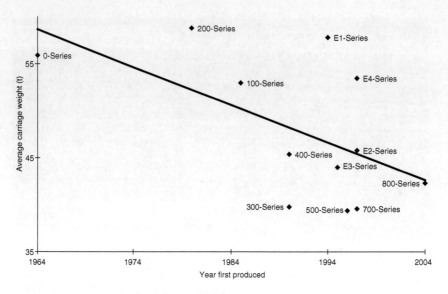

Figure 7.5 Reducing the shinkansen's weight.

Source: Data taken from Appendix 3.

Notes
E1-series and E4-series shinkansen are heavier due to being double-decker; JR East shinkansen tend to be heavier due to the snow-countermeasures.

when the individual has chosen to have their house constructed in that location does not strike me as being particularly fair. The shinkansen-operating companies are not alone in trying to address noise pollution as much work is also done at RTRI. Indeed RTRI's facility at Maibara (see Figure 5.9a) contains the world's largest and quietest wind tunnel so that research can be done into improving the design of rolling stock and also the design of the infrastructure. When you travel past the building on the shinkansen it is hard to gain an appreciation of just how huge it is. But more than the size, it is the low-noise quality that is most amazing as I discovered when I was taken inside and stood in places which created absolutely no echo and I could not hear the voice of someone stood only a few metres from me, separated only by special noise-reducing filters. That such a huge investment has been made is testament not only to how seriously the companies are taking the noise pollution issue, but also what an important issue it is.

Due to its immense size, the shinkansen also creates vibrations, which can be strong enough to feel like an earthquake and cause cracks to appear in walls (Groth 1996:216). Furthermore, the train can interfere with electrical signals. These problems led to the creation of around one hundred citizens' movements along just the Tōkaidō and Sanyō Shinkansen by the early 1980s (Groth 1996:215). The most well known of these groups was in Nagoya for residents living within 60 m of the Tōkaidō Shinkansen. While the group initially negotiated with JNR, the Ministry of Transportation and the Environment Agency, when this proved unsuccessful, a suit was filed against JNR in 1974 which led to compensation being awarded while the demand for injunction continued to be rejected (Groth 1996:217). The action was supported by both Dōrō and Kokurō, whose members sometimes slowed the shinkansen in the area to 110 km/h, which was still above the 70 km/h limit (just a third of the shinkansen's top speed in non-urban areas at that time) the citizens' movement was seeking (Groth 1996:217, 226, 236), while keeping the train on time by going faster in other areas (Mutō interview 2004). The Nagoya movement also planned to boycott the payment of NHK fees, which became a policy adopted by other groups opposed to shinkansen construction in later years. While this policy encouraged some people to join the movement merely as an excuse not to pay the fees (Groth 1996:239), that the shinkansen was interfering with the reception of TV signals would suggest that there was a valid basis for the policy itself. Yet, the movement was not without its ironies as they used trucks with loud speakers on (Groth 1996:240), and so created a noise pollution which is all too common in Japan, as the same methods are favoured by those running for political office, by right wing groups and the like. This form of noise pollution appears to contaminate every corner of Japan, other than, at least in theory, areas around schools, and is not merely restricted to large urban areas.

Anyone who has ever stood underneath a concrete elevated structure when a shinkansen passes will be aware of just how intrusive it can be. While research continues into reducing this form of pollution, it also reveals the heavy usage of concrete in Japan. The other notorious examples of this are the use of tetrapods, which apparently cover more than half Japan's coastline (Kingston 2004:131–2), and the concrete paving on many of Japan's riverbeds. The criticisms tend to be based on the concern of corruption and over-spending on construction (e.g. Kingston 2004:122–56), and overlooks both the important role that can be played by such interventions in a country that suffers so greatly from forces of nature,[5] and that this use of concrete is not limited to Japan as I have see tetrapods and other variants, as well as paved river banks, in many European countries too. However, there are also concerns that not all the concrete construction is of the appropriate standard. It was for this reason that a tunnel collapsed when the Tōhoku Shinkansen between Ueno and Tōkyō was being constructed (Hadfield 1995:159–60). In 1999, sections of concrete tunnelling on Sanyō Shinkansen began to fall due to the mixture not being correctly made. When part of the Sanyō Shinkansen collapsed in the wake of the Great Hanshin Earthquake it was found that some of the defects were caused by there being too much delay in the second stage of construction of the piers, while in some piers wooden blocks were found (Hadfield 1995:204).

Rather than just noise pollution and vibrations, I would suggest that the elevated sections are also a source of what could be termed visual pollution. Shinkansen lines cut through the local environment in an uncompromising manner. Even when a train is not passing, the infrastructure is clearly visible and can be a scar on the landscape. A trip along the Maruyama river near Kinosaki *onsen*, for example, would lose some of its beauty had the San'in Shinkansen been built. Only when a train passes is one really aware of the location of the conventional line despite being only a few metres from the river in places. The shinkansen's visual pollution is a problem that is not restricted merely to the elevated sections, but also applies to many of the shinkansen stations. A prime example of this is perhaps Kumagaya, where the platform is one of the highest structures in the city at about the equivalent of the seventh floor of an office building. Part of the problem of the rapid urbanization of Japan during the 1960s, combined with the 'concrete-ization' and the closeness of buildings, so that there is little opportunity to observe their exterior, is that not much attention appears to have been made to the 'wrapping' of many buildings. The result is summed up succinctly by Palin:

> All these towns seem to be alike, and considering how important a part aesthetics play in Japanese culture, remarkably unattractive. There is a constant feeling of being cramped. The houses are small and narrow, the streets have no pavements, the architecture of a shanty town in Ethiopia is more inventive.
> (Palin 1997:53–5)

While the elevated sections of the shinkansen may be unpleasing, what they pass can often be even worse, especially when one considers what the surrounding area used to be like (see Figures 7.6b, c and d). So, while Palin was cheered up by seeing the result of his favourite football team, Sheffield United, in the English-language newspaper *Japan Times* and is able to escape the drabness in due course due to being a visitor, for the Japanese the results of the country's developments and modernization cannot be as easily escaped. Two venues in many cities that allow escapism, particularly in terms of architecture, are *pachinko* parlours,[6] which also demonstrate the Japanese love-affair with noise, and love hotels.

In recent years there appears to have been a keener interest in the design of shinkansen stations. While the examples along the Yamagata Shinkansen may not be totally comparable due to their location and the unique way in which they were funded, the stations along the Kyūshū Shinkansen clearly reveal that the time of constructing dull monolithic cuboid concrete structures has been resigned to history. The Kyūshū Shinkansen stations have their own unique design. For example, the station Shin-Minamata has an open feel with sail-like panelling reflecting the importance of sailing ships in the city's development (see Figure 2.4g). Izumi station also reveals the importance of local characteristics with shapes in the roof similar to that of wings of cranes (see Figure 2.4d), which are found in abundance in the area. This change in the way in which stations are designed may be reflecting more than merely an increased interest in buildings' aesthetics. Shinkansen construction was previously included as part of '*toshi keikaku*' ('city planning'), which was a nationwide policy concerned with standardization, and station designs, although award winning, were the source of some criticism (Yamanouchi 2000:241). Today the concern is the much more locally focused '*machizukuri*' ('town-making'). Although '*machizukuri*' can be just seen as a buzz-word that hides the standardization that continues to take place (e.g. in one advertisement I have seen the Sōmushō using the term in promoting the policy of city, town and village mergers that have become a major issue in recent years) there is clearly an

Figure 7.6 (a) A 800-series shinkansen passes the patchwork of a concrete-reinforced cutting; (b) A 400-series joined with an E4-series shinkansen travel along an elevated section south of Ōmiya; (c) A 100-series-renewal shinkansen passing the once-beautiful coastline of the Seto Inland Sea, now blighted by oil refineries near Tokuyama station; (d) A 500-series shinkansen passing factories and rubbish near the foot of Mt Fuji.

underlying difference in the perception of what it means. That railways have an important role to play in this respect is best demonstrated by the fact that I have even come across JR employees who joined the company because of their interest and study at university of *machizukuri*. As mentioned in Chapter 3, the station is the *genkan* to a town or city. It is more than just a place to catch trains, it is a commercial and leisure focal-point of the city. Regeneration and new construction of shinkansen stations, therefore, can play an important role in the city's development – both economic and social (JR Kyūshū interview August 2003). Only in Japan, perhaps, has the station become what the nineteenth-century French poet Theophiele Gautier hoped for – 'the cathedral of humanity' (quoted in Yamanouchi 2000:240). One example in this respect is Kyōto station. The new station building is extremely impressive. I have come across many who are opposed to its design, which resembles a 'Borg cube' from *Star Trek*, saying that it does not have the traditionally Japanese cultural elements that one would expect from a building located in the historically important former capital. That it is one of the tallest buildings in the city, which has restricted the height of buildings so

180 *Mirror of Japan*

as not to detract from the beauty of the city's history, is a further issue. Yet, speaking to many locals, the view tends to be that as the station is in the southern part of the city, 'it is not in real Kyōto' so the design does not matter. That the huge building effectively creates a barrier between the main city and a *burakumin* suburb,[7] may be another bonus as far as some of the proud – to the point of snobbish – people of Kyōto are concerned.

As the shinkansen network continues to expand, so new technology is helping to ensure that routes are not only the most economic in terms of construction costs, but that all the environmental considerations can be taken into account. In this it is possible for routes to be planned so that they 'circumvent historical sites, hospitals and schools' and so they do not 'spoil the beauty of nature and the atmosphere of cities' (Nozue and Shirotori 1997).

Reflecting Japanese society

One of the things that makes Japan such an interesting country to explore is the intertwining of the modern with the traditional. Traditional Japanese cultural elements can be seen in the operations of the shinkansen, for example, JR Kyūshū's use of traditional bamboo and *igusa* rush rope curtains around the washrooms on its 800-series shinkansen (see Figure 7.7). Indeed JR Kyūshū even had a *haiku* competition

Figure 7.7 (a) An 800-series shinkansen near Shin-Minamata station; (b) Bamboo window covers, traditional patterns and wooden seats on the 800-series shinkansen.

Mirror of Japan 181

as part of its PR campaign to coincide with the start of the Kyūshū Shinkansen. One could even suggest that along the Sanyō Shinkansen, the sight of the sleek, modern 500-series shinkansen next to the classic 40-year old 0-series is another example. Yet the shinkansen has also been responsible for the spread of modernity and thus the breakdown of other traditions and ways of life. Today some traditional practices are maintained by modernity through so-called 'heritage tourism', but in reality much of this is little more than turning the country into a large theme-park rather than truly continuing the way of life. While it is likely changes would have occurred anyway, the shinkansen is likely to have sped them up. One particularly alarming change that was seen with the spread of the shinkansen was that as the Tōhoku Shinkansen was constructed, so the number of unwanted pregnancies and abortions in the prefectures where it was constructed rose (Hayashi 1987:83). Just as many in the Meiji period claimed that 'railways bring disease' (Mito 2002:29), so it seems in modern Japan the shinkansen has had its less desirable influences.

In considering the way in which the shinkansen reflects different aspects of Japanese society, let us begin with religion. Religion in Japan is a complex business. According to official statistics, there are about 107 million Japanese who follow Shintō, 95 million Buddhists, 2 million Christians and 11 million believers in other religions (Asahi Shimbun 2004:237). This is a total of 214 million. Yet the Japanese population is nearly half this at 127 million. The reason for this situation is that many Japanese 'believe' in more than one religion, with Shintō being mostly commonly associated with day-to-day living and weddings, and Buddhism being associated with death rituals. Without the moral codes of religions such as Christianity and Islam, it is often suggested that the Japanese are not religious people. Yet, Shintō, let alone the philosophical foundations of Confucianism, continues to play an important role. Buruma suggests that although on

> the surface of Japanese life has changed almost beyond recognition... enough remains under the concrete and glass façade of the economic miracle to amuse the gods. Despite all the changes Japan is a profoundly traditional country. Every new building has a shrine on the roof, dedicated to the fox Inari, guardian of rice-crops and export figures. In many ways the Japanese continue be a nation of farmers not quite sure what make of their new affluence.
>
> (Buruma 2001:16–7)

Even in the world of the shinkansen there is a place for Shintō. When I visited the offices of JR Tōkai's new Shinagawa shinkansen station, a small shrine and special shelf was clearly on display, albeit in an area that would not normally be seen by the public. Shintō rituals are also regularly carried out at the beginning of new construction work on shinkansen lines or stations, just as they would be when construction starts on new office buildings or houses. The importance of Shintō and its links with day-to-day life, which may be regarded as 'superstition' in some cultures, was revealed when during a tour of a tunnel that would connect Haruna with Annaka-Haruna shinkansen station, 30 women over the age of 16 were prevented from entering as the head of the construction site said that 'the deities of mountains are female and if women enter (the site), the mountain deity will be jealous'. The representative of Sata Construction Company said 'It was practice from olden days and we had to listen to workers on site who are concerned about accidents' (*Japan Today*, 28 August 2001). JR Kyūshū staff also visit a local shrine to receive blessings for its shinkansen and safe travel, with amulets placed in the cab (JR Kyūshū interview 2004).

Linked to religion is the issue of suicide, with the samurai practice of hara-kiri (also known as *seppuku*) and the *kamikaze*[8] pilots of the Pacific War being the symbols of this. In Japan, while someone may commit suicide to take responsibility for something having gone wrong, this does not necessarily mean that that individual had done anything wrong. Rather they are 'doing the right thing' due to the 'expectations of the social environment' for, as was discussed in relation to pollution, 'When things go wrong, it is rarely the responsibility of individuals' (Buruma 2001:124). With organizations and groups structured like families, the head often has to take responsibility for things beyond their control, since children have no responsibilities but are 'submissive' members (Buruma 2001:151). According to the Ministry of Health, Labour and Welfare, suicide was the sixth most common cause of death in 2002, with 29,920 deaths and a rate of 23.7 deaths per 100,000 people (Asahi Shimbun 2004:194),[9] about 50% higher than it was prior to 1998 (Asahi Shimbun 2004:35; see also Curtin 2004). Given that there was a rapid increase in the number of redundancies and companies filing for bankruptcy from 1998, it is likely that economic reasons account for much of this increase. Indeed, of the 9,530 suicides where notes were left explaining the reasons, 'economic and domestic problems' was the second largest category and accounted for 34.6%, after 'health problems' (38.6%), although there are also categories 'problems within the family' (9.4%) and 'problems at work' (5.8%) which appear to overlap this category (Asahi Shimbun 2004:35).

As far as suicide and the railways are concerned in Japan, it tends to be conventional railways where most incidents occur. The Chūō Line in Tōkyō has become particularly noted for the high number of these incidents. There appear to be four main reasons for this. First, it is a main commuter line, and so with economic reasons being a significant motive for suicides, many who are adversely affected by economic changes live along this route. Second, and linked to the previous reason, the route goes through a heavily populated area and so is used by a large number of passengers. Third, there are several stations which contain the character '*ji*' meaning a Buddhist temple and this religious association may appeal to some. Finally, as it is a heavily used line and due to the cramped nature of the platforms along it, there are many trains that do not stop at all stations but pass through at high speed, giving easy opportunity for one who wishes to commit suicide. Standing at the front of any commuter train that passes through such stations where passengers walk along the platform within a few centimetres of the edge, one begins to appreciate how horrific it must be for the driver of these trains when someone does go in front of them, whether deliberately or not, and even the concern that someone might do it, and that the issue should not be treated flippantly.

Countermeasures are in place to try to reduce the number of suicides on the railways. For example, railway police and JR officials stand on the platforms keeping an eye out for people who may be considering suicide. Suicides are much more infrequent on the shinkansen. Part of the reason for this may be due to one of the powers that the railways companies have following a suicide, namely that they may seek compensatory costs for loss of earnings and damage from the family of the person that has committed suicide (JR Tōkai interview September 2004). Clearly these costs have the potential to be extremely high, particularly on the shinkansen lines. On shinkansen lines it is also hard to access positions to commit suicide in front of a shinkansen as, with the exception of a few stations which have always had barriers,[10] there are no stations where the shinkansen passes next to a platform at high-speed. Tall barriers and fences, including on bridges that go over shinkansen lines, let alone the prevalence of tunnels and elevated sections, also make it hard to access the line directly. These barriers tend to be effective in keeping back both humans and animals.[11] A suicide on the Tōkaidō Shinkansen in 2004 demonstrated just what an impact such an incident can have. Although the location of the accident meant that the driver apparently did not see the victim, the accident caused a loud noise. Subsequently, a large hole, about 40 cm across, was found on the train. It was also discovered that the force was such that the victim was struck by two shinkansen, travelling in each direction, before the body was flung clear of the shinkansen line and struck by a conventional train on the nearby Tōkaidō Line. This incident led to 44 shinkansen being delayed and 38,000 passengers being inconvenienced (*Mainichi Shimbun*, 9 February 2004).

One of the greatest challenges facing Japan today is the change in the profile of its population. Japan is an ageing society. The 'Ageing index' of Japan, which measures the ratio of the population that are over 65 compared to those aged 0–14, has risen from 19.0 in 1960 to 38.7 in 1980 to 119.1 in 2000 (Asahi Shimbun 2004:30). With the average number of babies born per woman down to a record 1.32 in 2002, having been below 2.1, which is the level needed to maintain population size, since 1975 (Asahi Shimbun 2004:33),[12] the population of Japan is destined to fall. As the size of the working population shrinks, so it is probable that less business trips will be taken, which will impact the income of the JR companies. Consequently the companies are keen to encourage more people to go on domestic trips when they have free time. However, holidays have not commonly been taken in Japan, so a change in culture is needed. The government has been trying to help in this respect. For example, some national holidays are now on Mondays, rather than on the date that they should be, in the hope that this will encourage people to use the long weekend as a holiday and perhaps travel within Japan. Naturally the retired population could be a potential market for the JR companies. However, as there is a tendency to save money to help cover ones own medical and other expenses, it can be hard to attract such customers. Although various special tickets and discounts are available, they are not always as convenient as perhaps they could or should be. For example, one of the popular passes for older people does not allow travel on *Nozomi* services, which has greatly restricted their benefit with the change to a predominantly *Nozomi*-service-timetable in 2003.[13]

Recent attention on making Japanese society 'barrier-free' for the old and disabled has led to an improvement in access at conventional as well as shinkansen stations (which have tended to always have better facilities), since most will need to take at least one train to get to a shinkansen station. For while many of Japan's pavements have had yellow-studded markings for the visually impaired and brail has been made available as standard on many station signs (and even on home appliances in recent years), improving accessibility for those who had physical disabilities seems to have been a lower priority in Japan until recently.

With the working population set to fall, it has become more imperative that Japan fully develops opportunities for women to work. While approximately 51.1% of the population are female, only 48.5% of those aged 15 or over are in work. Furthermore, despite the existence of Equal Opportunity Laws, average wages are only 65.3% that of men's (Asahi Shimbun 2004:11). While clearly the statistics suggest a negative picture, one should not overlook the fact that the expectations, both of the women themselves as well as society, have been, and sometimes continue to be, different to that in some other societies (Condon 1991:295). Being a housewife in Japan is often regarded with higher status than the equivalent in some Western societies, recognizing the important role played in organizing the household finances – to the extent that traditionally the husband passed all of their salary to their wife, who then gave them pocket money – and in raising children and ensuring their educational progress was sufficient. This significant role tends to be hidden from official statistics and can be easily forgotten. Indeed, Hendry (1999:179) points out that 'perceptions of outsiders about the "meek and demure" Japanese housewife can be shown in practice to be quite false'.

Women's public status has sometimes been more limited. This is despite the fact that Shintō, which is linked with 'native identity', has matriarchal deities at its centre (Buruma 2001:37), and there was a history of female-worship and female clan leaders in early Japanese history. Indeed, there have been Empresses, though legislation from the nineteenth century currently prevents this. Yet this argument may overlook some of the importance of what occurs behind the scenes. Indeed, when the unmarried, divorced Koizumi became prime minister, one Japanese woman commented to me that she was unsure who would run the country! However important the role of women 'behind the scenes' is, there are many more who are wanting equal pay and to be able to fulfil their full potential, but Japan has been slow to both respond to this desire and to promote it further. With the government trying to make it easier for women to take time of work to have babies, and so address the problem of a falling population, policies appear to be pulling in opposite directions.

Considering the shinkansen, until recently it appeared as though the traditional roles were being maintained. Women working on the shinkansen were most likely to be performing one of two jobs, which appeared to have been age related. Older women were most likely to be seen cleaning the shinkansen. Although sometimes seen on-board the shinkansen, it is generally when shinkansen arrive at the terminal, unless it is to be cleaned at a depot, that they are met by a fleet of neatly presented cleaners. This process takes about five minutes, while all the chairs are rotated, new sheets are placed on the headrests, toilets and sinks are cleaned, seats are brushed, window sills are wiped and all the rubbish is removed either by bins at platform level, or by it being passed down to under the train where it is taken away by other staff or by a conveyer belt in the case of Tōkyō, for example. The other role commonly performed by women, usually younger women, on the shinkansen is to sell food, drinks and other products. These women tend to be employed by a subsidiary company of the main JR company, but are still dressed as smartly as the other crew on the train, also bow when entering and exiting the carriages, and the trolley sometimes even plays a jingle as it moves at the same time as the woman calls out examples of what she has to sell.

However, today women are involved in other areas of shinkansen operations. It is becoming increasingly common to see women conductors on the shinkansen. Yet without doubt, it is the driver's job that has the greatest prestige. Following an amendment in the Equal Opportunity Law that allowed for women to do the same hours of work as men (MLIT interview January 2001; JR Tōkai interview April 2004), it has been possible for women to become shinkansen drivers also. JR West was the first to have women drivers in 2000 (*Japan Times*, 19 August 2000; JR West interview 2003). However, as mentioned in Chapter 6, JR Tōkai is particularly adept at PR activities and on two occasions the story that women were driving Tōkaidō shinkansen made headline news. In the first instance, in 2000, the women were driving shinkansen to get their licence as is expected of most JR Tōkai employees. However, in the second case on 20 June 2003, the four women were beginning their work as main shinkansen drivers, which was described as heralding a new era by the Transport Minister (*Japan Times*, 21 June 2003). That the story made headlines, globally as well as in Japan, reveals the significant role that the shinkansen plays within society. That women are now driving the shinkansen should not, however, be seen so much as a sign that Japan is an equal society as that it is becoming an equal society. For, in the case of the shinkansen, there are still only a handful of women drivers,[14] but these women may be helping to inspire others to follow dreams of a fulfilling job. One of these drivers, Umeda Yasuko, was the first female conductor also and felt it her duty to get promoted to being a driver for the sake of the women that would follow her (interview April 2004). Another female driver (interview October 2001) told me that she does not feel there is any discrimination, but due to the nature of the work, which can mean not returning home at the end of the day, there is a concern about child-care and the like for women drivers who marry and have children.

It has been suggested to me by a JR East employee that the 2000 instance was nothing more than a publicity stunt by JR Tōkai. While there may be an element of truth about this, it perhaps overlooks the significance of the event. For example, the drivers were presented as a special story for children, in both Japanese and English, on the TBS website. Given its potential target audience, it is interesting to note that while the English title for the article was 'First Woman Shinkansen Operator', the translation of the Japanese title means 'My dream is to become a Shinkansen Driver' (TBS 2000). Clearly the women shinkansen drivers

186 *Mirror of Japan*

Figure 7.8 On the left hand side of the picture staff confirm their orders for their next shinkansen service. On the right hand side, one of JR Tōkai's first female shinkansen drivers, Umeda Yasuko, confirms the details of the service that she has just completed.

had a role to play and the media attention was key. Indeed the media attention was so great in 2003 that JR Tōkai delayed my interviewing one of the drivers for a few months as they were concerned that the continuing attention and pressure was creating too much stress. The media story has created the idea that the job can and is done by women. Although the reality is that there are not many women shinkansen drivers, unless one stands on the platform at the front of the train or hears the introduction of staff names by the conductor on leaving the first station, it is unlikely that passengers will be aware of the gender of the driver (Figure 7.8).

Changing behaviour

As Japanese society changes, the shinkansen has to adapt to these changes. For the purpose of this part of my research I observed the activities of passengers when using shinkansen. While differences were spotted according to the time of day, as will be discussed in the following section, Table 7.3 summarizes the activities observed on five of the shinkansen lines. Group 1 represents those activities that were carried out by a majority of passengers in a carriage; Group 2 are those which were carried out by a large number of passengers; Group 3 are those that

Table 7.3 Activities on shinkansen

Group	Tokaido	Sanyo	Tohoku	Joetsu	Hokuriku
1	Talking Sleeping Sitting quietly/watching view	Talking Sitting quietly/watching view Sleeping		Sleeping Talking Sitting quietly/watching view	Sitting quietly/watching view
2	Drinking (non-alcohol) Using mobile (not talk)	Reading book Drinking (non-alcohol)	Sleeping Sitting quietly/watching view Talking	Drinking (non-alcohol)	Talking Sleeping
3	Reading magazine Eating snack Reading book Listening to music	Reading magazine Using mobile (not talk) Drinking (alcohol) Looking at timetable/map Reading newspaper Eating *bentō* Listening to music	Using mobile (not talk) Reading book Reading magazine Drinking (non-alcohol) Reading newspaper Eating *bentō* Listening to music	Reading newspaper Eating snack Reading magazine Listening to music Reading documents Reading JR magazine Eating *bentō* Using mobile (not talk) Reading book	Reading JR magazine Reading newspaper Drinking (non-alcohol) Reading magazine Listening to music
4	Eating *bentō* Playing non-elec. games Drinking (alcohol) Reading newspaper Reading *manga* Using computer/PDA Putting on make-up Playing elec. games Reading documents Eating fast food Reading diary	Eating snack Reading *manga* Reading documents Writing/studying Reading diary Using computer/PDA	Eating snack Reading *manga* Drinking (alcohol) Playing non-elec. games Looking at timetable/map Using computer/PDA	Playing elec. games Putting on make-up Drinking (alcohol) Writing/studying Looking at timetable/map Reading *manga*	Using mobile (not talk) Drinking (alcohol) Reading book Eating snack Using computer/PDA Reading *manga* Kissing Writing/studying Talking on mobile phone

were carried out by some passengers; Group 4 are those which were only carried out by one or two passengers.

The most popular activities do not appear to vary greatly between the lines. Sitting quietly or looking at the view, for example, appears in Group 1 for four lines and Group 2 for the other studied. For some the shinkansen is a place to have time to think. Nakasone (2002:65), for example, finds the shinkansen to be a 'good place for generating ideas'. Unlike on the TGV or some British trains, the shinkansen is designed so that all rows line up with a window ensuring the person in the 'window seat' has a view. However, with the large number of tunnels, and the height of the noise-reducing-barriers on some lines, the view can be limited, so it is perhaps surprising that this activity is so popular on all lines. As companies continue to compete with airlines, so they may have to provide more entertainment themselves. Although providing individual TV screens may increase the weight of the train too much, having a portable-DVD-player hire service, as found on Eurostar services, would probably prove popular. Introducing such a service should also be helped by the fact that credit cards, the use of which ensures that if the machine is not returned, the passenger can be easily fined, are becoming increasingly used in what has been a largely cash society. That looking out of the window is a popular activity has not been lost on Japan's advertising companies. Komiya (2003:43–65) in his study of how the Japanese economy can be seen from the shinkansen, comments on the number of advertising boards that are placed in fields next to the Tōkaidō Shinkansen. Between Shin-Yokohama and Odawara, there are 71 (SuperTV 2004). One of these 4 m × 8 m boards costs about ¥600,000 per year, which is significantly cheaper than those found on buildings in big cities, and represents just a fraction of the huge amount, about ¥6 trillion in 2001, spent on advertising in Japan each year (Komiya 2003:52–5; SuperTV 2004). Even inside the shinkansen carriages there are some advertisements, though not the number as on conventional trains where it seems every conceivable space, including hanging down from the roof of the train, has been utilized.

Chatting with travel companions is also a popular activity as would be expected in any society. However, Japanese train travellers are also noted for sleeping, whether it be on conventional trains, where many have an uncanny knack of waking up just in time to alight at their station, or on shinkansen. Representatives from American railway companies were said to be 'surprised' when they travelled on the shinkansen in its early years and found people asleep, but 'considered it to be proof of people's confidence in high-speed transport' (Oikawa and Morokawa 1996:12). That the journey on a shinkansen can be spent so comfortably is also undoubtedly helped by the fact that carriages are pressurized so that passengers' ears do not 'pop' when passing through tunnels, which was a concern to the designers during initial tests (Oikawa and Morokawa 1996:11). Another reason why it is commonplace to see Japanese sleep on trains is due to the long hours that are worked. While the average in 2002 was 1,837 hours and was down from levels in excess of 2,100 hours during the 1980s, it is still a comparatively high figure internationally (Asahi Shimbun 2004:80). This culture of long working hours starts from a young age, with a popular adage amongst those taking entrance examinations being 'pass with five, fail with seven' in reference to the number of hours of sleep.

Eating and drinking is another popular activity on the shinkansen. However, there have been changes. It has become increasingly popular to purchase *bentō* (lunch box), particularly specialty ones that can only be bought at specific stations – known as *ekiben* – and drinks in plastic bottles. With this generally healthy form of 'fast food' becoming increasingly popular amongst travellers and non-travellers alike, dining cars and free water-dispensers have been withdrawn from some shinkansen. Removing the water-dispensers has the added advantage of further reducing waste, as disposable cups are also provided, but more importantly for the JR companies or their affiliated companies it has boosted the revenue from sales of food and drink at the station and on the train. While my survey found that it was generally non-alcoholic drinks that were consumed, alcoholic drinks are also consumed – seemingly more so on Sanyō Shinkansen than shinkansen on other lines. The average Japanese consumes 55.9 litres of beer a year, considerably less than that of the United States (82.4 litres), United Kingdom (97.1 litres), Germany (123.1 litres) or Czech Republic (155.5 litres), but more than China (17.5 litres) (Asahi Shimbun 2004:179). So popular is beer drinking on the shinkansen that Japan's beer companies have even produced special commemorative cans for significant milestones in the shinkansen's history. Beer is not the only alcohol the Japanese drink, with domestic production of many other drinks, such as *shōchū* (a low-grade alcoholic drink) and low-malt alcoholic drinks increasing, with sake and beer recently becoming seemingly less popular (Asahi Shimbun 2004:142).[15] As physical differences mean that Japanese tend to feel the effects of alcohol more than their Western counterparts, the amount of alcohol consumed during working hours tends to be limited. However, drinking of alcohol in adult social gatherings is often expected,[16] so one will see groups of people drinking on the shinkansen. Drinking alone, like eating alone, is also perfectly acceptable in Japan, whether it be at a restaurant or on the shinkansen.

Reading is another popular activity on the shinkansen, with newspaper reading tending be to most popular in the morning. Magazines and *manga* tend to vary greatly in content. Although *manga* is commonly translated as 'comic', this perhaps does not do justice to its content, which tends to be graphic, in all senses of the word. Indeed, although there are many types of magazines in Japan, those most commonly read on trains, due to their great number at kiosks on platforms, tend to be full of gossip, some intellectual articles by academics or social commentators, and soft-pornography. Apparently Japanese men think nothing of browsing these magazines while on trains, regardless of who is sitting next to them. *Manga*, which can have far more hard-core imagery, also seems to be widely accepted. As Buruma (2001:220) points out, 'If the Japanese are indeed a gentle, tender, soft and meek people with hardcore fantasies of death and bondage, few of these dreams appear to spill over into real life.' There is a certain poeticism about such material being read on trains, given the imagery of one of Japan's most famous novels, Kawabata's *Snow Country*, including the suggestive linking of trains and tunnels, although interpretations vary as to whether this was a reference to sexual intercourse or the desire to return to the mother's womb.[17] Given the popularity of reading, it is interesting to note that of the lines studied only JR East has a magazine that is available for passengers to read.[18] This magazine, as well as a further publication provided at the seat, also contain items – whether they be holidays, food, souvenirs or everyday items – that can be purchased either on the train or by mail-order.

One consequence of some of the given activities is rubbish. Bins, both on the shinkansen, but particularly on the platform, encourage customers to separate their rubbish out into appropriate bins to help speed up both the cleaning of the train, as well as speeding up the recycling process. The amount of plastic bottles being collected has become so great that JR East even has some of it recycled and made in to their company uniforms (Matsuda 2002:143).

Communication is important for Japanese people. Indeed, the need for face-to-face contact is one of the factors that has helped the shinkansen to be an economic success. Chapter 6 mentioned how it is possible for CTC to contact any shinkansen, or vice-versa, at any time. However, the means by which this is possible was not restricted to company use. Until recently anyone dialling '107' on a Japanese phone was able to telephone a shinkansen and request to be put in contact with a specific passenger on that service. Yet, it was noticeable that over the years the number of announcements for passengers to go to their nearest phone on the train as there is an incoming call for them had reduced. In the end the service was withdrawn. The reason being the prolific rise in usage of mobile phones. As of 2002, there were over 75 million mobile contracts, with little sign in the rise slowing significantly (Asahi Shimbun 2004:161). While it is the case that some people have more than one phone, the impression today is that everyone in Japan has a mobile phone. Mobile phones in Japan for some years have had extra features such as music players, internet access, personal organizers, cameras, games and such like. Consequently their usage on trains, particularly the shinkansen, is commonplace. Yet, the way in which they are used is somewhat Japanese. For, although often ignored on conventional trains, users are encouraged to turn off the ringing tone while on a train, to only talk on the phone by the doors and not at their seat (in the case of shinkansen), and not to use it in areas where the electrical interference could cause problems for those with pacemakers, for example, nearby. During my survey I only came across one single passenger, seemingly a self-important local politician concerned about the path of a severe typhoon approaching his prefecture, talking on a mobile phone at his seat.

While the mobile phone was initially only used by business people, today its usage is more widespread. Furthermore as the way in which it can be used, with internet usage by mobile phone being particularly prolific in Japan, so the JR companies have had to not only respond to the trend but have also looked for ways in which to exploit it. JR Tōkai have been particularly active in this regard and at present it is only on the Tōkaidō Shinkansen where there is no loss of connection on mobile phones when travelling through tunnels. It is not a surprise, therefore, that mobile phone usage for activities other than talking was highest on this line in my survey. Shinkansen tickets cannot only be reserved by mobile phone, but in the case of the Tōkaidō Shinkansen, the booking can be stored on the phone so that a paper ticket is no longer necessary and is cheaper than buying a non-reserved ticket (JR Tōkai interview October 2004). This clearly has advantages in reducing company costs and for the environment, particularly as more than one ticket is often needed for the shinkansen. For those who use paper tickets, new automated barriers on some lines mean that if the passengers have a seat reservation, their ticket no longer needs to be checked by the conductor on-board the train. With the use of special electronic cards becoming increasingly popular and gradually more usable nationwide, so automatic barriers are replacing inspectors at stations, with the continuous clicking of the hand-held hole-punch no longer heard.

The IT revolution has also led to other changes over the years on the shinkansen. Digital displays at the end of each carriage provide details on where the train will next be stopping as well as the latest news, weather and advertisements. On board *Hikari Rail Star* services there are also computer monitors in some locations so that passengers can access information about the train, train timetables, information about stations (including what *ekiben* are available at each station), and such like (see Figure 7.9c). Despite the existence of the various displays, the Japanese hunger for information and noise, together with the need to

Figure 7.9 (a) A 700-series *Hikari Rail Star* shinkansen passing a bridge near Okayama; (b) A *Hikari Rail Star* at Hakata station; (c) A computer information screen on-board a *Hikari Rail Star*; (d) A *Hikari Rail Star* at Shin-Ōsaka station.

assist those who cannot read the display, means that the information is also repeated in not just recorded announcements but also by the conductor on-board that shinkansen. The recorded announcements, and a few other announcements on certain services, are given in both Japanese and English (only JR Kyūshū currently uses other languages also (see Chapter 6)). Signs are also written in both languages. Yet this presents a problem. While generally one will not come across the unique combinations of English words and phrases favoured by many companies in Japan, in much the same way that many Western companies seem to be incapable of using correctly written *kanji* on their products, the shinkansen tends to have a mixture of English and American-English. Although English used to be taught at schools in Japan, since the 1950s, it has been American-English that has been used. However, the railway world still tends to be dominated by English words rather than their American-English equivalents. With many Japanese being unaware of the differences,[19] it is perhaps not surprising to find both systems being used side-by-side. For example, on the *Hikari Rail Star*, the information board will occasionally announce that the train is 'travelling at 285 km/h', while information about the train refers to where 'diapers' (rather than 'nappies') can be changed; on the *Hikari Rail Star* services, there is a announcement about the use of 'mobile phones' (the term used on *Hayate* services also), while a notice on seat backs refers to 'portable phones' and an information sheet in the 'Silence Car' refers to 'cellular phones' – these latter two phrases being the ones used on most other services; on all services 'carriages' are referred to as 'cars'. Interestingly the announcer on the Tōkaidō and Sanyō Shinkansen services has tended to have a British accent (the JR East services has an American accent), with JR Tōkai President Kasai apparently ensuring this was continued when changes were made in 2003 as he feels that it sounds better and is easier to hear (JR Tōkai interview August 2004). Another difference is that a female voice is used on the lines west of Tōkyō, while JR East uses a male voice.

Most announcements on the shinkansen are accompanied by jingles, which are commonplace on Japanese express trains on conventional lines as well. Jingles are often unique to a particular service. Some conventional line stations, notably those on the Yamanote Line, also have their own individual jingle. While station jingles may be helpful in alerting sleeping passengers to the arrival at their stop, those on trains help to make the start of announcements seem less abrupt. However, they also have a role in developing identity. In Japan the school or company song is taken much more seriously than it is in many Western countries. It is sung at ceremonies at the very least and sometimes everyday if the whole organization is assembled, such as when morning exercises are done with music. With the creation of the JR companies, each company had a new song. Railway songs, sung by employees and travellers alike at appropriate functions, also have a long history in Japan (Noguchi 1990:54), but have been largely confined to history, other than often forming part of the jingles on board trains. With the launch of the 'Ambitious Japan!' campaign, Kasai apparently hoped that a new age of railway songs could be begun, and the jingle on board Tōkaidō shinkansen was changed to include part of the song by Tokio (JR Tōkai interview April 2004).

Returning to my survey, while it is possible to see people playing traditional games, the innovations of Japanese companies have clearly led to greater ease in portability of entertainment, with many passengers using personal stereos of some kind, electronic game machines and the like. Of course increasing numbers of passengers have their own portable computer, but the battery life of these machines tends to be limited. While some seats on certain shinkansen do have access to power points, so far there appears to be limited use of this facility. However, as being able to use a computer for the entire journey is another area that the shinkansen has a clear advantage over planes, in due course it is likely that this service will have to be provided for all seats.

Although not included in my survey – as I took it for granted that this activity was done by most if not all the passengers in the appropriate carriages at some point in the journey – smoking is another popular activity in Japan, helped by the comparatively low cost compared to most other developed countries. The rate in Japan is estimated to be about 47% for men and 12% for women, whereas in the United Kingdom the figures are 28% and 26%, and in the United States they are 21% and 17% (British Heart Foundation 2004). However, as elsewhere, there has been an increasing awareness of the health hazards associated with smoking and as well as anti-smoking campaigns by the government, albeit seemingly fairly tame when compared to many EU countries and in the face of the huge amount of tobacco advertising that continues. On the whole the trend has been downwards (*Mainichi Shimbun*, 21 October 2004), and this has been reflected in the reduction in number of smoking carriages. Whereas there were no no-smoking carriages on the Tōkaidō Shinkansen until 1976, 11 of the 16 carriages are now no-smoking on the Tōkaidō and Sanyō Shinkansen. It is not all good news for non-smokers as those wishing to get to the forward most non-reserved no-smoking carriages on Tōkaidō and Sanyō Shinkansen, for example, either have to walk past the wall of smoke surrounding the 'Smoking Corner' on the platform, or through smoking carriages if they boarded the train near the middle. Smoking carriages are unlikely to be ever fully done away with as for those companies that have the most severe competition with airlines, for example JR West on the Tōkyō–Hakata route, the shinkansen is a comfortable alternative for smokers despite the greater journey time since planes are increasingly becoming all no-smoking.

On board comfort is one area where the shinkansen appears to have a great advantage over planes. Over the years the seat pitch has continued to increase on shinkansen, and while the official figures for the E2-series shinkansen, for example, show it to be 98 cm in standard class, the amount of leg room this equates to is about 49 cm. It is about 6 cm greater on the 700-series shinkansen, which reflects JR Tōkai's need to offer much better space and comfort than planes on the competitive Tōkyō–Ōsaka route.[20] Seats also recline, and most sets of seats can be rotated so that passengers can either face the direction of travel or larger groups can sit face-to-face. With individual table-trays on the back of seats, there are no tables, other than in a few special compartments, for example, those introduced on JR West's *Hikari Rail Star* services to try to make areas more convenient for business meetings or for families to relax together. Of the modern designs of shinkansen, it is the 500-series that has suffered some stigma as being uncomfortable. The reason for this is that the rounded shape of the body means that is marginally less space than on other shinkansen. However, I believe that, like the criticism of an apparent lack of space on *Concorde* in the past, such comments are in part an attempt for some people to be blasé rather than there really being any discomfort.

One of the popular myths about Japan is the idea that it is a classless society. While it may be true that Japan has greater income equality than most other countries (New Internationalist 2004),[21] to suggest that it is classless is overly simplistic (Ishida 1993:238; Sugimoto 1997:9–10). Yet some 90% of Japanese see themselves as middle class. The shinkansen appears to reflect this equality. Although trains in Japan have historically had first and second classes, in 1969 they were replaced by 'green' and 'ordinary' (or 'standard') classes respectively. Today, about 90% of shinkansen carriages tend to be 'ordinary'. Indeed, some JR staff that I have spoken to have suggested that there is no need for the 'green' class these days as the comfort in the 'ordinary' class has become so good, and that it tends to only be used sumō *rikishi* who appreciate the extra width of the seats[22] and by politicians. It is the latter group for whom they feel obligated to provide the service in particular. The Imperial Family also travel in the 'green' carriage, which is one reason for some windows being bullet-proof (Nishimoto 2003:50), and even have special waiting rooms and station entrances at certain stations which are regularly used – most notably at Tōkyō and at Nasu-Shiobara, close to where the Imperial Family have a residence. While there are undoubtedly others who do use the 'green' carriages, though I have certainly seen them used by politicians, sumō *rikishi* and professional football teams travelling to matches (something I could not imagine in Britain), there often many empty seats in 'green' carriages on trains that are completely full in the 'ordinary' carriages. In response to this, and because 'green' carriages exist on other services, JR West does not have 'green' carriages on its *Hikari Rail Star* services, but has rows of 2-by-2 seating in all reserved carriages, as would normally be found in 'green' carriages, rather than the usual 2-by-3 seating that is found in 'ordinary' carriages of shinkansen.

Luckily Japan has managed to remain relatively unscathed from some of the worst side-effects of modernity. Graffiti remains rare in Japan and will not be found on the shinkansen. Furthermore, Japan remains a safe country. In 2000, Japan's crime rate (the number of reported crimes per 100,000 people) was 1,925, which compares favourably to other developed countries such as the Untied States (4,124), France (6,421), Germany (7,625) and the Untied Kingdom (9,767) (Asahi Shimbun 2004:19). While I have come across Japanese people who have not reported crimes, and this may be more prevalent than in the other countries listed, the perception as one walks around Japan is that of being safe. Indeed, on the shinkansen I have seen times where on-board trolleys loaded with food and drink have been left for a few minutes, apparently without any concern that something will be stolen. However, one British acquaintance did have a suitcase stolen on-board the shinkansen. When this was reported at the station, he was told that this had never happened before. Although this may be an exaggeration, it does point to the degree of comfort that one can usually spend time in Japan.

Although there may be a decreasing school population, one common sight at stations and on shinkansen at certain times of the year are huge groups of students. As education in Japan is as much about becoming Japanese as it is learning information (Hood 2001:13–17), school trips, which help to bond groups together and form common memories, play an important role. These trips, while traditionally being linked to formal studies in their content so that visits were made to sites of historical importance, have become more and more like the excursions one would expect any group of tourists to take. Although the temples and shrines of Kyōto and the atomic bomb memorials at Hiroshima remain popular, Tōkyō Disneyland, Universal Studios in Ōsaka have also made it on to the itinerary of some schools. Despite the fall in the school-age population, class sizes tend to remain large as Japan does not have the same regard for low student–staff ratios as in many Western countries. As a consequence, when schools go on such a long-distance excursion, whole carriages, to the extent a whole train may only be for this purpose, are often reserved for school-children, easily identifiable as they have to continue wearing the uniform.

In conclusion, one can see that many different aspects of Japanese society are reflected in the operations of the shinkansen and the passengers who use it. However, the relationship does not appear to be one-way, as the shinkansen also appears to have a role in shaping the society around it too.

8   Conclusion

> As fast as winds,
> As calm as forests,
> As aggressive as fire,
> As steady as mountains.
> (Sun Tzu)

This study has considered the shinkansen from many different angles – its history, its importance as a symbol, the apparent desire of politicians and communities to have shinkansen built to their town or city, the economic impact of the shinkansen, the importance of the employees that operate it, and the way in which it reflects and responds to various challenges presented by the environment in which it operates. While it is normal to reflect upon some of the most salient points from a study in a concluding chapter, I intend to do that here largely in considering an issue that has only been touched upon in passing so far, namely the export of the shinkansen.

Perhaps the question I have been asked the most in relation to my research on Japan, let alone the shinkansen, is 'How do they manage to run their trains so well and why can't we do that here?' While the answer to the question may need to vary depending on where 'here' is, the previous chapters have already highlighted key information to help answer the first part of the question. Armed with such information, is the export of the shinkansen possible? While the answer to this question appears to be 'yes', given that it is being exported to Taiwan, what has happened in the case of Taiwan would suggest that a 'yes, but' answer may be more appropriate. This chapter will thus consider the 'Taiwan Shinkansen', following a similar structure to the book as a whole, before drawing conclusions about it and the exportability of the shinkansen to other countries, and some final conclusions about the shinkansen itself.

'The Taiwan Shinkansen' – project overview

Taiwan is located in the Pacific to the south west of Japan and 160 km east of China across the Taiwan straits. Following Japan's victory over China in the

Sino-Japanese War of 1895, Taiwan was annexed to Japan. Taipei, the capital (population: 2.65 million people), is at the north of island, with its other major cities, Kaohsiung (1.48 million), Taichung (0.94 million) and Tainan (0.73 million), all located on the west coast (total population: 22.5 million). That the populated areas are all on one side of the country means that comparisons with Japan, where most of the population is concentrated along the Pacific Coast, are commonplace.

The railways in Taiwan were established by the Japanese, and, as a consequence, used the standard Japanese gauge. The railways are now operated by the Taiwan Railway Administration (TRA). However, as was the case in Japan, the limitations of the gauge mean that it is difficult to significantly increase speed. In Taiwan, where economic development and prosperity has been relatively recent, this has meant that the airline industry has managed to gain a huge advantage. At present the airline industry dominates the Taipei–Kaohsiung route, which will be used by the 'Taiwan Shinkansen', the eleventh busiest route in the world (see Table 2.3). The line is initially to be 345 km in length and has 12 stations, similar to Tōkyō–Nagoya (342 km, which currently has 13 stations, though it initially opened with 8). With trains operating at up to 300 km/h, the journey time by train between Taipei and Kaohsiung will be cut from 4.5 h to just 80 minutes (THSRC interview April 2004). However, while the conventional railway continues to be operated by the TRA, the high-speed line will be operated by the Taiwan High-Speed Railway Corporation (THSRC), which is made up of five major Taiwanese corporations, including the Evergreen Group led by Evergreen Marine Corp. Not only are there no formal links between THSRC and TRA, the latter appears to be hostile to the new company (anonymous interview April 2004).

The project itself is the world's largest Build-Operate-Transfer (B-O-T) project. THSRC is responsible for the construction of the line, and will operate it for 35 years before it is transferred to the government. THSRC can also use land around the stations for 50 years. Both time limits may be extended depending on performance. Although underwritten by the Taiwanese government, all of the money has been privately raised. The total cost of the project is expected to be about NT$440 billion (about ¥1.6 trillion) (Okada 2001; THSRC interview April 2004). As THSRC had no railway expertise of its own initially and due to the limitations of Taiwan's own railway industry, specialist help and technology was to be imported. The two main bidders were a German-French consortium, Eurotrain, and a Japanese consortium. In 1998 the European consortium was considered the preferred bidder, but THSRC stopped short of signing a contract with them. In December 2000 a ¥332 billion contract was awarded to the Japanese Taiwan Shinkansen Corporation (TSC) to provide the rolling stock and core electrical and mechanical systems. For the first time the shinkansen was to be exported and was the first large-scale contract to have been won by any Japanese companies since the collapse of the 'bubble economy' in 1990 (Okada 2001).[1] TSC is made up of seven companies, including Hitachi, Mitsui, Toshiba and Mitsubishi Heavy Industries. However, other companies are also involved – most notably JR Tōkai and JR West – with Japan Railway Technical Service (JARTS) playing a co-ordinating role. The main advisor from JARTS is Shima Takashi, son of

Shima Hideo and grandson of Shima Yasujirō. Although having already worked on the design of the original Tōkaidō Shinkansen, Shima Takashi's involvement in the export of the shinkansen to Taiwan means that the Shima family has been central in each stage of the shinkansen's progress, from the initial concept in the Pacific War to its creation in the 1960s to its first export at the start of the twenty-first century. Although less common in the West, in Japan it is often the case that children will follow their father's occupation, particularly where the father is the head of the company (about three-quarters of the Japanese workforce are employed in small-sized (less than 100 employees) companies rather than in the large multinationals that have been focus of most studies on Japanese companies (Sugimoto 1997:80). Even in politics, about 37% of LDP members elected to the Diet in 1996 were related to former members of the Diet (Stockwin 1999:147).

Understandably the European consortium was shocked by its loss to TSC and took legal action against THSRC, with $73 million in compensation being awarded in 2004 (*Taipei Times*, 17 March 2004). However, as construction had already begun by 2000 when TSC won its contract, some of the standards, such as the size of tunnels, were to European specifications rather than Japanese. Perhaps due to the concern over the ongoing legal action and as an attempt to placate the Europeans, THSRC has continued to take advice and implement ideas from them. The result is that the system will be a hybrid of European high-speed railways and the Japanese shinkansen. Although it is common amongst the Japanese to still refer to the 'Taiwan Shinkansen', due to its hybrid nature, it is more accurate and appropriate to use the term Taiwan High-Speed Railway (THSR).

The Taiwan high-speed railway – the importance of symbolism

In preparation for my first visit to Taiwan I asked many people, particularly Japanese, what I should expect. I was told that it is much like Japan was 20–30 years ago. Based on what I experienced, this seems somewhat unlikely. Culturally, at least, Taiwan seems very different, not only to Japan, but also to other East Asian countries. One particular incident emphasized this difference to me. In Japan it is uncommon to be asked questions by a stranger, and if should it happen, conversations tend to revolve around experiences of visiting each other's countries and learning the languages of those countries. However, on my first day in Taiwan I was confronted by a stranger who proceeded to ask me various questions on global and Taiwanese political issues, and only asked about my background some 30 minutes or so into the conversation. One question was about what I thought about the relationship between China and Taiwan. In answering this question I began to appreciate what an important role the THSR may have to play.

The relationship between China and Taiwan is highly complex and certainly beyond the scope of this study. In short, the Chinese Nationalist Party, Kuomintang (KMT), fled to Taiwan in 1949 following the Communist Party taking control of the mainland. The result was two Chinas – the People's Republic of China, commonly just called China, and the Republic of China, commonly referred to as Taiwan. Although it may seem remarkable now, the KMT vowed to unify China by retaking the mainland. As time has gone by the likelihood of unification has been more likely in the other direction. In the meantime, Taiwan continues in a state of effective independence but with few formal diplomatic links with other countries, who are wary of doing anything that would hinder their relationship with China. Sadly, the result has been that although Taiwanese products have become well known globally, the voice and spirit of the Taiwanese people has remained less well known. Taiwan does not have a global symbol that reflects its essence. The often seen 'Made in Taiwan' on products even became a source of jokes in Britain, for example. The THSR may become a more positive symbol for the country. This study has already shown how the shinkansen reflects many aspects of Japanese society and culture. There is also no doubt that it is a well-known global icon. Furthermore, the shinkansen was symbolic and it helped to unite the Japanese people and give them pride in the developments that were being made during the 1960s. The THSR could do the same for the Taiwanese people.

In considering the imagery and symbolism of trains in Taiwan, it is worth mentioning the apparent importance of Japanese railways and the shinkansen. Since the railways were originally established by the Japanese, rolling stock has continued to be exported there. However, whether most Taiwanese are aware of these trains' origins is questionable. Yet, in a shop in Taipei's main station, many of the toy trains for children were based on the shinkansen – particularly the 300-series, although with a green rather than blue stripe and the words 'City Express' written on the side. Even the timetable in the station used a 'clipart' image of the 300-series to indicate express trains. Shelves in the shop were also covered with magazines imported from Japan covering topics such as travelling around Japan by train to the leading Japanese railway enthusiast magazines. Interestingly, though as the station is still wholly operated by the TRA perhaps not surprisingly, there was no sign of any pictures, models or advertisements for the THSR. Yet the most striking image of the shinkansen came in an enormous poster promoting World Citizenship Day 2004. The poster was mostly textual, but also included a map of Taiwan and its whole railway network. In front of this, and by far the most prominent part of the poster, was an image of a 0-series shinkansen, painted with a red stripe rather than blue (see Figure 3.5d). The representatives seemed uncertain as to why this image had been chosen other than 'a new railway will be opening'.

Chapter 3 considered the issue of the beauty of the shinkansen. For the shinkansen exported to Taiwan, it was decided that the 700-series would be used as the basic design, but that its 'duckbill-platypus' face would be replaced by something more attractive. The result is the 700T-series. While the outside looks different to the 700-series, its Japanese heritage is unmistakable to those who are familiar with the shinkansen. All of the 700T-series shinkansen are to be made in Japan by a variety of constructors, as is the normal practice in Japan for the construction of the shinkansen. Clearly, as discussed in Chapters 3 and 6, the issue of the symbolism of the train and the choice of its colour would also be important. The 700T-series uses the same basic scheme as for Japanese shinkansen, but has orange and black stripes rather than blue, green or red. When I asked THSRC why these colours were chosen, I was told that they are the company colours and are those liked by the company's president, Nita Ing. The choice of colours seems ironic given JR Tōkai's involvement in the project, since orange is their corporate colour but it is not used on Tōkaidō shinkansen. While the THSRC explanation for the colours seems plausible, it is also worth noting that their choice may be influenced by a need or desire to avoid overtly political colours – with blue and green being those of the two main parties in Taiwan (Deans interview 2004). That the main express train, which the 700T-series will effectively replace, is also coloured orange adds an element of familiarity to those used to thinking of long-distance trains being that colour.

The THSR may suffer initially from the lack of symbolism in one respect. Unlike with the construction of the Tōkaidō Shinkansen, Hokuriku Shinkansen and Tōhoku Shinkansen extension to Hachinohe, there is no event ensuring that not only there is a need to complete the work on time, but there is also nothing, as was particularly the case for the Tōkaidō Shinkansen and the staging of the first Olympics in Asia, that will create a link between the railway and a global event in which the people of the country can also take pride. Yet the THSR is expected to be symbolic in other way. Although the system will be a hybrid of European and Japanese technology and expertise, it is clearly being most associated with Japan. As a consequence, some, such as Chiang Ping-Kun, who was Chairman of the Council for Economic Planning and Development, have been suggesting that the railway will be a symbol of Taiwan–Japan relations (Okada 2001). This may well be the case, but such statements are likely to also be politically loaded in an attempt to help normalize Taiwanese relations and gradually bring the curtain down on Japan's, and other countries', two China policies.

The Taiwan high-speed railway – political involvement

When the European consortium was given preferential bidder status in 1998, it was a surprise to many. This was not due to any widespread belief or knowledge in Japanese expertise in this field, but rather that relations between Taiwan and Japan were so strong and significant. While it is true that European, particularly French, bids for other projects, for example the underground system in Taipei, had been successful in the past, these decisions too had not been due to purely technical or engineering considerations. The Taipei underground, for example, was linked to a purchase of French military weapons. That the construction of the underground has been plagued with problems, for example, some of the system is better suited to a small French city rather that a large metropolitan city such as Taipei, may have undermined further bids (Deans interview 2004). However, that the European consortium was given preferential status at one point would suggest that there was a possibility that these problems could have been overlooked.

Japan is popular in Taiwan, both at a cultural level and at a political level, although sometimes 'to say "I love Japan" is a way to say "I don't love China"'

(Deans interview March 2004). In terms of politics, it is the KMT that has particularly strong links with Japan and Japanese politicians. With the Japanese bid being supported by Liu Tai-Ying, chairman of the China Development Industrial Bank and ruling financial boss of the KMT, the European consortium's preferential status was a particular surprise (Okada 2001). However, although it is other issues, as will be discussed in following paragraphs, that apparently led to the change in THSRC's position, the possibility that there was political intervention cannot be overlooked. For the greatest supporter of Japan was President Lee Teng-Hui. With tensions between China and Taiwan rising, Lee was keen to develop closer economic and political ties with Japan (Okada 2001). Due to the size and symbolic nature of the project, the THSR may well have become a pawn in a much larger game of diplomatic strategy.

Lee is known for being a Japan-ophile, having publicly stated that he was Japanese until he was 22 and stressed the importance of relations between Taiwan and Japan in the 1996 Presidential Elections (Okada 2001). Although he stressed that as far as the THSR was concerned political considerations should be after cost and safety, in his book *The Road to Democracy* (1999), he was critical of the Japanese government's relative lack of dynamism in promoting the shinkansen in comparison to the approach taken by the French and German governments in relation to their products (Okada 2001). This lack of dynamism on the part of the Japanese may have been due to two factors. First, with the shinkansen now being operated by essentially private companies, despite the possible diplomatic benefits to the Japanese government, there may have been a lack of desire to become involved in the project. However, much more likely, was the problems created by needing to maintain a two Chinas policy. With China also considering the shinkansen as the basis for its own high-speed railway, the Japanese government and leading Japanese politicians seemed to have been more concerned about improving the chances of winning this even bigger project at the possible expense of not gaining the Taiwanese project. Although Japanese politicians tend to be divided into either pro-Taiwan or pro-China camps, the Japanese government's position has tended to be dominated by a pro-China stance. Despite Lee's words about the priority of considerations in choosing a partner for the THSR, he also made it clear that if the Japanese government tried to sell the shinkansen, Taiwan would have given political consideration to the choice. According to Okada (2001), 'Rarely had a top political leader shown such a firm stance in regard to a private sector venture'. Yet, Okada also suggests that rather than the Japanese government's stance in relationship to the THSR, it was the Japanese government's eagerness to promote the shinkansen to China that was Lee's concern.

The opportunity for the Japanese consortium to be victorious came in 1999 when THSRC announced that it was postponing the start of the project due to a lack of funds. There was even a suggestion that THSRC may seek to cancel its contract altogether (Okada 2001). However, Liu expressed that he was prepared to give financial assistance to THSRC on condition that they used the shinkansen. While Liu's links with the KMT cannot be overlooked, apparently he was also told that low-interest loans from the Japan Export and Import Bank (now Japan Bank for International Cooperation) would be forthcoming (Okada 2001). THSRC head, Ing, was apparently not happy about the KMT's attempted intervention and stresses that the decision to choose the shinkansen was due purely to technological and economic considerations (Okada 2001). In reality it may have been a combination of reasons that led to the Japanese success, but the position of the KMT and President Lee's involvement cannot be ignored. The result may be, as Lee wished, that the THSR will become a 'memento for closer bilateral economic cooperation' (Okada 2001) between Japan and Taiwan.

The Taiwan high-speed railway – economic rationale

The promotional material prepared by THSRC shows clearly how there is an expectation that the line will turn Taiwan into one huge metropolis. Taiwan's two largest cities, located at either end of the island, let alone those in between, will become within commuting distance of each other. While the market between Taipei and Kaohsiung is already large, the expectation is that the THSR will provide a stimulus to even greater economic development, that the size of the market travelling along the route will increase, and that the THSR will come to dominate it. THSRC project that about 200,000 people will use the line daily (Nehashi 2001; THSRC interview April 2004). As mentioned earlier, the project is the world's largest B-O-T Project and the costs have been enormous. In a situation that echoes Sogō's obtaining a loan from the World Bank, THSRC's loan from the Japan Export and Import Bank not only provided funds, but perhaps more importantly gave a level of support and confidence in the project, due to the externality of the creditor, that helped to gain further support for the project domestically.

While the TRA is introducing more comfortable trains, with wider seats than the 700T-series and TV screens, in an attempt to compete with the THSR, it is probable that it will have to focus on short to medium-distance travellers, which already account for 71% of its ticket sales, despite tickets being cheaper than the THSR (*Taipei Times*, 3 March 2004). However, THSRC clearly believe their main rivals are the airline companies. While the speed of the train and length of the line, taking the Tōkaidō Shinkansen as a model, would suggest that THSR is likely to dominate the market, forecasts have been relatively conservative. THSRC themselves suggest that their market share for the Taipei–Kaohsiung will be about 50% and a further 20% will be taken by the airline companies. The experience of Japan suggests that at some point in the future this latter share will drop to 0% and that the former will rise to close to 100%. However, in Taiwan cars and buses, which tend to have large reclining leather seats on long-distance journeys, are still popular and affordable options, and until the west coast becomes the crowded and affluent urban sprawl that is being forecasted to be one of the impacts of the THSR, it is likely that the conservative estimate on the THSR's market share is appropriate. However, to help THSR compete with the airlines, it will be undercutting them on price. The ticket price has been set at a rate of NT$3.459/km with a maximum adjustment of 20% allowed on top of that (THSRC interview April 2004). What may be significant is that there is no set minimum fare, with special offers being allowed. This does not happen to any significant degree in Japan. If JR Tōkai, for example, were able to drop its fares or offer discounts, there is little doubt that it could recover some of the market share it has lost over the past ten years to the airline industry on the Tōkyō–Ōsaka route. When one also considers the greater environmental record of rail travel over air travel, it is certainly time that prices be allowed to drop more.

However, the economic impact of the THSR is not limited to Taiwan. As a nod to the economic, let alone political, significance of the export of the shinkansen to Taiwan, the BBC (2004) used an image of a 700T-series shinkansen being lifted onto a cargo ship rather than a car, electronic goods or such like to go with an article about the importance of exports and continued improvement in the Japanese economy.

It is important that one remembers that the influence of the shinkansen has been more than merely financial. Railways are huge investments and it can take several decades to recuperate all of the costs. However, more than this, one has to remember that there are more important things than money. While a return to the wasteful spending of the JNR days cannot be justified, there is a need for great developments. As Clarkson (2004a:230) suggests, 'human thirst for improvement' is being 'extinguished' by 'the bean counters'. Under such conditions, Clarkson (2004a:229–30) maintains, Columbus probably would not have reached America, Armstrong would not have walked on the moon, Amundson would not have gone to the South Pole and Turing would not have invented the computer.

The world is full of many wonders of previous civilizations, yet the modern world's concern for profit may mean that such developments may not be seen again. Arguably, the shinkansen was such an important development. Not only did it become a powerful symbol for Japan, it also 'provided a shock and stimulus to the European countries, which had well-developed rail systems' (Yamanouchi 2000:5). As a consequence these railways have been further improved and today these railways, as well as the shinkansen itself, are influential in countries such as South Korea and Taiwan. However, while the Tōkaidō Shinkansen may have been such a significant development, it may be less true of the other lines. It is the fact that the shinkansen has faded into the normal and everyday that makes it such a good tool by which to study Japanese society. While it may cause some excitement at the local level, particularly in the case of the Kyūshū Shinkansen with its more locally flavoured design of train, for most Japanese I expect a new section or line opening is often met with a 'so what?' or 'didn't they already have a shinkansen?' type reaction.

However, the shinkansen can still play a role in this respect. Clarkson (2004a:90) believes that countries need funds to do 'nothing but pay for great public buildings, follies, laser shows, towers, fountains, airships, aqueducts. Big, expensive stuff designed solely to make us go "wow".' For, while this study has suggested what the economic impact has been of opening certain shinkansen lines to date, and what the impact may be of opening other lines, this is too narrow a means by which to measure what the shinkansen can do. The opening of the Tōkaidō Shinkansen gave the Japanese people hope and belief. This in itself can, and did, lead to economic benefits. But it can also make people optimistic and happy. In other words, it can raise the quality of life. It is for this reason, more than any other reason suggested in this study why I believe that not only should Japan build the Chūō Shinkansen, but that it *must*. While, as mentioned previously, I believe the term 'lost decade' is largely without merit, there is no doubt that Japanese national morale has suffered in recent years. What is needed is a symbol that says 'we can do anything'. The Tōkaidō Shinkansen did it before, the Chūō Shinkansen can do it this time.

The Taiwan high-speed railway – reflecting Taiwan

While economic and political considerations have already been discussed in relation to the decision to choose the Japanese bid, the one that is now most cited as being the key reason is the shinkansen's ability to cope with earthquakes. This is not a problem faced by France and Germany, and as a consequence their expertise and high-speed railways are not perceived to need to, or necessarily to be able to, cope with earthquakes. Taiwan is in a tectonically active location, although not to the degree of Japan. The concern about the THSR's ability to cope with earthquakes may not have been a high priority at the initial stage, and may be another factor that had allowed the European consortium to gain preferential bidder status. However, at the time when THSRC was struggling financially and was apparently facing increasing pressure from the KMT and President Lee to support the Japanese bid, nature intervened to ensure that the shinkansen was chosen. For on 21 September 1999 Taiwan was struck by a huge earthquake that left over 2,400 people dead. Not only did this focus the attention upon the need for the THSR to be able to cope with earthquakes, but the rapid reaction of the Japanese government and other organisations to provide help following the quake 'helped remove the "psychological barrier" that existed between Japan and Taiwan due to the absence of bilateral diplomatic relations' (Okada 2001).

While it is certainly the case that Taiwan faces many of the same natural challenges that Japan does, and that the shinkansen may be better prepared in this respect, various decisions that have been taken consequently suggest that one has to be cautious in assuming that the THSR will be able to boast the same exemplary safety record as the shinkansen in years to come. The shinkansen is a whole system. It is not just the train. It is the train combined with the other technical and engineering features, and the people who work upon it. To try to operate the shinkansen without using shinkansen-standards in other areas is a recipe for problems. Yet, it is this route that Taiwan has taken. For example, while the basic system is the shinkansen, due to suggestions from the European advisors, instead of the line being closed at night for daily maintenance to be performed, trains are going to be able to run either direction on each line, so that maintenance can be carried out during the daytime and while passenger operations are continued. However, doing this has increased the costs of the project as the ATC needs to be set up in both directions on both lines. It also raises questions about the level of safety, particularly for those working on the lines as trains pass, and performance, since trains will have to operate at reduced speeds in the areas where maintenance work is being carried out. This could prevent the THSR from gaining the image of being the reliable, consistently on-time service that the shinkansen enjoys in Japan. Additionally, although the use of some European specifications, such as larger tunnel cross-sections than would normally be used by the shinkansen, do not give rise to any concern, others, such as the use of points that are not of the design used by shinkansen in Japan, may do (anonymous interview April 2004).

There are other differences on the THSR which appear to reflect the different ways in which Taiwanese like to operate their modern railways. As on the underground, the trains will be driven automatically. The driver is there purely as a checking mechanism. JR companies consider the human element to be of considerable importance in maintaining high safety standards, and although the linear shinkansen is automated, the computer, such as ATC, is usually there to support the human driver in Japan rather than to be the primary driving system as will be the case in Taiwan. Similarly, the trains have had to be designed so that conductor can open and close the doors of the train at any door, rather than having their allotted position on the train. Again, not only does this potentially remove a safety check, as anyone who has ever stood on a shinkansen platform will be familiar with the sight of the conductors all having their heads out of the window of their room or the door of the cab as the train passes along a platform where the train is arriving or departing to check there are no problems,[2] but it also removes an element of certainty for passengers and other THSR employees alike about where conductors can be found.

These various differences from the shinkansen in Japan reflect not only THSRC's apparent desire or need to appease the Europeans, but also the way in which Taiwan tends to have a mixture of influences maintained. Whether the different ideas can coexist well together remains to be seen. However, despite lengths that have been gone to in trying to avoid problems, concerns appear to exist amongst many Japanese commentators about the future of THSR (anonymous interviews April 2004; Yomiuri Shimbun Chūbu Shakaibu 2002). As there is no history of high-speed railways in Taiwan, the THSRC employees that work on the trains, whether they be drivers, other train crew, or even maintenance staff, have been trained in Japan (THRSC interview April 2004). Clearly these training programmes work well in Japan. However, due to cultural, let alone linguistic, differences between Japan and Taiwan, it seems that some of those that I have interviewed, and others, for example Nehashi (2001) and Yomiuri Shimbun Chūbu Shakaibu (2002), appear to be concerned that the high standards taught in Japan will not be maintained. The result is that I have gained the impression that the Japanese are already beginning to emphasize that the THSR is *not* a shinkansen and are gradually distancing themselves from responsibility for the project.

The Taiwan high-speed railway – conclusions

There is much in common between Taiwan and Japan, and certainly more in common between those two countries than Japan and European or North American countries. Yet despite the similarities, important differences remain. While, as discussed previously, it is easy to over-emphasize differences at the expense of recognizing the mass of similarities, even small differences can be significant. After all, human DNA is suggested to be 96% the same as a chimpanzee and 60% the same as a fruit fly (Yahya 2004:48),[3] yet clearly the 4% and 40% respectively are highly significant. The THSR will only be about 70% shinkansen (Shima interview April 2004). The other 30% may be highly significant. The 30% need not be seen in a negative light, although it is certainly causing some Japanese concerns. The Taiwanese appear keen to use the shinkansen as a basis on which to develop the world's best performing high-speed railway, while being 'persistently cost-conscious' (Nehashi 2001). Given the high standards of the shinkansen, it will be a target that may be hard to achieve. It also has to be remembered that while the Tōkaidō Shinkansen was an immediate success, other lines have taken time to reach their full potential. Given the similarities between the THSR route and the Tōkaidō Shinkansen between Tōkyō and Nagoya, it is highly probable that the line will be successful, that the new towns planned along the route will develop, and the line will become used by long-distance travellers and commuters alike. However, given that Mishima's development as a commuter city, for example, has also been relatively recent, some of the THSR's success may not become truly visible until near the end of THSRC's 35-year operation period.

Whatever the real reason for the European consortium's loss of preferential status may have been, with the lack of other items, particularly military items, being offered the most probable (Deans interview March 2004), the end result has had benefits for both Taiwan and Japan. Due to the devaluation of the Euro, had THSRC chosen the European consortium the total cost would have ended up much greater than what has been spent on the shinkansen (THSRC interview April 2004). The decision by a foreign country to choose the shinkansen, the first time that this has happened, has given the opportunity for the shinkansen to establish itself as a high-speed railway system that can be exported. While there may be concerns amongst some Japanese about the way in which the THSR is being built and how it will be operated, distancing themselves from the project too much, while being the prudent approach, may also be an approach that prevents any further exports. The shinkansen needs to be more flexible if it is to be exported again.

Exporting the shinkansen to other countries

While the THSR is the first time that Japan has successfully exported the shinkansen, it is not the first time it has been attempted. In 1975 research was done about a possible shinkansen in Iran to link Tehran and Mashhad. The following year, a ten-year project called the North–East Corridor Improvement Project (NECIP) was begun in the United States to look at the possibility of the shinkansen being used in the United States. This project included research being done in Ohio, on a Los Angeles–San Diego route, and a Tampa–Orlando–Miami route (Nehashi 2001). However with all of these projects, the necessary funding could not be found and none of them came to fruition.

The next possibility was in South Korea with a high-speed line being constructed between Seoul and Busan. This was the first opportunity following the reform of JNR. However, coming at the start of the 1990s, it was a time when the shinkansen was still only just emerging from its period of the 'Dark Valley'. New research and development was being carried out in Japan into ways to improve the shinkansen, but the results were not yet on display. Europe, particularly France, appeared to have taken the position as the world's leading high-speed railway country. Given this fact, let alone the cultural, political and historical problems that existed, and still continue to exist, between South Korea and Japan, there was never any likelihood of the shinkansen being selected. Taiwan was the next battleground. Having appeared at first that the Europeans had won again, Japan managed to win the main contract.

The next country likely to make a decision on developing a high-speed railway line is China. Indeed China is hoping to develop a high-speed railway network, but will first concentrate on building a line between Shanghai and Beijing. The idea of such a line has been around for many years and when Beijing was selected to host the 2008 Olympics, it was expected that this would be the final impetus to make the Chinese government come to a decision. However, as the government began to suggest that it was going to choose the shinkansen, having already made it clear that Taiwan's decision to adopt the system was not going to be a hindrance to its possible adoption in China, with some even suggesting that Taiwan choosing the shinkansen was a prerequisite for China to adopt the shinkansen (Okada 2001), there were protests at a popular level. For, as in South Korea, there are still many who feel as though Japan has not fully atoned for its actions during the Pacific War. With the Shanghai–Beijing route going via Nanjing, scenes of some of the worst atrocities in China during the War, the scale of which remains a contentious issue in China and Japan to this day, there were protests from those opposed to the possibility of a Japanese railway being constructed in the city – although these protests may have been supported at higher levels as a negotiating tool. Just as it may have been cultural and political reasons as much as technical or financial reasons that won Japan the Taiwan project, so it would appear that cultural if not political reasons may intervene to prevent Japan from winning what may be the largest goal, given that the Shanghai–Beijing route alone is estimated to cost ¥2 trillion (Okada 2001). That victory would, according to some, allow the shinkansen to play a major role in East Asia. For example, Okada suggests

> With China's imminent accession to the World Trade Organization, economic integration will erode the political barrier that has existed across the Taiwan Strait since 1949. If Japan succeeds in winning the forthcoming Chinese deal following its victory in the Taiwan deal, the project could turn out to be not

only a bridge between Taiwan and China but also a model for multilateral and regional cooperation in East Asia, which still remains the legacy of the Cold War.

(Okada 2001)

Such an idea may seem a little extreme, but it certainly shows how the shinkansen is more than just a train.

In the future other countries may also try to develop high-speed railways. The proposal for at least one high-speed line has been around for many years in Britain, for example. This chapter has demonstrated how the decision may be influenced by more than technical matters. Even considering technical matters, it is not totally clear whether a European system, often spearheaded by the French TGV, is superior or whether it is the shinkansen. The problems that South Korea, which has a system based on the TGV, experienced would suggest that the TGV is far from faultless. Yet, the rigidity of the shinkansen system, and the concerns amongst the Japanese that appear to exist when the system is modified, as in Taiwan, would suggest that the shinkansen may not be best suited to some countries.

A further issue that some often cite as a problem with exporting the shinkansen, while suggesting that it is the reason why it is successful in Japan, is that the shinkansen is a 'dedicated track', with the associated costs of having to build everything new. There is nothing unusual about this in Japan for in practice, many lines are dedicated. While there is a possibility of transferring from one line to another, generally speaking the services are kept separate. This avoids the problem of bottle necks at major stations, for example, as the same trains always go to the same platforms, every day. The shinkansen has gained the reputation of being built on a dedicated track, but the origin for that, as was discussed in Chapter 2, was the fact that the standard Japanese railway gauge was narrower than the international standard, and the one used for the shinkansen. Had Japan already had standard gauge railways, it is highly likely that either Japan would be struggling to cope with having many types of trains (commuter, freight and express) on the same tracks, or a new line may have been built but using the existing stations rather than, as in many cases with the shinkansen, new ones being built. In fact, as this study has hinted at, it is a fallacy to describe the shinkansen as being 'dedicated'. Many lines – for example, the Sanyō Shinkansen and most of the JR East shinkansen network – have trains with significantly differing performances. Furthermore, south of Ōmiya, the track is used by trains that currently serve five different lines, with through-running to the Hokkaidō Shinkansen to increase this further in the future. The experience and problems with upgrading the West Coast mainline in Britain, as discussed in Chapter 5, also suggests that starting afresh may also be more cost-effective, let alone allow for a much greater increase in performance.

Based on what happened in all of the earlier examples, should a country such as Britain, for example, decide to build a high-speed line, it is clear that it will not be simply a matter of economics and comparisons of engineering information that will be key in the final decision being made. Britain, like other European countries, will always have to be mindful of the 'need' for trains to be able to operate on lines in other countries, even if this is likely to remain a theoretical need rather than a real need. By the same token, it should be noted that Japan would not necessarily want to export the shinkansen to all countries. The cultural differences between Japan and Taiwan have clearly raised concerns, a non-East Asian culture, particularly if it is a country where labour unions remain strong in the railway industry, given the experiences of JNR, is likely to discourage from Japan wanting to risk harming the shinkansen's reputation. This is a view that has been expressed to me in those terms by one JR company that has been asked on many occasions why they did not invest in British railways following privatization (anonymous interview). The establishment of THRSC in Taiwan, free from the history of TRA's activities and ways, reveals a way in which the shinkansen can be imported without being bound to the traditional ways of running the railways. Indeed, many employed in THRSC have joined the company from the airline industry (THRSC interview April 2004). This is not totally different to the situation of the shinkansen. For while the shinkansen was operated by JNR, it was overseen by a special 'Shinkansen Division' within JNR, but which was a 'nearly independent organization' (Yamanouchi 2000:117). Had this not been the case, given the problems of JNR during the 1960s and 1970s, it is questionable whether the shinkansen would have managed to have maintained its excellent safety and service record.

Final conclusions

This study has covered the history of the shinkansen, its symbolism, its relationship with politics, its economic performance, its culture and the way it both reflects and leads changes in Japanese society. A study such as this is long overdue. However, I hope that it is the beginning, not the end of the journey. Each chapter, and many of the subsections, are worthy of their own books. Now that the significance and importance of the shinkansen has been made clear, I hope that others, with their own specialist backgrounds, whether they be economists, historians, anthropologists or political scientists, will now take this study as a basis upon which they can use the shinkansen as a tool in their field of research.

The shinkansen is clearly much more than a means of transportation. It is a symbol that has captured the imagination of the Japanese people. From the live TV broadcast of its first journey between Tōkyō and Ōsaka to the extensive media coverage of events such as the retirement of 0- and 100-series shinkansen and the fortieth anniversary of the opening of the Tōkaidō Shinkansen, the shinkansen has become something that the Japanese people take pride in. It is a true national symbol. That the media coverage of the shinkansen has extended to stories such as the first women drivers further reveals the role that the shinkansen can and does play in not only reflecting Japanese society, but also inspiring changes within that society. So intertwined is the shinkansen with Japanese society that

210 *Conclusion*

Figure 8.1 (a) A 300-series shinkansen passing cherry blossom and office buildings in Tōkyō; (b) a 300-series shinkansen passing through the historically important town of Sekigahara; (c) a 700-series shinkansen passing Hamanako; (d) a 400-series shinkansen approaches Tōkyō station, with the new elevated section for the Chūō Line, built due to the expansion of shinkansen platforms, above; (e) a 0-series-renewal near Higashi-Hiroshima; (f) The Ōsaka shinkansen depot.

views along the shinkansen lines reflect the diversity of Japanese life; from maintaining links with its history and traditions to the hustle and bustle of modern cities (see Figure 8.1).

It is easy to be too dismissive about national symbols. The shinkansen is significant. It is a symbol which the Japanese people have chosen to embrace, although it is clearly one that the state has also been keen, particularly in the past, to support and develop. When Shima Hideo died on 18 May 1998, Yamanouchi (2000:259) described it as the end of an era. *The Times* obituary stated that he had 'helped to create one of the most potent symbols of Japan's postwar reconstruction and emerging industrial might' (*The Times*, 10 April 1998, 25). *The Times* was wrong. The shinkansen was Japan's *most* potent symbol then, and as it continues to evolve is likely to be so for many years to come.

Appendix 1
Chronology of significant dates in the history of the shinkansen

Year	Date	Period/event
		Imperial expansion
1939	October	The term 'shinkansen' is used in a report improving links between Tōkyō and Shimonoseki
1940	March	Approval given to the building of the shinkansen and *dangan ressha*
1941	August	Work began on Nihonzaka Tunnel
1942	20 March	Work began on Shin-Tanna Tunnel
1943	January	Work on Shin-Tanna Tunnel abandoned
1944	September	Abandonment of all construction of shinkansen with completion of Nihonzaka Tunnel
		Wilderness years
1945		No work on shinkansen during these years
1955		
1956	10 May	Study group to improve Tōkaidō Line set up
		Genesis
1957	25 May	Railway Technology Centre speak of Tōkyō to Ōsaka being cut to 3 hours by 'Super Express of Dreams'
	29 July	JNR sets up trunk line study group
	30 August	Trunk line study group set up within Ministry of Transport
1958	19 December	Tōkaidō Shinkansen authorized
1959	20 April	Ground-breaking ceremony for Tōkaidō Shinkansen at Shin-Tanna Tunnel
1960	5 May	World Bank investigatory group visit Japan
1961	2 May	World Bank agreed $80 million (¥28.8 billion) loan for construction of Tōkaidō Shinkansen
	18 October	Final route of Tōkaidō Shinkansen agreed
1962	23 June	Kamonomiya Test Track (12 km long) completed
	31 October	Research began on linear-motor propulsion for railways
	29 November	Testing of ATC system began
1963	30 March	Record speed of 256 km/h attained
	19 August	Interior design of trains finalized
1964	24 March	'Tōkaidō Mainline (Shinkansen)' decided on as official name for line

(Appendix 1 continued)

Appendix 1 Continued

Year	Date	Period/event
	15 June	Decision made to have signs say 'Shinkansen New Tokaido Line'
	1 July	The final rail is laid in Kawasaki, completing the 515 km route
	7 July	*Hikari* and *Kodama* names adopted for shinkansen services
	25 July	First test train runs between Tōkyō and Ōsaka, taking 10 hours
	24 August	First test of *Kodama* services between Tōkyō and Ōsaka
	25 August	First test of *Hikari* services between Tōkyō and Ōsaka, broadcast live on NHK
		The boom years
1964	1 October	Tōkaidō Shinkansen opens for service – *Hikari* taking 4 hours, *Kodama* 5 hours – 60 trains per day. 36,128 passengers carried on first day with all seats reserved
		Buffet – provided by Nihon Shokudō and the Imperial Hotel – and in-car food – provided by Tōkai Shahan – service began
		Japan Railway Construction Public Corporation (JRCC) formed to build new railways with aid of government finance
	14 October	Shinkansen awarded the 'Prime Minister's Award' for its successful establishment
1965	10 January	First running between Gifu-Hashima and Maibara in snow
	19 March	Total passengers carried reaches 10,000,000 after 170 days of operations
	20 May	Carriages 1–7 of *Kodama* services made non-reserved
	1 June	First use of public phones on shinkansen
	9 September	Construction of line between Shin-Ōsaka and Okayama authorized
	1 November	Time of fastest services between Tōkyō and Shin-Ōsaka cut to 3 hours 10 minutes
	10 November	0-series awarded the 'Blue Ribbon Award'
1966	1 January	JNR produced its vision for the next 20 years – included plans for Sapporo to Hakata to be linked by a shinkansen network
	10 March	Some *Kodama* services made completely unreserved
	12 June	Total shinkansen passenger numbers reached 50 million after 619 days of operation
1967	16 March	Construction of line between Shin-Ōsaka and Okayama started
	13 July	Total shinkansen passenger numbers reached 100 million after 1,016 days of operation
	25 November	Sprinklers installed at key sites to cope with snowfall
1968	25 November	JRCC announced plans for nationwide 4,750 km Shinkansen network
1969	1 March	Total shinkansen passenger numbers reached 200 million after 1,613 days of operation
	25 April	Mishima shinkansen station opened
	10 May	First class renamed as 'Green carriage' and second class as 'Ordinary carriage'
	6 August	Total distance travelled reaches 100 million km after 1,771 days of operation
	12 September	Okayama to Hakata Shinkansen extension authorized
	8 December	*Hikari* services increased from 12 to 16 carriage formations

Appendix 1 Continued

Year	Date	Period/event
1970	10 February	Construction of Okayama to Hakata extension began
	15 March	EXPO starts at Ōsaka
	13 May	National Shinkansen Development Law passed
	2 July	Total shinkansen passenger numbers reached 300 million after 2,101 days of operation
1971	18 January	Basic proposals for Tōhoku, Jōetsu and Narita Shinkansen announced
	1 March	Start of dedicated services for school trips
	1 April	Directive issued for work to start on Tōhoku and Jōetsu Shinkansen
	7 September	Total shinkansen passenger numbers reached 400 million after 2,533 days of operation
	16 October	Test running starts on Sanyō Shinkansen
	26 November	Construction of Tōhoku and Jōetsu Shinkansen began (official ceremony on the 28 November)
	3 December	Escalators introduced on Tōkyō platforms
	12 December	Construction of Dai-Shimizu Tunnel in Gunma and Niigata Prefectures began
1972	24 February	951 series prototype set world record speed of 286.0 km/h
	15 March	Sanyō Shinkansen between Shin-Ōsaka and Okayama opened Renewal of Tōkaidō Shinkansen began with heavier rails COMTRAC put into operation
	2 May	Plans for Hokkaidō, Tōhoku (north of Morioka), Hokuriku, Kyūshū (Kagoshima and Nagasaki) Shinkansen lines announced
	23 September	Total shinkansen passenger numbers reached 500 million after 2,915 days of operation
	14 October	Maglev tests began with ML100
1973	1 February	Night services on Tōhoku main line between Nagamachi and Higashi-Sendai replaced by buses until 30 June 1977 while Tōhoku Shinkansen was being constructed in that area
	21 February	Train derails at low speed at Ōsaka Depot
	20 April	Commemorative stone marking opening of shinkansen unveiled on platform at Tōkyō station
	27 April	Services disrupted by one day strike
	1 May	Break-through of Shin-Kanmon Tunnel for Sanyō Shinkansen connecting Honshū and Kyūshū (length 18.7 km)
	29 July	First prototype of shinkansen for Tōhoku and Jōetsu shinkansen, capable of 260 km/h, delivered
		The dark valley
1973	November	The *Seibi Keikaku* is established, prioritizing certain Shinkansen routes while putting other planned construction on hold
1974	1 February	Construction of Narita Shinkansen began
	1 March	Services disrupted by one day strike
	5 September	Restaurant carriages introduced on *Hikari* services
1975	28 January	Opening of Tōhoku Shinkansen deferred from 1977 to 1979
	10 March	Sanyō Shinkansen extension opened between Okayama and Hakata

(Appendix 1 continued)

Appendix 1 Continued

Year	Date	Period/event
	12 May	HM Queen Elizabeth II and HRH Duke of Edinburgh travel from Nagoya to Tokyo during state visit
	14 July	Fukushima Tunnel on Tōhoku Shinkansen line completed (11.7 km)
1976	25 May	Total Shinkansen passenger numbers reached 1 billion
	20 August	No smoking carriage (No. 16) introduced on Tōkyō to Shin-Ōsaka *Kodama* services
	19 November	Replacement of original 0-series stock began
1977	21 January	Doors open on Hiroshima-bound *Hikari* service between Hakata and Kokura. Train makes emergency stop. Faulty wiring blamed.
	1 May	Advertisements inside shinkansen carriages first introduced
	26 July	Maglev tests began with ML500 in Miyazaki
1978	18 July	Total distance travelled reaches 500 million km after 5,039 days of operation
1979	25 January	Completion of Dai-Shimizu Tunnel on Jōetsu Shinkansen (22.2 km – then the largest land tunnel in the world)
	20 March	Sixteen people killed in fire at Dai-Shimizu Tunnel construction site
	12 December	Unmanned maglev sets world record speed of 517.0 km/h
1980	25 January	Construction of Ōmiya to Tōkyō section of Tōhoku/Jōetsu Shinkansen began
	8 March	Nakayama Tunnel on Jōetsu Shinkansen flooded by water
	11 September	Legal ruling relating to train noise made in favour of JNR
	1 October	No-smoking carriage introduced on *Hikari* services
	31 October	Trail running on Tōhoku Shinkansen began
	5 November	Trail running on Jōetsu Shinkansen began
1981	15 May	World Bank loan fully repaid
	29 October	*Yamabiko*, *Aoba*, *Asahi* and *Toki* chosen as names for Tōhoku and Jōetsu Shinkansen services
	23 December	Nakayama Tunnel on Jōetsu Shinkansen finally completed (14.9 km)
1982	3 February	Names of Tōhoku and Jōetsu Shinkansen stations finalized
	23 June	Tōhoku Shinkansen opened between Ōmiya and Morioka – 20 trains daily
	15 November	Jōetsu Shinkansen opened between Ōmiya and Niigata – 42 trains daily
1983	23 June	Satellite based earthquake warning system introduced for Tōhoku and Jōetsu Shinkansen
	30 October	The 200-series wins Laurel Award
1984	3 April	Total shinkansen passenger numbers reached 2 billion
	1 July	No-smoking carriages increased to two on *Hikari* and *Kodama* services
1985	22 January	Route for Hokuriku Shinkansen announced
	14 March	Ueno to Ōmiya part of Tōhoku/Jōetsu Shinkansen opened Top speed of Yamabiko raised to 240 km/h Fastest service between Tōkyō and Ōsaka cut to 3 hours 8 minutes Mizusawa-Esashi and Shin-Hanamaki shinkansen stations opened

Appendix 1 Continued

Year	Date	Period/event
	1 April	No-smoking accommodation increased on all shinkansen to include reserved carriages
	24 June	Reformed six carriage 0-Series enters service between Hakata and Kokura
	1 October	The 100-series entered into service for passenger evaluation
1986	23 May	The 100-series wins Laurel Award
	1 November	Top speed on Tōkaidō Shinkansen raised to 220 km/h
		The 100 Series used for fastest services, with best time between Tōkyō and Shin-Ōsaka cut to 2 hours 56 minutes – other than 21:00 departure from Tokyo ('Cinderella Express') which was 2 hours 52 minutes – and between Tōkyō and Hakata cut to 5 hours 57 minutes
1987	4 February	Maglev MLU001 reached 400.8 km/h with passenger aboard

A new hope

Year	Date	Period/event
1987	1 April	JNR privatized and broken up
	16 October	Reserved seating on some *Kodama* services changed from 3 + 2 to 2 + 2
	11 November	Conversion of Ou Line between Fukushima and Yamagata to 'Mini-Shinkansen' began
1988	13 March	Seikan Tunnel between Honshū and Hokkaidō is opened after 24 years of construction. The 53.85 km tunnel (of which 23.35 km is under the sea) is the world's longest
		West Hikari services introduced on Sanyō Shinkansen
		Shin-Fuji, Kakegawa and Mikawa-Anjō stations opened on Tōkaidō Shinkansen
		Shin-Onomichi and Higashi-Hiroshima opened on Sanyō Shinkansen
1989	9 February	UrEDAS earthquake detection system introduced
	9 March	Introduction of news text on 100-series shinkansen
	11 March	Top speed on Sanyō Shinkansen raised to 230 km/h
		Introduction of *Grand Hikari* services
	2 August	Construction of Hokuriku Shinkansen began
	1 October	Twenty-five years since start of shinkansen services, a total of 2.75 billion passengers had been carried
1990	22 January	Tunnel collapses during construction of Tōhoku Shinkansen in Tōkyō city
	6 February	Ministry of Transport directs JRCC and JR Tōkai to begin surveys for Chūō Shinkansen. Maglev system to be considered
	10 March	Part of Jōetsu Shinkansen has top speed raised to 275 km/h
		Kurikoma-Kōgen station opened on Tōhoku Shinkansen
	1 April	Shinkansen extended from Hakata to Hakata-Minami
	23 June	Some double-deck carriages added to *Yamabiko* services on Tōhoku Shinkansen
	28 November	Work began on Yamanashi Maglev Test Line
	20 December	Gāla-Yuzawa ski resort station is opened
1991	30 January	Test running of 200-series and 400-series coupled together began
	20 June	Ueno to Tōkyō extension of Tōhoku/Jōetsu Shinkansen opened

(Appendix 1 continued)

Appendix 1 Continued

Year	Date	Period/event
	4 September	Construction of extension of Tōhoku Shinkansen from Morioka to Aomori began
	7 September	Construction of Kyūshū Shinkansen began
	17 September	Construction of Hokuriku Shinkansen between Karuizawa and Nagano began
	1 October	Shinkansen network sold by Settlement Corporation to the Honshū JR companies
	6 December	*Nozomi* adopted as name for new service on Tōkaidō/Sanyō Shinkansen to start in March 1992
	9 December	*Tsubasa* adopted as name for Yamagata shinkansen services
1992	13 March	Construction of Akita 'Mini-Shinkansen' began
		The golden age
1992	14 March	The 300-series *Nozomi* service began with top speed of 270 km/h, cutting the Tōkyō to Shin-Ōsaka journey time to 2 hours 30 minutes
		JR East's STAR21 test train completed
	April	JR West's experimental 500–900 'WIN350' train delivered
	10 June	Tōkyō's No.1 Shinkansen depot at Shinagawa closed
	1 July	Yamagata Shinkansen service began
1993	18 March	The 300-series *Nozomi* began service between Tōkyō and Hakata with fastest time of 5 hours 3 minutes
	27 May	The 300-series wins Laurel Award
	26 October	First shares of JR East sold on stock market
1994	15 July	E1-series introduced – the world's first completely double-decker high speed train
		JR West began construction of 500 Series prototype
	1 October	Thirtieth anniversary of opening of Shinkansen – 2.8 billion passengers carried with no fatalities
1995	17 January	Sanyō Shinkansen suspended between Kyōto and Okayama following Great Hanshin Earthquake. Kyōto – Shin-Ōsaka and Himeji – Okayama services restarted three days later
	8 April	Services between Shin-Osaka and Himeji recommenced. Complete normal service restored on 2 May
	April	E2-series and E3-series prototypes completed
	2 May	JR Tōkai's 300X's testing began
	22 October	Takasaki – Echigo-Yuzawa services suspended for morning (on 25 and 29 October also) while testing carried out of high-speed points for Hokuriku Shinkansen at Takasaki
	1 December	New all-stations stopping *Nasuno* service began on Tōhoku Shinkansen between Tōkyō and Nasu-Shiobara
	27 December	A 17-year old killed at Mishima station after being trapped in door and dragged along platform
1996	31 January	First 500-series set delivered to JR West
	16 March	No-smoking accommodation on *Nozomi* and *Hikari* services increased to ten carriages
	25 March	Permission granted for second Tōkaidō Shinkansen terminal to built in Tōkyō
	30 March	Tazawako Line closed for conversion to 'Mini-Shinkansen'
	26 July	The 300X sets Japanese record of 443.0 km/h

Appendix 1 Continued

Year	Date	Period/event
	29 September	Trial running began on Hokuriku Shinkansen between Karuizawa and Takasaki
	8 October	JR West shares listed on Tōkyō stock market
	17 November	Chūō Line terminal at Tōkyō station moved to make space for Hokuriku Shinkansen
1997	21 February	JR East announced that Yamagata Shinkansen would be extended to Shinjō by 2000
	22 March	Services on Akita Shinkansen began
		The 500-series began passenger operations on Sanyō Shinkansen
		New 4-carriage 0-series formation operates as *Kodama* between Hakata and Kokura
	3 April	Tests began on Yamanashi Test Line
	16 April	*Asama* announced as name for Hokuriku Shinkansen service
	26 May	Ground-breaking ceremony held at Shinagawa for new Tōkyō Terminal
	12 July	New Kyōto Station building – the biggest in Japan – opened
	29 September	JR East moves headquarters to Shinjuku
	1 October	Services on Hokuriku Shinkansen began
		Aoba and *Toki* service names withdrawn
		Tanigawa service began between Tōkyō and Takasaki/Echigo Yuzawa
	8 October	JR Tōkai shares listed on Tōkyō stock market
	29 November	Maximum speed of 500-series increased to 300 km/h on Sanyō Shinkansen, service extended to run between Tōkyō and Hakata, cutting time to 4 hours 49 minutes
	12 December	Manned MLX01 Maglev train set record of 531 km/h
	20 December	E4-series shinkansen with maximum of 1634 passengers began services
	24 December	Unmanned MLX01 Maglev train set record of 550 km/h
1998	12 March	Ministry of Transport approved three new Shinkansen projects – Hachinohe to Shin-Aomori, Nagano to Jōetsu, Funagoya to Shin-Yatsushiro
	March	Installation of automatic barriers completed all Tōkaidō stations other than Shin-Ōsaka
	6 October	Total JR East Shinkansen passenger numbers reached 1 billion
	30 October	Final running of a 0-series *Hikari* service
	30 November	Test running of gauge-changing shinkansen began at RTRI in Tōkyō
	8 December	Journey times on Akita and Tōhoku Shinkansen reduced. E2-series becomes main set on Tōhoku Shinkansen
	16 December	JR East announced new station on Jōetsu Shinkansen for Honjō to be built – most of the cost being met by the local community
	17 December	Maglev trains successfully passed each other at combined speed of 966 km/h
1999	February	Second General Control Centre for the Tōkaidō Shinkansen opened in Ōsaka
		First 200-series-renewal train set introduced

(Appendix 1 continued)

Appendix 1 Continued

Year	Date	Period/event
	13 March	The 700-series *Nozomi* services entered service
		Asa station opened on Sanyō Shinkansen
	14 April	Manned 5 carriage Maglev train set reached 552 km/h
	18 September	Last of JR Tōkai's 0-series withdrawn from service
	2 October	Dining carriage services on 100-series withdrawn
		All *Nozomi* services between Tōkyō and Hakata became operated by 500- and 700-series
	4 December	Yamagata Shinkansen extended to Shinjō with new E3-series being introduced
2000	10 March	All dining carriage services withdrawn from Sanyō and Tōkaidō Shinkansen
	11 March	JR West introduced 8 carriage 700-series *Hikari Rail Star* service
	September	The 700-series wins Laurel Award
	11 September	Tōkaidō Shinkansen services severely delayed by typhoon – 74 services suspended with 52,000 passengers on board. The worst shinkansen delays ever at 22 hours 21 minutes
	2 December	Massage service introduced on double-decker carriages of 200-series shinkansen
2001	7 May	E4-series introduced on Jōetsu Shinkansen
	3 September	T4 Doctor Yellow began operations
	1 October	Number of no-smoking carriages increased to 11 on *Nozomi* services
		East-i introduced
	November	GCT01 'Free Gauge Train' started tests between Ogori and Shin-Shimonoseki
	21 November	E2-1000-series entered service
	1 December	Green carriages on *Komachi* services made all no-smoking
2002	31 January	The 300X withdrawn
	4 February	New 6 carriage 100-series introduced on *Kodama* services on Sanyō Shinkansen
	14 May	*Hayate* announced as new service between Tōkyō and Hachinohe
	17 May	New 0-series-renewal introduced on Sanyō Shinkansen
	12 August	New 100-series-renewal entered service on Sanyō Shinkansen
	1 December	Tōhoku Shinkansen extension between Morioka and Hachinohe opened
		Asahi services renamed to *Toki*
2003	6 April	Modified E2-shinkansen records speed of 362 km/h in test running on Jōetsu Shinkansen
	16 September	Last 100-series withdrawn from Tōkaidō Shinkansen
	1 October	Shinagawa Shinkansen station opened
		Nozomi becomes main service on Tōkaidō Shinkansen
2004	30 January	Unveiling of first 700T-series shinkansen to be exported to Taiwan
	13 March	Honjō-Waseda station opens on Jōetsu Shinkansen
		Kyūshū Shinkansen services start
	1 October	Tōkaidō Shinkansen celebrated fortieth anniversary since opening
	23 October	Jōetsu Shinkansen derailed by Chūetsu Earthquake

Sources: Semmens (2000:113–18); Fossett (2004); Aoki *et al.* (2000:205); Suda (2000:168–74); Yamanouchi (2002:171–4); Hoshikawa (2003:167–9).

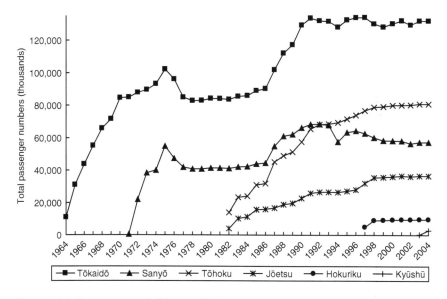

Figure A1.1 Passengers carried by the shinkansen.

Notes
Data is for fiscal years rather than calendar years. Data for Tōkaidō and Sanyō shinkasen between 1971 and 1986 are estimates as only total figures for the two lines together were kept until the reform of JNR. Figures for 2004 are estimates based on previous years and data received from JR Kyūshū for period 1 April–30 September.

Appendix 2
Shinkansen lines and stations

The tables that follow provide details about stations on each of the shinkansen lines. Each table includes the following information:

- Station name
- Map code – as shown on the map in this book.
- City or town served – Locations are cities unless stated otherwise. Two cities/towns are shown for one station only where more than one city/town name appears in the name of the station. Many stations serve other cities and towns also, and some are closer to these cities than the centre of the town or city they nominally serve. Cities in bold type indicate the prefectural capital.
- Population – Populations are as of 31 March 2003 and was taken from http://www.towninf.co.jp/. The population of Japan was 127,435,000.
- Date opened – Two dates are shown where additional services began at a terminal at a later date.
- Distance from origin – Both the operating distances (*eigyō kiro*) and the real distances are given. The difference between the two is due to main shinkansen lines not following exactly the same route as the original conventional line (where applicable). Generally the real distance is shorter than the operating distance due to the improved construction methods – for example, in the form of tunnels – that are used. Fares, which are based in part on a rate per distance travelled, are usually calculated on the basis of the operating distance. Some sections of lines have a special 'fare distance' which are shown in brackets on the table.
- Distance from previous station – Based on real distance. Some distances between stations may appear different to their distance from the origin due to rounding of figures.
- Number of trains per day – Details for number of trains taken on 1 October 2004. This was a Friday, when some extra services run. The figures includes the station where the train terminated. D = Down (usually from Tōkyō) and U = Up.
- Number of services – The total number of a particular type of service (e.g. *Nozomi*) that ran on the census date.
- Number of patterns – The number of different patterns for a particular service type. The higher this number, the less passengers can rely on using the service name solely as the basis for checking whether a service will stop at their desired station.

Other sources: JR Railway Timetables, Kōsoku Tetsudō Kenkyūkai 2003:97.

Tōkaidō Shinkansen

Station name	Map code	City or town served	Population	Date opened	Distance from origin (Operating)	Distance from origin (Real)	Distance from previous station	N D	N U	H D	H U	K D	K U	Tot. D	Tot. U	Total
Tōkyō	T1	**Tōkyō**	8,083,980	01.10.1964	0	0	—	83	82	30	30	36	39	149	151	300
Shinagawa	T2			01.10.2003	6.8	6.8	6.8	55	61	16	16	36	39	107	116	223
Shin-Yokohama	T3	**Yokohama**	3,466,875	01.10.1964	28.8	25.5	18.7	67	62	14	14	36	39	117	115	232
Odawara	T4	Odawara	198,269	01.10.1964	83.9	76.7	51.1	0	0	6	6	36	39	42	45	87
Atami	T5	Atami	42,582	01.10.1964	104.6	95.4	18.8	0	0	2	3	36	39	38	42	80
Mishima	T6	Mishima	111,373	25.04.1969	120.7	111.3	15.9	0	0	7	5	37	40	44	45	89
Shin-Fuji	T7	Fuji	237,620	13.03.1988	146.2	135.0	23.7	0	0	0	0	34	34	34	34	68
Shizuoka	T8	**Shizuoka**	703,255[a]	01.10.1964	180.2	167.4	32.4	0	0	16	17	36	35	52	52	104
Kakegawa	T9	Kakegawa	80,563	13.03.1988	229.3	211.3	43.9	0	0	0	0	33	32	33	32	65
Hamamatsu	T10	Hamamatsu	575,943	01.10.1964	257.1	238.9	27.6	0	0	13	11	34	32	47	43	90
Toyohashi	T11	Toyohashi	357,554	01.10.1964	293.6	274.2	35.3	0	0	7	12	32	30	39	42	81
Mikawa-Anjō	T12	Anjō	162,008	13.03.1988	336.3	312.8	38.6	0	0	0	0	32	30	32	30	62
Nagoya	T13	**Nagoya**	2,117,904	01.10.1964	366.0	342.0	29.2	84	83	32	32	33	31	149	146	295
Gifu-Hashima	T14	Hashima	66,813	01.10.1964	396.3	367.1	25.1	0	0	17	16	16	15	33	31	64
Maibara	T15	Maibara[b]	31,822	01.10.1964	445.9	408.3	41.2	0	0	17	17	16	15	33	32	65
Kyōto	T16	**Kyōto**	1,386,372	01.10.1964	513.6	476.3	68.0	84	83	30	26	16	15	130	124	254
Shin-Ōsaka	T17	**Ōsaka**	2,490,172	01.10.1964	552.6	515.4	39.1	84	83	30	26	16	15	130	124	254
Number of services								84	83	32	32	40	42	156	157	313
Number of patterns (in Tōkaidō region only – some extra variations when through-running past Shin-Ōsaka)								4	4	12	11	8	8	24	23	47

Notes

N = *Nozomi*, H = *Hikari*, K = *Kodama*, Tot. = Total.

a Shizuoka city merged with Shimizu city in 2003 – prior to merger the population of Shizuoka city was about 469,000.

b Maibara town merged with Santō, Iubki and Sakata in 2005 – prior to the merger the population of Maibara town was about 12,345.

Sanyō Shinkansen

Station name	Map code	City or town served	Population	Date opened	Distance from origin Operating	Distance from origin Real	Distance from previous station	Number of trains per day N D	N U	H D	H U	K D	K U	Tot. D	Tot. U	Total
Shin-Ōsaka	S1	**Ōsaka**	2,490,172	01.10.1964/15.03.1972	0	0	—	40	43	41	40	20	20	101	103	204
Shin-Kōbe	S2	**Kōbe**	1,483,670	15.03.1972	36.9	32.6	32.6	40	43	41	40	20	20	101	103	204
Nishi-Akashi	S3	Akashi	291,422	15.03.1972	59.7	54.8	22.2	0	0	15	14	20	20	35	34	69
Himeji	S4	Himeji	476,939	15.03.1972	91.7	85.9	31.1	8	9	41	40	20	20	69	69	138
Aioi	S5	Aioi	33,638	15.03.1972	112.4	105.9	20.1	0	0	14	13	20	20	34	33	67
Okayama	S6	**Okayama**	624,841	15.03.1972	180.3	160.9	55.0	40	43	40	39	39	36	119	118	237
Shin-Kurashiki	S7	Kurashiki	434,466	10.03.1975	205.5	186.7	25.8	0	0	0	0	36	36	36	36	72
Fukuyama	S8	Fukuyama	407,456[a]	10.03.1975	238.6	217.7	31.0	14	14	26	26	36	38	76	78	154
Shin-Onomichi	S9	Onomichi	93,091	13.03.1988	258.7	235.1	17.4	0	0	0	0	36	36	36	36	72
Mihara	S10	Mihara	81,250	10.03.1975	270.2	245.6	10.6	0	0	0	0	36	36	36	36	72
Higashi-Hiroshima	S11	Higashi-Hiroshima	119,344	13.03.1988	309.8	276.5	30.9	0	0	0	0	35	35	35	35	70
Hiroshima	S12	**Hiroshima**	1,118,767	10.03.1975	341.6	305.8	29.3	35	38	26	26	45	43	106	107	213
Shin-Iwakuni	S13	Iwakuni	106,142	10.03.1975	383.0	350.0	44.2	0	0	0	0	31	32	31	32	63
Tokuyama	S14	Shūnan[b]	156,608	10.03.1975	430.1 (434.5)	388.1	38.2	2	2	8	8	31	31	41	41	82
Shin-Yamaguchi[c]	S15	Ogōri	22,524	10.03.1975	474.4 (478.8)	429.2	41.0	8	10	18	16	32	32	58	58	116
Asa	S16	Sanyō town	22,610	13.03.1999	509.5 (513.9)	453.3	24.1	0	0	0	0	31	32	31	32	63
Shin-Shimonoseki	S17	Shimonoseki	246,282	10.03.1975	536.1 (540.5)	477.1	23.8	0	0	3	3	32	33	35	36	71
Kokura	S18	Kitakyūshū	997,398	10.03.1975	555.1 (559.5)	497.8	20.7	21	23	25	24	46	46	92	93	185
Hakata	S19	**Fukuoka**	1,315,007	10.03.1975	622.3 (626.7)	553.7	55.9	21	23	25	24	46	46	92	93	185
Number of services								40	43	41	40	65	61	146	144	290
Number of patterns (in Sanyō region only – some extra variations when through-running past Shin-Ōsaka)								6	7	7	7	11	12	24	26	50

Notes

N = *Nozomi*, H = *Hikari*, K = *Kodama*, Tot. = Total.
[a] Fukuyama city merged with Chūkaku city in 1998 – prior to merger the population of Fukuyama city was about 380,000.
[b] Tokuyama city merged with Shinnanyō city and Kumage county in 2003 – prior to merger the population of Tokuyama city was about 103,000.
[c] Originally this station was called Ogōri, but was renamed on 01.10.2003.

Tōhoku Shinkansen

Station name	Map code	City or town served	Population	Date opened	Distance from origin		Distance from previous station	Number of trains per day																
					Operating	Real		H		K		Y		MaxY		T		N		MaxN		Tot.		
								D	U	D	U	D	U	D	U	D	U	D	U	D	U	D	U	Total
Tōkyō[a]	Th1	**Tōkyō**	8,025,538	20.06.1991	0	0	—	16	15	15	14	26	25	22	20	17	16	15	17	0	1	111	108	219
Ueno[a]	Th2			14.03.1985	3.6	3.6	3.6	15	15	14	13	26	25	21	19	16	15	15	17	0	1	107	105	212
Ōmiya[a]	Th3	**Saitama**[b]	1,038,100	23.06.1982	30.3	31.3	27.8	16	14	15	13	26	25	22	20	17	16	15	17	0	1	111	106	217
Oyama	Th4	Oyama	153,900	23.06.1982	80.6	80.3	48.9	0	0	0	0	10	8	3	3	0	1	15	17	0	1	28	30	58
Utsunomiya	Th5	**Utsunomiya**	445,780	23.06.1982	109.5	109.0	28.8	0	0	0	0	24	23	21	19	16	15	15	16	0	1	76	74	150
Nasu-Shiobara	Th6	Kuroiso	59,650	23.06.1982	157.8	152.4	43.4	0	0	0	0	10	8	3	3	0	0	15	16	0	1	28	28	56
Shin-Shirakawa	Th7	Nishi-Gōmura[c]	19,028	23.06.1982	185.4	178.4	26.1	0	0	0	0	11	10	3	2	0	0	6	7	0	0	20	19	39
Kōriyama	Th8	Kōriyama	331,602	23.06.1982	226.7	213.9	35.5	0	0	0	0	25	24	20	17	14	13	6	7	0	0	65	61	126
Fukushima	Th9	**Fukushima**	288,632	23.06.1982	272.8	255.1	41.2	0	0	0	0	25	24	23	20	17	16	0	0	0	0	65	60	125
Shiroishi-Zaō	Th10	Shiroishi	40,514	23.06.1982	306.8	286.2	31.1	0	0	0	0	6	5	17	14	0	0	0	0	0	0	23	19	42
Sendai	Th11	**Sendai**	991,169	23.06.1982	351.8	325.4	39.2	17	16	16	15	29	27	23	20	0	0	0	0	0	0	85	78	163
Furukawa	Th12	Furukawa	73,136	23.06.1982	395.0	363.8	38.5	1	2	1	1	16	15	2	2	0	0	0	0	0	0	20	20	40
Kurikoma-Kōgen	Th13	Shiwahime town	7,524	10.03.1990	416.2	385.7	21.8	1	1	1	1	16	16	2	2	0	0	0	0	0	0	20	20	40
Ichinoseki	Th14	Ichinoseki	61,847	23.06.1982	445.1	406.3	20.6	1	2	1	1	17	16	2	2	0	0	0	0	0	0	21	21	42
Mizusawa-Esashi	Th15	Mizusawa Esashi	60,431 33,985	23.06.1982	470.1	431.3	25.1	1	1	1	1	17	16	2	2	0	0	0	0	0	0	21	20	41
Kitakami	Th16	Kitakami	92,470	23.06.1982	487.5	448.6	17.3	1	2	1	1	16	15	2	2	0	0	0	0	0	0	20	20	40
Shin-Hanamaki	Th17	Hanamaki	72,672	23.06.1982	500.0	463.1	14.5	1	1	1	1	17	16	2	2	0	0	0	0	0	0	21	20	41
Morioka	Th18	**Morioka**	281,245	23.06.1982	535.3	496.5	33.3	18	16	16	15	17	16	2	2	0	0	0	0	0	0	53	49	102
Iwate Numakunai	Th19	Iwate Town	17,296	01.12.2002	566.4	527.6	31.1	9	8	0	0	0	0	0	0	0	0	0	0	0	0	9	8	17
Ninohe	Th20	Ninohe	27,581	01.12.2002	601.0	562.2	34.6	13	11	0	0	0	0	0	0	0	0	0	0	0	0	13	11	24
Hachinohe	Th21	Hachinohe	244,075	01.12.2002	631.9	593.1	30.9	18	16	0	0	0	0	0	0	0	0	0	0	0	0	18	16	34
Number of services								18	16	16	15	29	27	23	20	17	16	15	17	0	1	118	112	230
Number of patterns[d]								6	6	5	5	9	9	7	8	4	5	2	3	0	1	33	37	70

Notes

H = *Hayate*, K = *Komachi*, Y = *Yamabiko*, MaxY = *MAX Yamabiko*, T = *Tsubasa*, N = *Nasuno*, MaxN = *MAX Nasuno*, Tot. = Total.

[a] Figures for number of trains stops does not include Joetsu and Hokuriku shinkansen stopping at this station.
[b] Ōmiya city merged with Urawa and Yono to become Saitama city in 2001.
[c] The population of Shirakawa is 47,319.
[d] Total number of *Yamabiko/MAX Yamabiko* patterns: 14 down and 14 up, total number of *Nasuno/MAX Nasuno* patterns: 2 down and 3 up.

Jōetsu Shinkansen

Station name	Map code	City or town served	Population	Date opened	Distance from Tōkyō[a] Operating	Distance from Tōkyō[a] Real	Distance from previous station	Number of trains per day To D	To U	MaxTo D	MaxTo U	Ta D	Ta U	MaxTa D	MaxTa U	Tot. D	Tot. U	Total
Tōkyō[b]	Th1	**Tōkyō**	8,025,538	20.06.1991	0	0	—	11	11	19	19	2	3	17	17	49	50	99
Ueno[b]	Th2			14.03.1985	3.6	3.6	3.6	8	11	19	18	2	3	17	17	46	49	95
Ōmiya[b]	Th3/J1	**Saitama**[c]	1,038,100	23.06.1982/15.11.1982	30.3	31.3	27.7	10	11	19	18	2	3	17	17	48	49	97
Kumagaya[d]	J2	Kumagaya	155,626	15.11.1982	64.7	67.9	36.6	4	2	4	6	2	3	17	17	27	28	55
Honjō-Waseda[d]	J3	Honjō	59,082	13.03.2004	86.0	89.0	21.1	3	2	4	1	2	3	11	14	20	20	40
Takasaki[d]	J4	Takasaki	242,359	15.11.1982	105.0	108.6	19.6	9	7	16	17	2	3	17	17	44	44	88
Jōmō-Kōgen	J5	Tsukiyono town	11,368	15.11.1982	151.6	150.4	41.8	3	2	4	5	2	0	15	15	22	22	44
Echigo Yuzawa	J6	Yuzawa town	8,968	15.11.1982	199.2	182.7	32.2	9	11	19	17	2	0	15	15	43	43	86
Urasa	J7	Minami-Uonuma[e]	43,262	15.11.1982	228.9	212.3	29.6	4	6	13	12	0	0	0	0	17	18	35
Nagaoka	J8	Nagaoka	191,212	15.11.1982	270.6	245.1	32.9	11	11	20	20	0	0	0	0	31	31	62
Tsubame-Sanjō	J9	Tsubame	43,989	15.11.1982	293.8	268.7	23.6	9	11	20	20	0	0	0	0	29	31	60
Niigata	J10	Sanjō	85,510	15.11.1982	333.9	300.8	32.1	12	12	20	20	0	0	0	0	32	32	64
		Niigata	515,192					12	12	20	20	0	3	17	17	51	52	103
Number of Services																		
Number of Patterns[f]								9	8	8	12	1	1	3	3	21	24	45

Notes

To = *Toki*, MaxTo = *MAX Toki*, Ta = *Tanigawa*, MaxTa = *MAX Tanigawa*, Tot. = Total.

a Technically the terminal of the Jōetsu Shinkansen is Ōmiya at present, but in practice most trains use the Tōhoku Shinkansen between Ōmiya and Tōkyō.
b Figures for number of trains stops does not include Tōhoku and Hokuriku shinkansen stopping at this station.
c Ōmiya city merged with Urawa and Yono to become Saitama city in 2001.
d Figures for number of trains stops does not include Hokuriku shinkansen stopping at this station.
e Urasa served Yamato town (population approximately 15,600) until it merged with Muika town to become Minami-Uonuma city on 1 November 2004.
f Total number of *Toki*/*MAX Toki* patterns: 12 down and 16 up, total number of *Tanigawa*/*MAX Tanigawa* patterns: 3 down and 3 up.

Hokuriku Shinkansen

Station name	Map code	City or town served	Population	Date opened	Distance from Tōkyō[a]		Distance from previous station	Number of trains per day (Asama only)		
					Operating	Real		D	U	Total
Tōkyō[b]	Th1	**Tōkyō**	8,025,538	20.06.1991/ 01.10.1997	0	0	—	33	33	66
Ueno[b]	Th2			14.03.1985	3.6	3.6	3.6	30	30	60
Ōmiya[b]	Th3/J1	**Saitama**[c]	1,038,100	23.06.1982	30.3	31.3	27.7	33	33	66
Kumagaya[d]	J2	Kumagaya	155,626	15.11.1982	64.7	67.9	36.6	8	9	17
Honjō–Waseda[d]	J3	Honjō	59,082	13.03.2004	86.0	89.0	21.1	6	5	11
Takasaki[d]	J4/H1	Takasaki	242,359	15.11.1982/ 01.10.1997	105.0	108.6	19.6	25	26	51
Annaka Haruna	H2	Annaka Haruna town	48,461 22,349	01.10.1997	123.5	127.1	18.5	14	11	25
Karuizawa	H3	Karuizawa town	17,455	01.10.1997	146.8	150.4	23.3	33[e]	30	63
Sakudaira	H4	Saku	67,025	01.10.1997	164.4	168.0	17.6	30	28	58
Ueda	H5	Ueda	121,809	01.10.1997	189.2	192.8	24.8	33	29	62
Nagano	H6	**Nagano**	359,100	01.10.1997	222.4	226.0	33.2	34	32	66
Number of services								35	33	68
Number of patterns								10	11	21

Notes

a Strictly speaking the terminal of the Hokuriku Shinkansen is Takasaki. However, in practice trains all use the Jōetsu Shinkansen between Takasaki and Ōmiya and the Tōhoku Shinkansen between Ōmiya and Tōkyō.
b Figures for number of trains stops and passenger numbers does not include Tōhoku and Jōetsu Shinkansen stopping at this station.
c Ōmiya city merged with Urawa and Yono to become Saitama city in 2001.
d Figures for number of trains stops and passenger numbers does not include Jōetsu Shinkansen stopping at this station.
e As one special service from Karuizawa to Nagano was a continuation of a service from Tōkyō, the number of down departures was left at 33. On an ordinary day, the number of departures/arrivals at Karuizawa and all stations to Nagano would be one less.

Kyūshū Shinkansen

Station name	Map code	City or town served	Population	Date opened	Distance from origin		Distance from previous station	Number of trains per day (Tsubame only)		
					Operating	Real		D	U	Total
Shin-Yatsushiro	K1	Yatsushiro	106,269	13.03.2004	151.3[a]/0.0	130.0/0.0	—	23	32	55
Shin-Minamata	K2	Minamata	30,545	13.03.2004	42.8	42.1	42.1	20	23	43
Izumi	K3	Izumi	39,537	13.03.2004	58.8	58.1	16.0	20	23	43
Sendai	K4	Satsuma-Sendai[b]	104,979	13.03.2004	91.5	90.8	32.7	24	35	59
Kagoshima-Chūō	K5	**Kagoshima**	596,635[c]	13.03.2004	137.6	126.1	35.3	25	35	60
Number of services								25	35	60
Number of patterns								4	5	9

Notes
a Distance from Hakata using planned route for extension of Kyushu Shinkansen.
b Sendai city merged with Satsuma county to become Satsuma-Sendai on 12 October 2004 – population of Sendai prior to merger was about 73,454.
c Kagoshima merged with Yoshida town, Sakurajima town, Kire town, Matsumoto town and Kōriyama town on 1 November 2004 – population of Kagoshima city prior to merger was about 553,400.

Yamagata Shinkansen

Station name	Map code	City or town served	Population	Date opened	Distance from origin		Distance from previous station	Number of trains per day (Tsubasa only)		
					Operating	Real		D	U	Total
Fukushima	Th9/y1	**Fukushima**	288,632	23.06.1982/ 01.07.1992	0	0	—	17	17	34
Yonezawa	y2	Yonezawa	92,057	01.07.1992	40.1	40.1	40.1	17	17	34
Takahata	y3	Takahata town	26,871	01.07.1992	49.9	49.9	9.8	7	8	15
Akayu	y4	Nanyō	35,937	01.07.1992	56.1	56.1	6.2	13	15	28
Kaminoyama Onsen	y5	Kaminoyama	36,917	01.07.1992	75.0	75.0	18.9	13	15	28
Yamagata	y6	**Yamagata**	250,517	01.07.1992	87.1	87.1	12.1	18	17	35
Tendō	y7	Tendō	63,316	04.12.1999	100.4	100.4	13.3	8	10	18
Sakuranbo Higashine	y8	Higashine	45,754	04.12.1999	108.1	108.1	7.7	8	10	18
Murayama	y9	Murayama	29,389	04.12.1999	113.5	113.5	5.4	8	10	18
Ōishida	y10	Ōishida town	9,398	04.12.1999	126.9	126.9	13.4	8	10	18
Shinjō	y11	Shinjō	41,404	04.12.1999	148.6	148.6	21.7	9	10	19
Number of services								18	17	35
Number of patterns								8	6	14

Akita Shinkansen

Station name	Map code	City or town served	Population	Date opened	Distance from origin		Distance from previous station	Number of trains per day (Komachi only)		
					Operating	Real		D	U	Total
Morioka	Th18/a1	**Morioka**	281,245	23.06.1982/ 22.03.1997	0	0	—	15	15	30
Shizukuishi	a2	Shizukuishi town	19,723	22.03.1997	16.0 (17.6)	16	16	4	4	8
Tazawako	a3	Tazawako town	12,714	22.03.1997	40.1 (44.1)	40.1	24.1	14	14	28
Kakunodate	a4	Kakunodate town	14,600	22.03.1997	58.8 (64.7)	58.8	18.7	14	14	28
Ōmagari	a5	Ōmagari	39,022	22.03.1997	75.6 (83.2)	75.6	16.8	15	15	30
Akita	a6	**Akita**	312,845	22.03.1997	127.3 (134.9)	127.3	51.7	15	15	30
Number of services								15	15	30
Number of patterns								3	3	6

Special shinkansen stations

Station name	Map code	City or town served	Population	Date opened	Distance from origin		Distance from previous station	Number of trains per day		
					Operating	Real		D	U	Total
Hakata-Minami[a]	SS1	**Fukuoka** (Minami ward)	1,315,007 (240,354)	01.09.1990	Hakata: 8.5	8.5	8.5	25	25	50
Gāla-Yuzawa[b]	SJ1	Yuzawa town	8,968	20.12.1990	Echigo-Yuzawa: 1.8	1.8	1.8	12	13	25

Notes

a Services (*Kodama*) operate between Hakata and Hakata-Minami only.
b All services (*Tanigawa/MAX Tanigawa*) operate on the Jōetsu Shinkansen between Tokyo and Gāla-Yuzawa. Numbers of trains per day taken from January 2004 timetable (station only open during the Winter). 'Gāla' is written in *katakana* in Japanese, and so if transcribed back into roman letters becomes 'Gāra'.

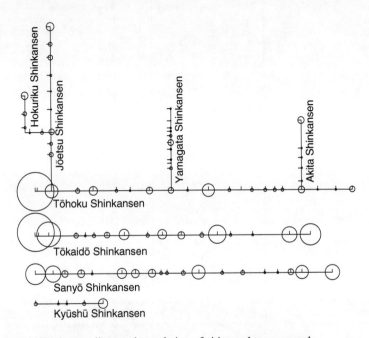

Figure A2.1 Shinkansen lines and population of cities and town served.

Notes
Small lines show the location of stations. Circles represent the population of the city or town served, as detailed in the tables given. The special shinkansen stations are not included.

Appendix 3
Shinkansen types

The tables that follow provide details about stations on each of the shinkansen lines.
Notes:

- The formation information shows how many cars have motors (M) and how many are 'trailers' (T). Each train has a 'registration number' (printed on the front/rear window and entrance to the driver's cab) made up of the formation letter and a number indicting which number it is in order of that particular formation (e.g. a 700-series C12 would be the twelfth train in the C-set formation of the 700-series). Each carriage of a set also has a registration number made up of a code relating to its series number and the order in which it was manufactured.
- Length of carriage – figures in brackets is the length of the end carriage.

All details assume longest formation unless stated otherwise (using formation code in brackets afterwards). Details in square brackets are for services that have been withdrawn.

Sources: Fossett (2004), East Japan Railway Company (2001); Umehara (2002:112–69); Hitachi (2003); JR Kyūshū interviews (2003); Haraguchi 2003a:172–3; Kōsoku Tetsudō Kenkyūkai (2003:204–5).

	0-series	100-series	200-series
Lines operated on	Sanyō (Tōkaidō)	Sanyō (Tōkaidō)	Tōhoku, Jōetsu
Year first produced	1964	1985	1980
Manufacturers	Nippon Sharyō, Kawasaki Sharyō, Kinki Sharyō, Kisha, Hitachi	Nippon Sharyō, Kawasaki Heavy Industries, Hitachi, Kinki Sharyō, Tōkyū Sharyō	Nippon Sharyō, Kawasaki Heavy Industries, Hitachi, Kinki Sharyō, Tōkyū Sharyō
Formations	R (6M) 1985–, (H, K, N, R, S (12M) 1964–1966),[a] (N_H (16M) 1974–2000) (Q (4M) 1997–2001) (S_K (12M) 1989–2000) (Y_K (16M) 1989–1999)	K (6M) 2002–, P (4M) 2000–, (G (10M2T) 1986), (G (12M4T) 1988–2003), (V (12M4T) 1989–2002), (X (12M4T) 1986–2000)	F (12M) 1982–, H (14M2T) 1991–, K (10M) 1997–, (E (10M) 1982–1992) (G (10M) 1987–1988) (G (8M) 1988–1999) (H (12M1T) 1990–1991) (K (8 M) 1992–1997)
Construction	Steel	Steel	Welded aluminium alloy
Total weight (when empty)	894 tons	848 tons (G), 850 tons (V)	705 tons
Total length	400.3 m	402.1 m	300.3 m (F)
Size of carriage (width/height/length)	3.383 m/3.975 m/25.000 m (25.15 m)	3.380 m/4.000 m/25.000 m (26.05 m)	3.385 m/4.000 m/25.000 m (25.15 m)
Maximum number of passengers	1,153 in standard, 132 in green: 1,285 total	1153 in standard, 168 in green: 1321 total (G) 1063 in standard, 222 in green: 1285 total (V)	833 in standard, 52 in green: 885 total (F)
Seat pitch	940 mm (980 mm from 1981) in standard 1160 mm in green	1040 mm in standard 1160 mm in green	980 mm in standard 1,160 mm in green
Maximum acceleration	1.0 km/h/s	1.6 km/h/s (G)–1.4 km/h/s (V)	1.5 km/h/s
Maximum braking	2.84 km/h/s	2.6 km/h/s	2.6 km/h/s
Top speed (operational/other)	220 km/h /–	230 km/h / 276 km/h (G) 270 km/h (V)	240 km/h (275 km/h Jōetsu down)/ 276 km/h
Total power output	11,840 kW	11,040 kW	11,040 kW
Power/weight ratio	12 kW/t	13 kW/t	16 kW/t

Note

a There was also a preproduction model – C (6M), with a capacity of 463 passengers. Different letters between 1964 and 1966 represented the manufacturers – H = Hitachi, K = Kisha, N = Nippon Sharyō, R = Kawasaki Sharyō, S = Kinki Sharyō.

	300-series	400-series	500-series
Lines operated on	Tōkaidō, Sanyō	Yamagata/Tōhoku	Sanyō, Tōkaidō
Year first produced	1990	1990	1996
Manufacturers	Nippon Sharyō, Kawasaki Heavy Industries, Hitachi, Kinki Sharyō	Kawasaki Heavy Industries	Kawasaki Heavy Industries, Hitachi
Formations	JR Tōkai: J (12M6T) 1992–, JR West: F (12M6T) 1993–	L (6M1T) 1995–, (L (6M) 1992–1995)	JR West: W (16M) 1997–
Construction	Aluminium alloy	Steel	Aluminium alloy (brazed honeycomb sections and hollow extrusions)
Total weight (when empty)	637 tons	318 tons	630 tons
Total length	402.1 m	148.7 m	404 m
Size of carriage (width/height/length)	3.380 m/3.650 m/25.000 m (26.05 m)	2.947 m/3.970 m/20.500 m (23.10 m)	3.380 m/3.690 m/25.000 m (27.00 m)
Maximum number of passengers	1,123 in standard, 200 in green: 1,323 total	379 in standard, 20 in green: 399 total	1,124 in standard, 200 in green: 1,324 total
Seat pitch	1,040 mm in standard	910 mm in standard (980 mm in reserved)	1,020 mm in standard
	1,160 mm in green	1,160 mm in green	1,160 mm in green
Maximum acceleration	1.6 km/h/s	1.6 km/h/s	1.6 km/h/s
Maximum braking	2.6 km/h/s	2.6 km/h/s	2.7 km/h/s
Top speed (operational/other)	270 km/h / 296 km/h	240 km/h on Tōhoku, 130 km/h on Yamagata/345 km/h	300 km/h / 320 km/h
Total power output	12,000 kW	7,200 kW	18,240 kW
Power/weight ratio	19 kW/t	23 kW/t	27 kW/t

	700-series	800-series	E1-series
Lines operated on	Tōkaidō, Sanyō	Kyūshū	Jōetsu (Tōhoku)
Year first produced	1997	2003	1994
Manufacturers	Kawasaki Heavy Industries, Hitachi, Nippon Sharyō, Kinki Sharyō	Hitachi	Kawasaki Heavy Industries, Hitachi
Formations	JR Tōkai: C (12M4T) 1999–, JR West: B (12M4T) 1999–, E (6M2T) 2000–	U (6M) 2004–	M (M6T6) 1994–
Construction	Aluminium alloy hollow extrusions	Aluminium alloy	Steel
Total weight (when empty)	634 tons	254 tons	693 tons
Total length	404.7 m	154.7 m	300.1 m
Size of carriage (width/height/length)	3.380 m/3.650 m/25.000 m (27.35 m)	3.380 m/3.650 m/25.000 m (27.35 m)	3.380 m/4.485/25.000 m (26.05 m)
Maximum number of passengers	1,123 in standard, 200 in green: 1,323 total (B/C) 571 (E)	392 in standard, 0 in green: 392 total	1,133 in standard, 102 in green: 1,235 total
Seat pitch	1,040 mm in standard, 1,160 mm in green	1,040 mm in standard	980 mm in standard 1,160 mm in green
Maximum acceleration	1.6 km/h/s on Tōkaido, 2.0 km/h/s on Sanyō	2.5 km/h/s	1.6 km/h/s
Maximum braking	2.7 km/h/s	2.7 km/h/s	2.69 km/h/s
Top speed (operational/other)	270 km/h (Tōkaidō), 285 km/h / (Sanyō) 338 km/h	260 km/h / 290 km/h	240 km/h
Total power output	13,200 kW (6,800 kW (E))	6,600 kW	9,840 kW
Power/weight ratio	19 kW/t (21 kW/t (E))	26 kW/t	14 kW/t

	E2-series	E3-series	E4-series[a]
Lines operated on	Tōhoku, Jōetsu, Hokuriku	Yamagata, Akita, Tōhoku	Jōestu, Tōhoku, Hokuriku[a]
Year first produced	1995 (E2-1000-Series from 2001)	1995	1997
Manufacturers	Kawasaki Heavy Industries, Hitachi, Nippon Sharyō, Tōkyū Sharyō	Tōkyū Sharyō, Kawasaki Heavy Industries	Kawasaki Heavy Industries, Hitachi
Formations	J (8M2T) 2002–, N (6M2T) 1997–, (J (6M2T) 1997–2002), (J52 (6M2T) (E2-1000) 2001–2002)	L (5M2T) 1999–, R (4M2T) 1998–, (R(4M1T) 1997–1998)	P (4M4T) 1997–
Construction	Aluminium alloy extrusions	Aluminium alloy	Aluminium alloy extrusions
Total weight (when empty)	366 tons (N)	220 tons (R (4M1T))	428 tons
Total length	201.4 m (N)	148.7 m	201.4 m
Size of carriage (width/height/length)	3.380 m/3.700/25.000 m (25.70 m)	2.945 m/4.080/20.500 m (23.10 m)	3.380 m/4.485/25.000 m (25.70 m)
Maximum number of passengers	579 in standard, 51 in green: 630 total (N) 763 in standard, 51 in green: 814 total (J)	247 in standard, 23 in green: 270 total (R to 1998), 338 in R from 1998, 402 in L	763 in standard, 54 in green: 817 total (1,634 when in 16-car formation)
Seat pitch	980 mm in standard 1,160 mm in green	910 mm in standard (980 mm in reserved) 1,160 mm in green	980 mm in standard 1,160 mm in green
Maximum acceleration	1.6 km/h/s	1.6 km/h/s	1.65 km/h/s
Maximum braking	2.69 km/h/s	2.69 km/h/s	2.69 km/h/s
Top speed (operational/other)	275 km/h (260 km/h on Hokuriku) / 315 km/h	247 km/h on Tōhoku, 130 km/h on Yamagata and Akita / 315 km/h	240 km/h
Total power output	7,200 kW (N)	6,000 kW (L)	6,720 kW
Power/weight ratio	20 kW/t	20 kW/t	16 kW/t

Note
a Special services only.

Glossary and abbreviations

Amakudari	Literally 'decent from heaven'. Process by which bureaucrats, normally, move into private companies, often related to their previous work upon 'retirement'
ANA	All Nippon Airways
ATC	Automatic Train Control
Bogie	The mechanism linking wheels to the carriages of trains ('truck' in American-English)
Chūbu	The area of Japan taking in Shizuoka, Aichi, Gifu and Nagano prefectures. Fukui and Ishikawa are also sometimes included
Chūgoku	The area of Japan taking in Okayama, Tottori, Shimane, Hiroshima and Yamaguchi prefectures
COMTRAC	Computer-Aided Traffic Control System
Conventional Line (*Zairaisen*)	Non-shinkansen Line in Japan
CTC	Centralized Train Control
Dangan Ressha	The original 'bullet train' planned during the Pacific War
Development Law	Law that aimed to have Shinkansen across the whole country
Diet	Japanese parliament
Dōrō	National Railway Locomotive Union
EJRCF	East Japan Railway Culture Foundation
EMU	Electric Multiple Unit trains
Gauge	The distance between the two rails of a railway line
Green Carriage	Equivalent to First Class (terminology used from 1969)
Hokuriku	The area of Japan taking in Niigata, Toyama, Ishikawa and Fukui prefectures
Honshū JR Companies	JR Central, JR East, JR West (Honshū is the main island of Japan)
JAL	Japan Airlines

Glossary and abbreviations 237

Japan Rail Pass	A pass which allows tourists to travel at significantly reduced cost on most JR trains
JARTS	Japan Railway Technical Service (*Kaigai Tetsudō Gijutsu Kyōryoku Kyōkai*)
JCP	Japanese Communist Party
JEIB	Japan Export and Important Bank, became Japan Bank for International Cooperation
JMS	Japan Meteorological Agency Scale for measuring earthquakes
JNR	Japan National Railways
JNRSC	JNR Settlement Corporation – *Nihon Kokuyū Tetsudō Seisan Jigyōdan*
JRCC	Japan Railway Construction Public Corporation (*Nihon Tetsudō Kensetsu Kōdan*) became JRTT in 2003
JR Central	Central Japan Railway Company – usually known as JR Tōkai due to its Japanese name, *Tōkai Ryokaku Tetsudō*
JR companies	The seven Japanese Railway companies created following the break-up and privatisation of Japan National Railways
JR East	East Japan Railway Company – *Higashi Nihon Ryokaku Tetsudō*
JR Freight	Japan Freight Railway Company – *Nihon Kamotsu Tetsudō*
JR Hokkaidō	Hokkaidō Japan Railway Company – *Hokkaidō Ryokaku Tetsudō*
JR Kyūshū	Kyūshū Railway Company – *Kyūshū Ryokaku Tetsudō*
JR Rengō	Japan Railway Trade Unions Confederation
JR Shikoku	Shikoku Railway Company – *Shikoku Ryokaku Tetsudō*
JR Sōren	Japan Confederation of Railway Workers' Union
JR Tōkai	Popular name for JR Central
JRTT	Japan Railway Construction, Transport and Technology Agency
JR West	West Japan Railway Company – *Nishi Nihon Ryokaku Tetsudō*
Junkyū	Semi-Express Service
Kaisoku	Rapid service
Kanji	Chinese characters used in the Japanese language
Kansai	The region of Japan that includes the prefectures of Mie, Shiga, Kyōto, Ōsaka, Hyōgo, Nara and Wakayama

Glossary and abbreviations

Kantō	The region of Japan that includes the prefectures of Ibaraki, Tochigi, Gumma, Saitama, Chiba, Tōkyō and Kanagawa
Karōshi	Death from overwork
Kinki	Another name for Kansai
KMT	Kuomintang (Nationalist party of Taiwan)
Kokurō	National Railway Workers' Union
Kyūkō	Express Service
LCA	Life Cycle Assessment
LDP	Liberal Democratic Party
Loading gauge	The maximum width locomotives and carriages can be (often wider than the gauge, and in Japan significantly so)
Local train (*futsū densha*)	Train that stops at all stations
MITI	Ministry of International Trade and Industry (became METI – Ministry of Economy, Trade and Industry in 2001)
MOFA	Ministry of Foreign Affairs
Narrow gauge	Gauge on many of Japan's conventional lines – 1067 mm
National Shinkansen	*Zenkoku Shinkansen Tetsudō Seibi Hō* – 1970
NRM	National Railway Museum (York, England)
Onsen	Hot springs – usually for bathing in, although some are just for observing due to their intense heat, for example
Ordinary Class	Equivalent to Second Class (terminology used from 1969)
Pantograph	The mechanism that collects electricity from the overhead cables
RTRI	Railway Technical Research Institute, *Tetsudō Sōgō Gijutsu Kenkyūjo*, also referred to as JR Sōken
Seibi Shinkansen	The Shinkansen lines to be constructed, which were prioritized in the 1973 *Seibi Keikaku*
Seiji eki	'Political station'
SHC	Shinkansen Holding Corporation
Standard gauge	Gauge on many railways internationally and the one used on the shinkansen – 1435 mm
Super-*Tokkyū*	Track which has the loading gauge for shinkansen, but the gauge and trains are conventional
Tatemae	The 'professed intention' (as opposed to the 'real intention' (*honne*))
Tetsurō	Railway Labour Union
Three island JR Companies	JR Hokkaidō, JR Shikoku, JR Kyūshū

THSR	Taiwan High-Speed Railway
THSRC	Taiwan High Speed Rail Corporation
Tōkai	The region of Japan from Yokohama to Nagoya (Kanagawa, Shizuoka and Aichi prefectures) Also sometimes includes Mie prefecture
Tokkyū	Limited Express Service
TRA	Taiwan Railway Administration
TSC	Taiwan Shinkansen Corporation
UrEDAS	Urgent Earthquake Detection and Alarm System

Notes

1 Introduction

1 These words, originally said by Saichō (767–822), founder of the Tendai Sect of Buddhism, appear on a memorial stone in Niihama city, where Sogō Shinji was born.
2 The figure does not include the Yamagata and Akita 'mini-shinkansen'.
3 In January 2001, the Ministry of Education, Science, Sports and Culture merged with the Agency for Technology to become the Ministry of Education, Culture, Sports, Science and Technology (abbreviated to MEXT).
4 Kasumigaseki is the area where most of the Ministry buildings are situated – equivalent to Whitehall in London.
5 The Chatham House Rule states that 'participants are free to use the information received, but neither the identity nor the affiliation of the speakers, nor that of any other participant may be revealed.' More information about the Rule is available on the 'About Chatham House' page via www.chathamhouse.org.uk.
6 I did conduct some research on the topic and its relationship to regional differences in Japan during 1998 after completing my thesis and presented a paper on the topic ('Regionalism in Japan: The Roles of Sport and Infrastructure', Japan Politics Group Colloquium, University of Kent/Chaucer College, 3 September 1998), but had to put the research on hold while I wrote my first book.
7 If Robin Hood did exist, it is suggested that he was from Loxley, part of modern-day Sheffield, rather than Nottingham. The famous tree in Sherwood Forest would only have been a sapling at the time so could not have been his base, which would suggest that there is at least some element of fiction about his story. Indeed, I have heard it suggested that his surname was taken from the hood on his cloak, and Robin may well have been a pun on the word 'robbing'.
8 In Japanese, my name is often rendered as 'Fuudo' rather than 'Fuddo' if I do not insist on the spelling being 'Hood as in Robin Hood'. More commonly, I have used this approach when spelling my name over the phone, for example, in northern England, where the standard pronunciation of 'Hood' is often transcribed as 'Hudd', as in 'Huddersfield'.
9 Naturally I have always felt a connection to '*the Hood*' due to our common name. However, this connection runs deeper as a recruitment office suggested that my father, who once visited the ship with my grandfather, became one of the crew of *HMS Hood*. Luckily my father was more interested in flying aircraft.
10 As well as the 7 JR companies, there are 15 large scale private companies, 6 medium-sized companies, 11 publicly owned railway companies (mostly underground and trams), 111 small-to-medium sized private companies, 12 non-JR freight companies, 9 monorail type operators, 10 'new type of transport system' companies, 16 steel-rope railway companies, 1 rail-less train company, 13 companies where their lines have not started operations, and the new Tōkyō Metro. The above list has taken out overlaps.

Notes 241

It should also be noted that some companies only own lines and do not operate the trains, for example.

2 From bullet train to low flying plane

1. These are the words that greet passengers – both on the digital display and on the audio playback – on the Tōkaidō Shinkansen.
2. The original Shimbashi station became a freight station, Shiodome, in 1914 and was subsequently closed following the reform of JNR. During subsequent construction work in the area, remains of the original building were discovered and a reconstruction of the original building was opened as a museum by the East Japan Railway Culture Foundation in 2003.
3. There are some that claim that the first line to be built was in Nagasaki, and that it was completed in 1868 (see Semmens 2000:1).
4. 14 October is now celebrated as being '*Tetsudō no hi*' (Railway Day). Although not a public holiday, many railway companies put on special services or events and offer discount fares on this day.
5. Although the figures may appear rather odd in metric measurements, one has to remember that railway engineering was developed within Britain and so Imperial measurements were used. Hence the standard gauge was 4 foot 8½ inches, and the 'narrow gauge' used in Japan was 3 foot 6 inches.
6. The dimensions (length × width × height) of a typical British train carriage are 23.0 m × 2.73 m × 4.00 m. This compares to a carriage on a Japanese conventional line being 20.5 m × 2.95 m × 4.05 m (Semmens 2000:13). The width to gauge ratios are 1.90 and 2.76 respectively.
7. Although 'road' is the normal translation for the suffix 'dō', it would be inappropriate for many of these routes to be described as roads. Most, even when in big cities, remained narrow as they were usually travelled along on foot or horseback. As the use of horse-drawn carriages or equivalents did not develop until the Meiji Period, when time came to build paved or tarmac roads, there was often little space to do so. The legacy of this can be seen today with the narrow streets of Japan's cities.
8. Due to time zone differences, the date of the attack on Pearl Harbor was 9 December in Japan.
9. See the Channel 4 (UK) television series *Speed Machines* (first broadcast in 2003).
10. The principal was abolished in 2001 and replaced with a requirement that trains are able to stop as quickly as possible in the case of an emergency (MLIT interview September 2004).
11. The commemorative plaque at Tōkyō station (see Figure 2.1a), however, does refer to the 'Tokaido New Line', which may be a reflection of the problems of translating 'Shinkansen' into English (see Chapter 3).
12. The Tanna and Shin-Tanna tunnels run between Atami and Kannami (east of Mishima) at the head of the Izu Peninsula. When the 7.8 km Tanna Tunnel opened in 1934 after 16 years of construction (and the loss of 67 lives), it meant that trains no longer needed to negotiate the steep Gotemba route and journey times were cut by 30 minutes (Aoki *et al.* 2000:55).
13. 'EXPO 70' was an 'international' event that was 'more than an anything an ostentatious exhibition of Japan's new-found wealth, and a monument to Japanese materialism'. EXPO attracted 64 million people – about 97% of which were Japanese (see Horsley and Buckley 1990:108).
14. Name to be confirmed.
15. The Nixon Shocks of 1971 related to the decision of President Nixon to visit China without first notifying Japan, as had been the norm, and then to impose an import duty on Japanese goods (Storry 1987:280).

242 Notes

16 A shinkansen derailed coming out of the depot near Shin-Ōsaka on 21 February 1973 after a driver error.
17 Karuizawa is a popular tourist destination as well as the site of one of the Imperial summer retreats. It is also unique in Olympic history as the only city to have host both Summer (horse jumping in 1964) and Winter (curling in 1998) Olympic events.
18 The location of Shin-Aomori would potentially mean that at least one change of train, and potentially a change in direction of travel, would be needed if a change to conventional trains were needed.
19 According to the Shintō legend, there were eight islands created by the deities: Awaji, Honshū, Shikoku, Kyūshū, Oki, Tsushima, Iki and Sado.
20 The term *kurai tanima* ('dark valley') is used to refer to the period from 1931 to 1941 when the liberalism and freedom of the previous decade was replaced by the build up to the Pacific War (Storry 1987:182–213).

3 Ambassador of Japan

1 '*Kyō-mo shinkansen-o goryō itakakimashite arigatou gozaimasu*' – literally 'Thank you very much for using the shinkansen again today'.
2 On other lines passengers are welcomed to 'The Tōhoku Shinkansen', 'The Kyūshū Shinkansen', etc.
3 For example 'hara-kiri', 'tsunami' and 'karaoke' often become 'harry-karry', 'tsoon-army' and 'karry oky' respectively.
4 The dictionary, *Jisupa*, also describes 'shinkansen' as 'a railway linking major cities at speeds above 200 km/h', which is the definition used in the *Seibi Shinkansen* Law (Yamanouchi 2000:248), and '*chōtokkyū*' as 'a train faster than *tokkyū*' in its Japanese–Japanese section.
5 This programme was designed not only to help improve the level of English taught in schools, but more significantly to improve international understanding, in part by giving young non-Japanese the opportunity to experience Japan (Hood 2001:60–2).
6 Other than *Nozomi*, there have been some sub-groups of names that have been decided by JR companies. '*Hikari Rail Star*' is a variation of '*Hikari*' service and is only found on the Sanyō Shinkansen. As the nature of this service is essentially the same as other '*Hikari*', in terms of the stations it stops at, JR West decided that it would too confusing to have an additional name. However, as the train itself is different, JR West wanted to be able to differentiate with the standard '*Hikari*' service. The company created the '*Rail Star*' name internally (source: JR West interview 2003). Similarly JR East gave the prefix 'MAX' (Multi Amenity eXpress) to various services that stop at the same stations as those without the 'MAX' prefix, but that use double-decker (E1-series or E4-series) shinkansen.
7 The top 10 names were: *Hayato, Satsuma, Mirai, Sakura, Tsubame, Shiranui, Minami, Hibiki, Hayate* and *Hayabusa* (Kyūshū Railway Company 2003). Hayate was ninth despite having already being adopted for JR East's shinkansen service to Hachinohe the previous year.
8 Due to the way in which *kanji* were imported from China, unlike in China, Japanese can, and often does, have multiple readings for *kanji*. This is perhaps the most difficult aspect of the Japanese language. *Hiragana* (and *katakana*) are phonetic alphabets, so can be used to avoid any confusion or problems. However, with the exception of Saitama city which is sadly always written in *hiragana*, it is not natural for place names to be written in this way and it appears somewhat childish (as it takes many years for children to master the main 1,950 *kanji*).
9 Ōmiya has long been a regional headquarters, a station where trains start/terminate, and the location of one of the training centres for JNR (now JR East). It will become the location of the new railway transport museum in 2006.

10 The new cover of the Japan Rail Pass uses the famous *ukiyoe* picture of a tsunami and Mt Fuji. No pictures or graphics of any trains are used.
11 This included one image where the shinkansen was incorrectly identified as an underground train in the caption, one taken from inside the shinkansen, and two pictures where it would not be immediately obvious to many that the train was a shinkansen rather than a conventional train.
12 The shinkansen accounted for 40% of all pictures and 74% of video clips when searching for 'high speed train'.
13 In a bizarre piece of editing, despite travelling from Tōkyō, the view of Atami appears after Fuji. Then, after alighting at Kyōto, the shinkansen is clearly identifiable as a 700-series, but then the character is standing next to back of a 100-series on a different platform.
14 The film differs greatly from the original Ian Fleming novel. Although much of the filming took place in Himeji (filmed prior to opening of Sanyō Shinkansen) and in Kyūshū, Bond does make a journey from Tōkyō to Kōbe by car, when shinkansen would undoubtedly have been the better choice given the apparent time limitations and resources at his disposal.
15 The voice over to the image of the shinkansen mistakenly spoke of 'trains travelling at 300 miles per hour'.
16 There are three categories of Japanese postage stamps; definitive (*futsū kitte*) which merely 'provide proof of purchase', commemorative (*tokushu kitte*) and home-town stamps (*furusato kitte*) which aim to promote 'regional development through themes highlighting local customs and scenes' (Dobson 2002:25–6).
17 Sanrio does not have any characters that are primarily violent as being '*nakayoshi*' (friends) is their main concept (Sanrio interview 2003).
18 Sanrio manages to take the shinkansen to a whole new level by allowing them to apparently be able to fly and swim in one of the drawing books published.
19 The reason why Kitty is drawn with no mouth is that 'she speaks with her heart'. On top of this, there was a concern that if a child returned from school unhappy and saw a smiling Kitty doll, they would no longer feel that Kitty understood their feelings and was their friend (Sanrio interview 2003).
20 The English translation of Katō's novel includes one interesting phrase that highlights the confusion of terms; 'Bullet train on the Shinkansen' (Rance and Kato 1980: 151).
21 The shinkansen was transported from Fukuoka to Yokohama via Kōbe by ship, before being taken to Southampton on a different ship. It was then taken by road to York, causing traffic chaos on the M1 motorway when people slowed to have a look. On one side of the train was a banner announcing 'Bullet Train to York (PS: This is the only time you will ever overtake one)'. The problems did not end on the road, however. The shinkansen had to be placed on a short stretch of the East Coast Mainline (which makes it the widest train ever to have travelled on British railways) in order to get it into the NRM, but the couplings did not fit, one of the signs (of about 30) that had to be moved was missed, and a concrete curb was found to be in the way. In the end it just managed to squeeze through a doorway (BBC TV 2001).

4 Whose line is it anyway?

1 Quote and information taken from 'The Boscombe Valley Mystery' in *The Adventures of Sherlock Holmes* by Sir Arthur Conan Doyle.
2 The 'home station' of Satō was in fact Tabuse on the Sanyō mainline rather than Iwakuni (Umehara 2002:51–2).
3 Japan top five prefectures by area are: Hokkaidō (83,454 km^2), Iwate (15,278 km^2), Fukushima (13,782 km^2), Nagano (13,585 km^2) and Niigata (12,582 km^2) (Asahi Shimbun 2004:44).

4 The ten most populated prefectures in 1884 were: (1) Niigata (1,583,400), (2) Hyōgo (1,442,600), (3) Aichi (1,364,400), (4) Hiroshima (1,256,600), (5) Ōsaka (1,168,900), (6) Tōkyō (1,152,500), (7) Fukuoka (1,133,800), (8) Chiba (1,107,500), (9) Nagano (1,040,100), (10) Kumamoto (1,000,000) (Hiraishi 2002:55).
5 Evidence of this can still be seen today in the way the track is constructed around Nagaoka Station.
6 Until the opening of the Kyūshū Shinkansen, with its 35‰ incline that has meant that the 800-series has needed to be more powerful than the 700-series on which it is largely based, this incline was the greatest one faced on the shinkansen network at 28‰, nearly double the 15‰ found on most other lines (there is a section of 20‰ on the Tōkaidō Shinkansen). On the conventional lines such inclines are not unusual, with the greatest being and 80‰ on the Hakone Tōzan Railway and 90‰ on the Ōigawa Line.
7 Due to the way in which Japan modernised, electricity frequencies are different in East and West Japan. In 1895 Tōkyō Electric purchased its original equipment from Germany and so used 50 Hz, whereas Ōsaka Electric purchased its equipment from the US and so used 60 Hz. The division still exists with the Fuji river being one of the main boundaries. The Tōkaidō Shinkansen, which crosses this boundary, uses 60 Hz by having its own dedicated power supply and substations. The Hokuriku Shinkansen is the only other shinkansen line that crosses the boundary. Due to advances in shinkansen design, the train itself has been designed to use either frequency and so overcome the problem which would have otherwise meant the costly construction of substations to provide 50 Hz within the 60 Hz region.
8 Although Urasa was within Tanaka's constituency, Nagaoka (another station on the Jōetsu Shinkansen) is actually closer to where Tanaka lived.
9 For example, the famous loyal dogs Hachikō and Sābu outside Shibuya station in Tōkyō and Nagoya station. Examples of inanimate objects would be footballs in the football-mad city of Shimizu (now part of Shizuoka city) or the statue of a *gyōza*, a local food specialty, outside Utsunomiya station.
10 Saigō was a central figure in the changes that occurred in the Meiji Period. Respected as a true samurai, he was the inspiration for the character of Katsumoto in *The Last Samurai* (2003). Sakamoto was a samurai who brought together the progressives of Satsuma (modern-day Kagoshima), including Saigō, and Chōshū (modern-day Yamaguchi) to bring an end to the Tokugawa Period. Ōtomo no Yakamochi was a leading editor and poet of Man'yōshu, Japanese poetry.
11 Gāla-Yuzawa (JR East) and Hakata-Minami (JR West) opened before the Ueno-Tōkyō section of the Tōhoku Shinkansen. However, in both cases the track already existed and it was only the station that was constructed.
12 Marunouchi is Tōkyō's business district – equivalent to 'The City' in London or Wall Street in New York.
13 Narita Airport is served by both JR and Keisei. At the station for Terminal 2, for example, each company has only one platform and one line, one of which would have been for shinkansen from Tōkyō (the JR side) and the other for shinkansen to Tōkyō (the Keisei side) had the Narita Shinkansen been built. A new faster rail link to Tōkyō is currently under construction.
14 Organized crime syndicates.
15 NHK's revenue primarily comes from monthly fees paid by households that have a television – in much the same way as the television license is used in the UK to fund the BBC. However, NHK has no legal recourse for collecting fees should residents refuse (interview with NHK official in April 2003), unlike the UK situation, which means they are prone to be more flexible in the face of threats of non-payment from large groups.

5 The bottom line

1 This section between Iwate-Numakunai and Ninohe stations also has the longest shinkansen rail at a staggering 60.4 km in length (Kōsoku Tetsudō Kenkyūkai 2003:225).

2 As with the Seikan Tunnel for JR Hokkaidō, JR Shikoku has a problem with the Seto Ōhashi bridges connecting Honshū and Shikoku, which were built to super-*tokkyū* standards and would have to still take some conventional trains even if it carried shinkansen in the future. However, it is now felt that full shinkansen are unlikely to ever be seen on Shikoku, and this shift can be seen by looking at the Shikoku's other connection to Honshū. At Naruto, the road bridge that connects Shikoku to Awaji-shima, clearly has a portal and other construction beneath it which could been used for a railway line, specifically shinkansen. However, when the Akashi Bridge was built linking Awaji-shima to Honshū, and so completing the Shikoku's second link to Honshū, no such construction was done.

3 Announcements tend to refer to waiting for a *Nozomi* service to pass (*tsūka*), for example, rather than using the expression *taihi*.

4 There are few shinkansen examples of *kankyū ketsugō*, whereby faster services stop on the opposing side of a platform to a slower service allowing passengers to transfer to the other service for the next leg of its journey without a significant wait.

5 For example, a passenger arriving at Kakegawa station having just missed *Kodama* 450 at 10:04 to Tōkyō, will have a 36 minute wait until the next service, *Kodama* 400, as there are only two services per hour. This service gets to Tōkyō at 12:23. However, if they alight at Shizuoka and wait on the platform for about 12 minutes, they can catch *Hikari* 266, which arrives in Tōkyō at 12:13. The difference between *Kodama* and *Hikari*, even over this relatively short distance of about 210 km is some 22 minutes.

6 Using the earlier example, had the passenger not already bought their shinkansen ticket at Kakegawa, they could take a local train to Shizuoka leaving at 10:13, and be able to catch *Hikari* 266 from Shizuoka, saving themselves ¥840 in the process.

7 For example, if a passenger were to go from Mikawa-Anjō to Tōkyō as quickly as possible, a change of train is also necessary. Going via Nagoya and using the *Nozomi* service, although meaning they pass by Mikawa-Anjō again, could be as much as 40 minutes quicker than if they take the direct route. Similarly on the Tōhoku Shinkansen, as the *Hayate* service only has limited stops, for passengers at stations, such as Shin-Hanamaki, the quickest way to Tōkyō may actually be to travel first away from Tōkyō so that they can get to a station where the *Hayate* service can be boarded.

8 There is a ¥200 supplement or reduction for certain times of the year.

6 The need for training

1 The official series numbers for East-i and the T4 Dr Yellow are E926 and 923 respectively. 9xx-series shinkansen are the numbers reserved for experimental and other non-passenger shinkansen.

2 Some early designs of shinkansen were not pressurized in the areas between carriages, leading to some passengers getting covered by the contents of the toilets when the train entered a tunnel (Yamanouchi 2000:131).

3 East-i is predominantly white, with a red stripe – using a colour scheme similar to that found on ambulances in Japan. Prior to its introduction, JR East also had a 'Dr Yellow' (T3) similar to T2, although it had the familiar Tōhoku/Jōetsu green-stripe down its side, rather than the blue stripe found on Tōkaidō and Sanyō shinkansen.

4 On the Yamagata Shinkansen to Yamagata, 11 of 91 (12%) were closed when the line was converted for mini-shinkansen usage; 45 of 79 (57%) were closed on the extension to Shinjō; 41 of 99 (41%) were closed on the Akita Shinkansen. Other crossings were converted to using a bridge or elevated section (6, 43 and 33 respectively) (Mini-Shinkansen Shippistu Gurūpu (2003:13).

5 The original system used to take the train down to 10 km/h below the ATC limit.

6 For example, the Yurikamome Line in Tōkyō.

7 Such as the Central Line and Docklands Light Railway on the London Underground.

8 For example, see BBC 2003; Mainichi Interactive 2003. Articles tended to suggest that the driver was merely 'asleep'. The BBC article also incorrectly claimed that shinkansen operate on 'automatic pilot'.
9 However, as some shinkansen, such as 0-series and 100-series shinkansen used on the Sanyō Shinkansen, are now used for shorter distances than in the past, it means that they take longer to reach their regular inspections. The most visible side of this is that the outside of the train is re-painted less regularly and so is not as clean as the shinkansen which do the longer journeys between Hakata and Tōkyō on a daily basis.
10 Ironically the shinkansen which the Queen travelled on only arrived at Tōkyō on time due to skill of driver keeping speed just below the ATC point after it had been delayed leaving Nagoya (Oikawa and Morokawa 1996:12–13).
11 On some conventional lines and underground lines there are also signs on the platform to show passengers where the best place is to board the train so that they will be close to the platform exit at the station they wish to alight at.
12 Although a more common sight on conventional trains than shinkansen, passengers who wish to stay on a busy train until another station, will often step off onto the platform, join the head of the queue of passengers waiting to board, and then board the train again once the alighting passengers have all got off.
13 If using a carriage other than the front or rear carriage, it is often the case now that the box in which passengers are waiting does not always exactly line up with the door of the carriage. This is not due to a driver error, but rather that doors are in marginally different positions on different series of shinkansen, which also vary slightly in length (see Figure 6.4e). The mark on the platform by the cab door at either the front or the rear carriage is the only way to see how accurately the driver has stopped the train.
14 The population aged 65 and over is the most prone to car accidents in Japan – accounting for 37.8% of 8,326 deaths in 2002 (Asahi Shimbun 2004:216).
15 Drivers are not required to take any medical examination before they go to work, even for alcohol or drugs, and the companies rely upon the employee's good judgment, which they hope have been developed through the extensive training programme (interviews with all four shinkansen-operating companies).
16 Knipprath (2003:10) notes within the studies of Japanologists, homogeneity is 'implied' and though 'some authors incidentally warn the reader that they do not necessarily imply homogeneity, they further do not pay (much) attention to within-country differences'.
17 While it is still uncommon to walk or run on escalators in Japan, the standard system as used in Tōkyō is to stand on the left. However, in Ōsaka, it is normal practice to stand on the right. The reason for this is said to be due to this being the international norm and dates back to Ōsaka's hosting of 'EXPO 70', although some say that it predates EXPO and is done as a desire not to conform to Tōkyō's ways.
18 Shinkansen drivers, like other train crew, do not have *meishi*, but do have a name badge on their uniform.
19 Differences between perceptions of blue and green, as well as other colours, existed in standard Welsh also, although modern colloquial Welsh tends to use the same definitions as in English (see Hendry 1999:26–7).
20 Other than the small coloured 'JR' logo that appears on shinkansen carriages, JR Kyūshū is the only company that now has all of its shinkansen sets using its corporate colour in a prominent position.
21 JR Tōkai have also moved their headquarters to the building from one near Tōkyō station that was out of sight of the station itself.
22 In the past, Shinagawa had been terminal for some long distance trains – particularly school excursion trains – so to have a shinkansen station, which can be used as a terminal for some trains, seems quite natural for some of that generation that I have spoken to.

Notes 247

23 Korean and Chinese announcements are not used on the *tokkyū Relay-Tsubame* service between Shin-Yatsushiro and Hakata. This has apparently led to some negative feedback from visitors from those countries (JR Kyūshū interview September 2004).

7 Mirror of Japan

1 In Japanese literature, the classic work *The Pillow Book* by Sei Shōnagon contains many entries in relation to snow and its beauty – although, revealing her aristocratic background, she considered its falling on the 'houses of common people', particularly when the moonlight fell on it, as being something 'regrettable' and 'unsuitable'.
2 Although commonly referred to as the Kōbe Earthquake, Great Hanshin Earthquake or Great Hanshin-Awaji Earthquake, its official title is the Southern Hyōgo Prefecture Earthquake.
3 The JMS scale is as follows: 0 – *Mukan* – when nothing occurs; 1 – *Bishin* – a faint tremor, which most will not feel; 2 – *Keishin* – a light tremor, which will cause some shaking; 3 – *Jakushin* – a weak tremor, which causes houses to shake; 4 – *Chūshin* – a medium tremor, which would be felt by most people; 5 – *Kyōshin* – a strong tremor, which would cause some damage; 6 – *Resshin* – a violent tremor, which causes people to fall over and will destroy up to about 30% of housing; 7 – *Gekishin* – a severe tremor, which destroys over 30% of housing. Fractions, such as 4.1, are also possible (source: Hadfield 1995:28–9).
4 The other 700 were on the Sanyō Shinkansen.
5 It has been suggested that tetrapods accelerate coastal erosion (Kingston 2004:132), which was one of the things they were supposed to be preventing.
6 *Pachinko* is a popular pastime in Japan. It is played on a vertical machine by inserting ball-bearings into a tray which are then projected at a speed determined by the turning of a knob by the player. Should the ball-bearing enter a particular hole, the player is rewarded by having more ball-bearings dropped into their tray. The aim is to win as many as possible. As most gambling for money is not permitted in Japan, the ball-bearings can strictly only be exchanged for prizes. However, there are often places near parlours where these prizes can be exchanged for cash.
7 *Burakumin* were traditionally those working in jobs related to death and dead bodies – for example, butchery, tanning and the leather industry. They have faced discrimination for many centuries (see for example Neary, I. (1997) 'Burakumin in Contemporary Japan' in Weiner, M. (1997) Japan's Minorities, London: Routledge, 50–78).
8 Kamikaze literally means 'divine wind' and is reference to strong wind which, on two occasions, saved Japan from almost certain defeat at the hands of the Mongols in 1274 and 1281 when all hopes seemed lost. The term was resurrected in the Pacific War as the manner in which the planes were used was a similar last hope, though such tactics are often used successfully in the Japanese board game *shōgi* (a variation of chess).
9 The most common causes of death in Japan in 2002 were: Cancer (304,286 deaths; a rate of 241.5 per 100,000), Heart disease (152,398; 120.9), Cerebrovascular disease (129,589; 102.8), Pneumonia (87,385; 69.3), Accidents (38,593; 30.6), Suicide (29,920; 23.7), Senility (22,675; 17.9), Renal insufficiency (18,171; 14.4), and Liver disease (15,465; 12.3) (Asahi Shimbun 2004:194).
10 For example, Shin-Yokohama, Atami and Shin-Kōbe.
11 Animals getting onto lines can be a problem in Japan. I have been held up by deer on the San'in Line – ironically near Yōka, which means 'Eight Deer' – on two occasions. In early October 2004, deer stopped a train on the JR Kosei Line in Shiga and wild boar stopped another train on another section of the same line (*Mainichi Shimbun*, 12 October 2004). In the case of the shinkansen, there are not many animal incursions onto the track. When the Tōkaidō Shinkansen was being tested, it was found that the

number of birds being hit by the train reduced significantly after the first few test runs 'as if the birds had become aware of the danger' (Yamanouchi 2000:115). However, if an animal does get through and is struck, due to the speed and force, not only is the animal's body thrown clear of the line, it sometimes requires analysis to be done on any remaining hair to find out what animal it was (JR Tōkai interview July 2001).

12 The rate fell below 2.1, to 1.58, for the first time in 1966. However, as this was a particular Year of the Horse (*hinoeuma*), with girls born in that year said to have a bad disposition and so find it hard to find partners (Hendry 1999:27; Asahi Shimbun 2004:33), many parents either delayed having children or may have attempted to have the registration of the birth altered in some way.

13 The same problem exists for those tourists travelling around Japan on the JR Rail Pass.

14 As of January 2004 there were 4 female shinkansen drivers at JR Tōkai, with a further 27 holding a shinkansen driving license, 10 at JR West, and none at JR East. The number of shinkansen drivers at each of these companies was about 600, 550 and 450 respectively. Even on the conventional railways there are not many female drivers with 6 of about 1,200 drivers at JR Tōkai, 22 of about 4,950 drivers at JR West and 7 of about 6,500 drivers at JR East (MLIT interview January 2004).

15 Domestic production of alcoholic drinks has increased to 9,518,000 kl in 2001 from 4,793,000 kl in 1970. The size in absolute terms and relative terms for the various types of drinks in 2001 and 1970 were: Sake (680,000 kl (7.1%) in 2001, 1,257,000 kl (26.2%) in 1970), *shōchū* (804,000 kl (8.4%), 219,000 kl (4.6%)), beer (4,813,000 kl (50.6%), 3,037,000 kl (63.4%)), low-malt alcoholic drinks (2,374,000 kl (24.9%), 0), whisky (112,000 kl (1.2%), 144,000 kl (3.0%)), wine (90,000 kl (0.9%), 0), others (645,000 (6.8%), 137,000 (2.9%)) (Asahi Shimbun 2004:142).

16 It should be noted that in relation to activities such as drinking and smoking, 'adult' in Japan means over 20 years old. Although legally adulthood starts at 18, many Japanese still regard 20, which is even expressed using a special term ('*hatachi*') rather than adding the usual suffix '*sai*' to the number, which, as 21 was in many Western countries, as the real start of adulthood.

17 My thanks to David Williams of the Cardiff Japanese Studies Centre for his insight into this topic – see also Noguchi 1990:45. Kawabata's use of the image of a train is somewhat different to Gill's (2003:58) description of the shinkansen as looking 'like God's suppository and is twice as fast' in a mistake-ridden article that mixes humour with serious ideas. For example, he suggests that all the toilets on the shinkansen are 'pee-in-your socks squat jobs', which is not the case. There are also urinals and 'Western-style' toilets on all shinkansen. The toilets can also be used at any time, rather than only in stations, as is the case on many British trains.

18 JR Kyūshū also has a magazine on its shinkansen and *tokkyū* services.

19 That Japanese people become confused about the differences is also not surprising since the education system does not educate in this respect. The English textbook which I used when teaching on the JET Programme, when introducing the word 'colour', merely stated that it was the same as 'color', with no explanation as to the difference and that they should not be interchanged.

20 The seat pitch on planes varies from airline to airline, as well as being depending on the plane. The typical seat pitch for a Boeing 747–400 is about 79–82 cm in economy class (see www.seatguru.com; tokyo.cool.ne.jp/higachanel/zaseki.htm).

21 According to the New Internationalist (2004) Japan ranked second, to Hungary, in terms of equality of income and consumption as measured on the GINI index.

22 The width of the sitting part of shinkansen seats in 'ordinary' class varies from series to series, but is 43 cm on the 700-series (the middle seat of the three-seat sets is a few centimetres wider), whereas the seats in 'green' class are about 47.5 cm (JR Tōkai interview). For comparison the width on a Boeing 747 seat is about 42.5 cm (see www.seatguru.com; tokyo.cool.ne.jp/higachanel/zaseki.htm).

8 Conclusion

1 The 'Bubble Economy' started in the late 1980s with overly optimistic valuations in the price of land and stock, following the easing of monetary and fiscal policy after the signing of the Plaza Accord.
2 These are the only windows which open on a shinkansen. All other windows need to be kept closed to ensure that the correct pressure is maintained and to allow the air-conditioning system to work.
3 The similarities in DNA of humans and fruit flies, combined with the fruit flies' short life-cycle, has led to experiments being done on fruit flies in checking that the magnets and other equipment on the linear shinkansen are not likely to cause any health problems for humans (RTRI interview January 2001).

Bibliography

To save space, and as many sources that I have read often cover similar material, I have generally only listed materials in the main text that are most accessible to readers. Similarly, this bibliography only contains those materials which have been cited in the text (particularly in the case of materials received from the JR companies and related organizations, as it would not be possible to list all of the more than 200 documents I obtained) and a selection of other significant works that have influenced the direction of this book or which may be useful reading. Only interviews of those listed in the text by name without any biographical information are included.

Bullet-In (the Japanese Railway Society publication), various dates.
Japan Railfan Magazine (*Tetsudō Fan*), Kōyūsha, various dates.
Japan Railway & Transport Review, East Japan Railway Culture Foundation, various dates.
Japan Railways timetables, various dates and publishers.
Tabi to Tetsudō, Tetsudō Jānarusha, various dates.

Abe, M. (2001) *Ren'ai Shinkansen*, Tōkyō: Tokuma Shoten.
Aoki, E., Imashiro, M., Kato, S. and Wakuda, Y. (2000) *A History of Japanese Railways 1872–1999*, Tōkyō: East Japan Railway Culture Foundation.
Asahi Evening News (1963) *Japan's Railway Engineering*.
Asahi Shimbun (2004) *Japan Almanac 2004*, Tōkyō: Asahi Shimbun.
Babb, J. (2000) *Tanaka: Profiles in Power*, Essex: Pearson Education Limited.
Barthes, R. (1983) *Empire of Signs*, trans. R. Howard, London: Jonathan Cape.
BBC (2003) 'Bullet Train Driver Nods Off', http://news.bbc.co.uk/1/hi/world/asia-pacific/2804349.stm accessed on 5 March 2003.
—— (2004) 'Japan Trade Surplus Keeps Rising', http://news.bbc.co.uk/1/hi/business/4040663.stm accessed on 25 November 2004.
BBC Television (1999a) *Great Railway Journeys: Tokyo to Kagoshima*.
—— (1999b) *Disaster: JAL 123 – A Japanese Tragedy*.
—— (2001) *York's Oriental Express*.
Befu, H. (2001) *Hegemony of Homogeneity*, Melbourne: Trans-Pacific Press.
Bell, P. (2001) 'Content Analysis of Visual Images' in Leeuwen, T. van and Jewitt, C. (eds) *Handbook of Visual Analysis*, London: Sage Publications.
Belson, K. and Bremner, B. (2004) *Hello Kitty*, Singapore: John Wiley & Sons (Asia).
Bradshaw, B. and Lawton Smith, H. (eds) (2000) *Privatization and Deregulation of Transport*, London: Macmillan Press.

British Heart Foundation (2004) 'Prevalence of Smoking' available from http://www.heartstats.org/temp/Tabsp4.9spweb04.xls accessed 10 October 2004.
Brunn, S.D. (2000) 'Stamps as iconography: celebrating the independence of new European and Central Asian states', *Working Paper*.
Burchell, A. (2000) 'Regulation of transport: An overview' in Bradshaw, B. and Lawton Smith, H. (eds) *Privatization and Deregulation of Transport*, London: Macmillan Press.
Buruma, I. (2001 reprint of 1984) *A Japanese Mirror: Heroes and Villains in Japanese Culture*, London: Phoenix.
Button, K.J. and Pitfield, D.E. (eds) (1985) *International Railway Economics*, Aldershot: Gower Publishing.
Cardiff University (2002) *Japanese Railway Conference Report*, Cardiff: Cardiff University/Ubiqus Reporting.
Central Japan Railway Company (2003a) 'Superconducting Maglev Technology Takes Another Leap Forward', Press release.
—— (2003b) 'Ambitious Japan!', Press Release.
—— (2003c) *Central Japan Railway Company Annual Report 2003*.
—— (2003d) *Central Japan Railway Company Data Book 2003*.
—— (2004a) *Central Japan Railway Company Annual Report 2004*.
—— (2004b) *Central Japan Railway Company Data Book 2004*.
Charrier, P. (2003) 'The South Manchuria Railway Company and the Visualization of Manchuria, 1932–1937', BAJS Conference presentation, University of Sheffield, 16 April 2003.
Chesnau, R. (2002) *Hood: Life and Death of a Battlecruiser*, London: Cassell.
Chūō Shinkansen Ensen Gakusha Kaigi (ed.) (2001) *Rinea Chūō Shinkansen-de Nihon-wa Kawaru*, Tōkyō: PHP Kenkyūjo.
Clarkson, J. (2004a) *The World According to Clarkson*, London: Penguin.
—— (2004b) *I Know You Got Soul*, London: Penguin.
Cole, S. (1987) *Applied Transport Economics*, London: Kogan Page.
—— (2004) (Researcher, Wales Transport Research Centre, University of Glamorgan), interview 29 October 2004.
Condon, J. (1991) *A Half Step Behind*, Rutland: Charles Tuttle.
Curtin, S. (2004) Suicide in Japan, Social Trend articles available via www.glocom.org/special_topics/social_trends/listindex.html suicide accessed on 8 November 2004.
Deans, P. (2004) (Director, Contemporary China Institute, School of Oriental and African Studies), interview 19 March 2004.
Department for Transport (2004) *Channel Tunnel Rail Link – Facts and Figures*.
Dobson, H. (2002) 'Japanese postage stamps: propaganda and decision making', *Japan Forum*, Vol. 14, No. 1, 21–39.
Dugan, S. (2003) *Men of Iron*, London: Channel Four Books.
East Japan Railway Company (2001) 'All Double Deck Shinkansen E4', pamphlet received from East Japan Railway Company.
—— (2004) Annual Report 2004.
East Japan Railway Culture Foundation (1995) *Iwasaki and Watanabe Collection – The Locomotions of Meiji Era*, CD ROM.
—— (1997) 'National Railways of Japan: Historical Statistics' – CD Rom and maps.
—— (2001) 'East Japan Railway Culture Foundation', pamphlet received from East Japan Railway Culture Foundation.
Ebihara, K. (1997) *Shinkansen – Kōsoku Dairyō Unsō no Shikumi*, Tōkyō: Seiyamadō Shoten.

Ericson, S.J. (1996) *The Sound of the Whistle: Railroads and the State in Meiji Japan*, Cambridge, MA: Harvard University Press.
Ford, R. (2003) 'New trains – value for money but not cheap', *Modern Railways*, Vol. 60, No. 655, 16–17.
Fossett, D. (2004) various pages accessed via http://www.h2.dion.ne.jp/~dajf/byunbyun/ on 1 September 2004.
Frewer, D. (2002) 'Japanese postage stamps as social agents: some anthropological perspectives', *Japan Forum*, Vol. 14, No. 1, 1–19.
Fullers (2003) email contact from company.
Gill, A. (2001) 'Mad in Japan', *The Sunday Times Magazine*, 9 September 2001, 56–67.
Goldthorpe, J. (1993) in Ishida H. (1993) *Social Mobility in Contemporary Japan*, London: Macmillan Press.
Groth, D. (1986) *Biting the Bullet: The Politics of Grass-Roots Protest in Contemporary Japan*, Unpublished PhD Thesis, Stanford University.
—— (1996) 'Media and political protest: the bullet train movements' in Pharr, S. and Krauss, E. (eds) (1996) *Media and Politics in Japan*, Honolulu, HI: University of Hawaii Press, 213–41.
Hadfield, P. (1995) *Sixty Seconds That Will Change the World*, London: Pan Books.
Haraguchi, T. (2003a) 'Kikan Ronsō no Temmatsu' in Hoshikawa T. (ed.) *Shinkansen Zenshi*, Tōkyō: Gakken.
Harris, N.G. and Godward, E.W. (eds) (1992) *Planning Passenger Railways*, Glossop, Derbyshire: Transport Publishing Company.
Hayashi, K. (1987) *Jūdai no Ninshin: Sono Shosō to Mondaiten*, Tōkyō: Jiyū Kikaku.
Hendry, J. (1993) *Wrapping Culture*, Oxford: Oxford University Press.
—— (1999) *Other People's Worlds*, New York: New York University Press.
—— (2003) *Understanding Japanese Society*, Third edition, London: RoutledgeCurzon.
Hiraishi, K. (2002) *Shinkansen to Chiiki Shinkō*, Tōkyō: Kōtsū Shimbunsha.
Hirota, R. (2004) 'Air-Rail links in Japan: present situation and future trends', *Japan Railway and Transport Review*, Vol. 39, 4–14.
Hirota, T., Hirota, I. and Saka, M. (2004) *Kaigyō 40 Nen Shinkansen no Subete*, Tōkyō: Yamakei.
Hitachi (2003) pamphlets and papers received about 800 series.
HMS Hood Association (2002) 'Hood's Legacy', http://hmshood.com/association/HoodsLegacy.html accessed 31 January 2002.
Hokkaidō Shinkansen Kensetsu Sokushin Kiseikai (2003) *Kita no Daichi-ni Shiknansen*, Sapporo: Hokkaidō Shinkansen Kensetsu Sokushin Kiseikai.
Hokkaidō Sōgō Kikakubu Shinkansen Taisakushitsu (ed.) (2000) *Hokkaidō Shinkansen*, Sapporo: Hokkaidō Sōgō Kikakubu Shinkansen Taisakushitsu.
Hood, C.P. (2001) *Japanese Education Reform: Nakasone's Legacy*, London: Routledge.
—— (2002) 'Images of Japan/Japan 2001 Survey – Report', www.hood-online.co.uk/academic/articles/report.pdf
Horsley, W. and Buckley, R. (1990) *Nippon: New Superpower*, London: BBC Books.
Hosaka, S. (1982) *Hadaka no Shinkansen*, Tōkyō: Tōyōdō.
Hoshikawa, T. (ed.) (2003) *Shinkansen Zenshi*, Tōkyō: Gakken.
Hosokawa, B. (1997) *Old Man Thunder: Father of the Bullet Train*, Colorado: Sogo Way.
Imai, M. (1997) *Gemba Kaizen*, New York: McGraw-Hill.
Ishida, H. (1993) *Social Mobility in Contemporary Japan*, London: Macmillan Press.
Ishikawa, T. and Imashiro, M. (1998) *The Privatisation of Japanese National Railways: Railway Management, Market and Policy*, London: The Athlone Press.

Ja, K.K. and Konami, H. (2002), 'The Impact of Shinkansen Construction on Regional Development', http://www.konamike.net/hiro/papers/ShinkansenKL2002.htm accessed 10 June 2003.
Japan Airlines (2004) 'World Top 11 City-Pairs', SORA Data File.
Japan Railfan Club (2004) 'Blue Ribbon and Laurel Awards' accessed via http://www.jrc.gr.jp/ on 2 October 2004.
Japan Railway Construction Public Corporation (2002) 'Japan Railway Construction Public Corporation', pamphlet received from East Japan Railway Company.
Jōetsu City (2001) *21 Seiki ni Nokoru Norimono Shinkansen*, Joetsu, Niigata: Joetsu City.
Kagoshima-ken Kyūshū Shinkansen Kensetsu Sokushin Kyōryoku Kai (2002) *Kyūshū Shinkansen Kagoshima Rūto*, Kagoshima: Kagoshima Prefectural Government.
Kakuhira, I., Fujita, Y. and Miyamoto, T. (2001) 'Development of the New Shinkansen Multipurpose Inspection Train', pamphlet received from Central Japan Railways Company.
Kakumoto, R. (2001) *Shinkansen Kaihatsu Monogatari*, Tōkyō: Chūō Kōron Shinsha.
Kasai, Y. (1999) 'A consideration on the sell-off of JR shares: JNR Reform: a paradox of privatising a public corporation', unpublished paper received from Central Japan Railway Company.
—— (2001) *Minkan no Kokutetsu Kaikaku*, Tōkyō: Tōyō Keizai.
—— (2003) *Japanese National Railways – Its Break-up and Privatization*, Folkestone: Global Oriental.
Kawashima, R. (1992) *Shinkansen Jijō Daikenkyū*, Tōkyō: Sōshisha.
—— (1999) *Shinkansen-wa Motto Hayaku Dekiru*, Tōkyō: Chūō Shoin.
—— (2001) *Tetsudō Mirai Chizu*, Tōkyō: Tōkyō Shoseki.
—— (2002) *Shinsen Tetsudō Keikaku Tettei Gaido*, Tōkyō: Sankaidō.
—— (2003) *Maboroshi no Tetsudō Rosen-o Ou*, Tōkyō: PHP Kenkyūjo.
—— (2004) *Nihon no Tetsudō Meisho 100-o Aruku*, Tōkyō: Kōdansha.
Kikuchi, Y. (2004) *Japanese Modernisation and Mingei Theory: Cultural Nationalism and Oriental Orientalism*, London: RoutledgeCurzon.
Kimata, M. (2001) 'Railways in the 21st Century (Tokaido Shinkansen)', http://www.jef.or.jp/en/jti/200103_008.html accessed on 15 January 2003.
Kingston, J. (2001) *Japan in Transformation, 1952–2000*, Harlow: Pearson Education.
—— (2004) *Japan's Quiet Transformation*, London: RoutledgeCurzon.
Kirimura, K., Tsujimura, T. and Mifune, N. (1997) 'Railway Materials Tending to be Increasingly Environment-Conscious in Japan', http://www.rtri.or.jp/infoce/wcrr97/E143/E143.html accessed on 10 June 2003.
Kishi, Y. (2004) 'Shikoku region', *Japan Railway and Transport Review*, July 2004, No.39, 43–51.
Kitakami, A. (1998) *Shinkansen Saishū Shirei*, Tōkyō: Kadokawa.
Kluckhohn, C. (1949) *Mirror for Man*, New York: McGraw-Hill.
Knipprath, H. (2003) 'The model of enquiry in Japanese Studies on the determinants of cognitive achievement: A methodological critique', Seminar presentation at the 10th International Conference of the European Association for Japanese Studies, 29 August 2003.
Kobayasahi, K. (2002) *Konkurīto ga abunai*, Tōkyō: Iwanami Shoten.
Komiya, K. (2003) *Shinkansen kara keizai ga mieru*, Tōkyō: Jitsugyōnonihon.
Konno, S. (1984) 'Strengths and weaknesses of the bullet train', *Japan Echo*, Vol. XI, No. 3, 73–80.
Kōsoku Tetsudō Kenkyūkai (2003) *Shinkansen – Kōsoku Tetsudō no Gijutsu no Subete*, Tōkyō: Sankaido.

Kotani, K. (1999) *Shinkansen 'Nozomi 47 gō' Shōshitsu*, Tōkyō: Tokuma Shoten.
Kyūshū Keizai Chōsa Kyōkai (2001) *'Kyūshū Shinkansen no kaigyō ni Mukete'*, Kyūshū Keizai Chōsa Kyōkai, Vol. 55, No. 8, 19–45.
Kyūshū Railway Company (2002) *Kyūshū Railway Company 2002–3 Kaisha Annai*.
—— (2003) 'Kyushu Shinkansen', email received on 7 April 2003.
—— (2004) *Kyushu Railway Company 2004*.
Leeuwen, T. van (2001) 'Semiotics and Iconography' in Leeuwen, T. van and Jewitt, C. (eds) *Handbook of Visual Analysis*, London: Sage Publications.
—— (2003) (Director, Centre for Language and Communication, Cardiff University) interview 15 January 2003.
Leeuwen, T. van and Jewitt, C. (2001) *Handbook of Visual Analysis*, London: Sage Publications.
Littlewood, I. (1996) *The Idea of Japan: Western Images, Western Myths*, London: Secker & Warburg.
Lonsdale, S. (2001) *Japanese Design*, London: Carlton Books.
Mainichi Interactive (2003) 'Bullet Train Driver Slept on the Job for 8 Minutes', accessed online on 27 February 2003.
Marzuki, N. (2002) 'Look East to Japan', http://www.jef.or.jp/en/jti/200211_015.html accessed on 15 January 2003.
Matsuda, M. (2002) *Making the Impossible Possible*, Tōkyō: Japan Productivity Center for Socio-Economic Development.
Mini-Shinkansen Shippistu Gurūpu (2003) *Mini Shinkansen Tanjō Monogatari*, Tōkyō: Seizandō.
Mito, Y. (2002) *Teikoku Hassha*, Tōkyō: Kōtsū Shimbunsha.
Mizutani, F. (1994) *Japanese Urban Railways*, Aldershot, Hampshire: Ashgate Publishing.
—— (2004) (Professor, Graduate School of Business Administration, Kobe University), interview 5 April 2004.
MLIT (2004a) *'Shinkansen ni kansuru keikaku to sono hyōka'*, document received from Ministry in 2004 – first date of publication unknown.
—— (2004b) *Heisei 15-Nendo Kōkyō-Jigyō-Kankeihi*.
—— (2004c) *Heisei 16-Nendo Kōkyō-Jigyō-Kankeihi*.
—— (2004d) *Seibi-Shinkansen-Kankei-Yosan-no-Suī*.
Morris-Suzuki, T. (1998) *Re-Inventing Japan: Time, Space, Nation*, London: ME Sharpe.
Mutō, H. (2004) (former Chairman of Kokurō), interview 16 April, 2004.
Nagasaki Shinkansen Kensetsu Kiseikai (2003) *Nagasaki Shinkasen*, Nagasaki: Nagasaki-ken Kikakubu Shinkansen Kensetsu Suishinshitsu.
Nagatomo, T., Miyauchi, T. and Tsuchiya, H. (1997), 'Preliminary Investigation for Life Cycle Assessment (LCA) of Shinkansen Vehicles', http://www.rtri.or.jp/infoce/wcrr97/E142/E142.html accessed on 10 June 2003.
Naitō, H. (2003) 'Memory of Old Bullet Train Tunnel', http://www.asahi-net.or.jp/~tu6a-nitu/jrs2/members/naito/tunnel/tunnel.htm accessed on 5 March 2003.
Nakabō, K. (2000) *Watashi no jikenbo*, Tōkyō: Shūeisha Shinsho.
Nakamura, A. (2004) 'JR Central's Career-track Workplace Training Programme', *Japanese Railway Society Bullet-In*, July–September 2004, 15–17.
Nakano Tsuneo (2004) (Professor of Accounting History, Graduate School of Business Administration, Kōbe University), interview 5 April 2004.
Nakao, K. and Itō, H. (1998) *Shinkansen no Mania no Kiso Chishiki*, Tōkyō: Ikarosu Shuppan.

Nakasone, Y. (2002) *Japan – A State Strategy for the Twenty-First Century*, trans. by Connors, C. and Hood, C.P. and Nishikawa, T., London: RoutledgeCurzon.
Nehashi, A. (2001) 'Shinkansen Technology for Export to Taiwan', http://www.jef.or.jp/en/jti/200103_010.html accessed on 15 January 2003.
New Internationalist (2004) 'Measure of Equality', Jan/Feb 2004, available from http://www.newint.org/issue364/facts.htm accessed 20 October 2004.
NHK (2001) *Project X: Shūnen-ga Unda Shinkansen*, DVD.
Nihon Gikai Gakkai (1999) *Kōsoku Tetsudō Monogatari*, Tōkyō: Seizando.
Nihon Tetsudō Kensetsu Kōdan (1996) '*Hokuriku Shinkansen*'
—— (2002a) '*Kyūshū Shinkansen*', pamphlet received from Kyūshū Railway Company.
—— (2002b) '*Atarashii Eki, Atarashii Deai*', pamphlet received from Kyūshū Railway Company.
—— (2002c) '*Kyūshū Shinkansen Gaiyōzu*', pamphlet received from Kyūshū Railway Company.
—— (2003) '*Kyūshū Shinkansen: Funagoya-Shin-Yatsushiro*', Nagano: Nihon Tetsudō Kensetsu Kōdan Hokuriko Shinkansen Kensetsukyoku.
Nishimoto, H. (2003) *Nihon no Tetsudō*, Tōkyō: Natsume-sha.
Nishimura, K. (1987) *Hikari 62 gō no satsui*, Tōkyō: Shinkosha.
—— (2001) *Jōetsu Shinkansen Satsujin Jiken*, Tōkyō: Kōdansha.
Nishio, G. (2001) (Honorary member, Japan Railway Engineers' Association), interview 13 April 2001.
Noble, R. and Tremayne, D. (1998) *Thrust*, London: Transworld Publishers.
Noguchi, P. (1990) *Delayed Departures, Overdue Arrivals: Industrial Familialism and the Japanese National Railways*, Honolulu, HI: University of Hawaii Press.
Noguchi, T. and Fujii, T. (2000) 'Minimizing the Effect of Natural Disasters', *Japan Railway and Transport Review*, No. 23, 52–9.
Nozue, N. and Shirotori, T. (1997), 'Decision Support System for Planning Routes Considering Economic and Environmental Assessment Using Remote Sensing Data', http://www.rtri.or.jp/infoce/wcrr97/E177/E177.html accessed on 10 June 2003.
Ogawa, O. (2001) *Shinkansen 100 kei noritsubushi, tabetsukushi monogatari*, Tōkyō: Seizando.
Ogbonna, E. and Harris, L.C. (2002) 'Managing organisational culture: insights from the hospitality industry', *Human Resource Management Journal*, Vol. 12, No. 1, 33–53.
Oikawa, Y. and Morokawa, H. (1996) *Shinkansen*, Tōkyō: Hoikusha.
Oka, N. (1985) 'The unhappy birth of a tunnel', *Japan Quarterly*, Vol. 32, No. 3, 324–9.
Okada, T. (2001) 'Former President Lee's Support Decisive Factor behind Taiwan's Decision to Opt for Japan's Shinkansen in Island-Spanning Rail Project', http://www.jef.or.jp/en/jti/200103_011.html accessed on 15 January 2003.
Okumura, F. (2001) 'Development of Superconducting Maglev in Japan', http://www.jef.or.jp/en/jti/200103_009.html accessed on 15 January 2003.
Palin, M. (1997) *Full Circle*, London: BBC.
Passin, H. (1975) 'Intellectuals in the decision-making process' in Ezra F. Vogel (ed.) *Modern Japanese Organization and Decision-Making*, Berkeley, CA: University of California Press, 1975, 281.
Pearce, W.A. (2004) 'Japanese train names used outside Japan', *Japanese Railway Society Bullet-In*, April–June 2004, 39–41.
Pharr, S. and Krauss, E. (eds) (1996) *Media and Politics in Japan*, Honolulu, HI: University of Hawaii Press.
Planet Ark Environmental Foundation (2004) *The Recycling Olympics*, available from http://www.planetark.com/nrw/04RecyclingReport.pdf

Bibliography

Powers, D. (2004) (Japan Interface and former BBC correspondent to Tōkyō), interview 9 January 2004.
Railway Technical Research Institute (RTRI) (2001a) 'Railway Technical Research Institute', pamphlet received from Railway Technical Research Institute.
—— (2001b) '*Shōrai no Tetsudō to Shakai-o Sōzo Suru*', pamphlet received from Railway Technical Research Institute.
—— (2001c) 'Yamanashi Maglev Test Line', pamphlet received from Railway Technical Research Institute.
—— (2001d) 'RTRI's Large-Scale Low-Noise Wind Tunnel', pamphlet received from Railway Technical Research Institute.
—— (2001e) 'Gauge Change Train', pamphlet received from Railway Technical Research Institute.
Railway Technical Research Institute and East Japan Railway Culture Foundation (RTRI and EJRCF) (eds) (2001) *Japanese Railway Technology Today*, Tōkyō: East Japan Railway Culture Foundation.
Rance, J. and Katō, A. (1980) *Bullet Train*, London: Souvenir Press.
Sakurai, K. (1994) *Shinkansen ga Abunai*, Tōkyō: Kenyūkan.
Sanrio (2002) *Sanrio: Company Brochure*, Tōkyō: Sanrio.
Sargent, J. (1973) 'Remodelling the Japanese archipelago: the Tanaka plan', reprint from *The Geographical Journal*, Vol. 139, Part 3, October 1973.
Sasaki, K., Ohashi, T. and Ando, A. (1997) 'High-speed rail transit impact on regional systems: does the *Shinkansen* contribute to dispersion?', *The Annals of Regional Science*, Vol. 31, January 1997, 77–98.
Satō, Y. (2000) 'Shinkansen operators strive to boost traffic', *IRJ*, May 2000, 23–5.
—— (2004) *Shinkansen Tekunorojī*, Tōkyō: Sankaido.
Semmens, P.W.B. (1997) *High Speed in Japan: Shinkansen – The World's Busiest High Speed Railway*, Sheffield: Platform 5.
—— (2000) *High Speed in Japan: Shinkansen – The World's Busiest High Speed Railway*, Second Edition, Sheffield: Platform 5.
Shimahara, N.K. (1986) 'The cultural basis of student achievement in Japan', *Comparative Education*, Vol. 22, No. 1, 19–26.
Shimizu, K. (2004) 'Derailments mars shinkansen safety myth', *Japan Times*, 19 November 2004.
Shin Tetsudō Shisutemu Kenkyūkai (ed.) (2003) *Tetsudō – Nazenaze Omoshiro Dokuhon*, Tōkyō: Sankaidō.
Shinohara, T. and Takaguchi, H. (1992) *Shinkansen Hatsuansha no Hitorigoto*, Tōkyō: Pan Research Publishing.
Shoji Ken'ichi (2004) (Professor, Transport Economics and Policy, Graduate School of Business Administration, Kōbe University), interview 5 April 2004.
Smith, R. (2004) 'Third sector railways: the Japanese model', *Train Times*, Association of Community-Rail Partnerships, No. 35, Summer 2004, 9–10.
Soloman, B. (2001) *Bullet Trains*, Osceola, FL: MBI Publishing.
Stockwin, J.A.A. (1999) *Governing Japan*, Oxford: Blackwell.
Storry, R. (1987) *A History of Modern Japan*, Middlesex: Penguin Books.
Strategic Rail Authority (2004) *West Coast Main Line – Progress Report April 2004*.
Suda, H. (1998) *Tōkaidō Shinkansen*, Tōkyō: Taishō Shuppan.
—— (2000) *Tōkaidō Shinkansen*, Tōkyō: JTB Publishing.
Sudō, H. (2004) (Professor of Sociology, University of Kitakyūshū), interview 10 April 2004.

Suga, T. (1997) 'Battle of shinkansen funding – the never ending story', *Japan Railway and Transport Review*, April 1997, 13.
Sugimoto, Y. (1997) *An Introduction to Japanese Society*, Cambridge: Cambridge University Press.
Sugiura, K. (2001) *Gekitotsu! Tōkaidōsen: 'Nozomi' tai kōkū shatoru*, Tōkyō: Sōshisha.
Sumita, S. (2000) *Success Story: The Privatisation of Japanese National Railways*, London: Profile Books.
Super TV (2004) *Shinkansen 100 no Himitsu*.
Takahashi, D. (2000) *Shinkansen-o Tsukutta Otoko: Shima Hideo Monogatari*, Tōkyō: Shogakukan.
—— (2003) '*Sogō Shinji to Tōkaidō Shinkansen Keikaku no Shidō*' in Hoshikawa, T. (ed.) *Shinkansen Zenshi*, Tōkyō: Gakken.
Tanaka, K. (1972) *Nippon Rettō Kaizō-ron*, Tōkyō: Nikkan Kōgyō Shimbunsha.
—— (1973) *Building a New Japan: A Plan for Remodelling the Japanese Archipelago*, Tōkyō: Simul Press.
TBS (2000) 'Sōrai no Yume –wa, Shinkansen no Untenshi', http://www.tbs.co.jp/catchat/news/main/000802/02news.html accessed on 18 August 2000.
Tokorozawa, H. (2003a) '*Shinkansen ga Nihonjin o Kaeta?*' in Hoshikawa, T. (ed.) *Shinkansen Zenshi*, Tōkyō: Gakken, 160–1.
—— (2003b) '*Senmei, Ekimei Kobore Hanashi*' in Hoshikawa, T. (ed.) *Shinkansen Zenshi*, Tōkyō: Gakken, 164–5.
Umehara, J. (2002) *Shinkansen no Nazo to Fushigi*, Tōkyō: Tōkyōdō Shuppan.
Vicom (2001) *Shinkansen Densha*, DVD.
—— (2002) *Yume no Chōtokkyū*, DVD.
West Japan Railway Company (2003) *Company Profile 2002*.
—— (2004a) *JR West's Business Report*, Fiscal 2003.
—— (2004b) *Dēta-de Miru JR Nishi Nihon 2004* (available via http://www.westjr.co.jp/company/data/)
Wyckoff, D.D. (1976) *Railroad Management*, Lexington: D.C. Heath and Co.
Yahya, H. (2004) 'The Secrets of DNA' available from http://www.harunyahya.com/dna05.php accessed 1 August 2004.
Yamagata Shinkansen Shinjō Enshin Suishin Kaigi (2003) *Yamagata Shinkansen Shinjō Enshin Purojekuto no Kiroku*, document received from Yamagata Prefecture Government.
Yamanouchi, S. (2000) *If There Were No Shinkansen*, JR East (translation of Yamanouchi, S. (1998) *Shinkansen-ga nakattara*, Tōkyō: Tōkyō Shimbunsha).
—— (2002) *Tohoku, Jōetsu Shinkansen*, Tōkyō: JTB Publishing.
Yasubuchi, S. (2004) (Executive Director, Investment Banking Department, UBS Securities Japan) interview 14 April 2004.
Yomiuri Shimbun Chūbu Shakaibu (ed.) (2002) *Umi o Wataru Shinkansen: Ajia Kōsoku Tetsudō Sōsen*, Tōkyō: Chūō Kōron.

Index

Note: Numbers in italics indicate photographs.

9/11 14, 136
1922 Railway Construction Law 20, 85
2002 World Cup 25, 86, 116, 127, 133; see also World Cup
0-series shinkansen 41–2, 54–8, 62, 69, 156, 174, 181, 246; abroad 64–5, 199; photographs of *36*, *41*, *210*; significant dates 212, 214, 215, 217, 218; technical information 135, 137, 165, 176, 232
100-series shinkansen 42, 56, 58, 59, 124, 156, 209, 243, 246; photographs of *12*, *28*, *157*, *179*; significant dates 215, 218; technical information 176, 232
200-series shinkansen 42, 56, 58, 156, 157, 169; photographs of *158*, *159*; significant dates 214, 215, 217, 218; technical information 165, 173, 176, 232
300-series shinkansen 42, 43, 56, 58, 59, 199; photographs of *10*, *57*, *74*, *123*, *142*, *160*, *175*, *210*; significant dates 216; technical information 124, 176, 233
400-series shinkansen 42, 58, 156; photographs of *98*, *111*, *179*, *210*; significant dates 215; technical information 124, 176, 233
500-series shinkansen 42, 55, 56, 58, 65, 121–2, 156, 160, 174, 181, 194; photographs of *121*, *179*; significant dates 216, 217; technical information 95, 122, 176, 233
700-series shinkansen 42, 55–6, 58, 116, 122, 132, 156, 199, 243; photographs of *55*, *97*, *110*, *160*, *175*, *210*; significant dates 218; technical information 142, 172, 173, 176, 193, 234, 248; see also Hikari Rail Star

700T-series shinkansen 56, 199, 200, 202, 203, 218
800-series shinkansen 42, 56, 58, 62, 152, 180; photographs of *179*, *180*; significant dates, technical information 176, 234, 244

acceleration 123–4, 232–5
accidents *see* safety; *specific accident location names*
activities on the shinkansen 186–95
advertisements 50–1, 59, 109, 114, 116, 157–60, 164, 178, 181, 185–93, 199, 214; for new shinkansen lines 84
airport *see* air travel; *individual names of airports*
air travel 21, 35, 37, 77, 101, 115, 125–6, 146, 164, 173, 203; *see also* airport
Akashi Bridge 96, 245
Akita 32, 38, 40, 94, 96, 228
All Nippon Airways *see* ANA
Amakudari 93, 236
Ambitious Japan 61, 158, *160*, 192
ANA 101, 122, 126, 236
Annaka-Haruna 114, 181, 225
anthem 66–7
Aoba 48–9, 214, 218
Aomori 38, 84, 216; *see also* Shin-Aomori
Asahi 48–9, 214, 218
Asama 49, 128, 156, 217, 225
Atami 81, 105, 125, 221, 241, 243, 247
ATC 26, 134–8, 142–6, 149, 205, 211, 236, 245, 246
Automatic Train Control *see* ATC
Awaji 242, 245

Index

barrier-free 184
Barthes, Roland 66
baseball 152, 157
beauty 53–9, 116, 152, 157, 162, 164, 177, 180, 199, 247
Beijing 207
bentō (and ekiben) 187, 189, 191
Biwako-Rittō 29
Blue Ribbon Award 56, 212
Boeing 747 *see* Jumbo jet
bogie 137, 236
Boom Years period 43, 76, 212–13
Bownas, Geoffrey 27
bubble economy 197, 249
Buddhism 173, 181, 240
Bullet train 1, 18–22, 44–6, 59–65, 236, 243; *see also* Dangan Ressha
business cards *see* meishi

Cardiff 128
Centralized Train Control *see* CTC
Channel Tunnel 95
China 19, 60–9, 107, 157–9, 189, 196–201, 207–8, 241–2, 247
Chinese Nationalist Party *see* KMT
Chitose Airport 37
chōtokkyū 45, 47, 242
Chūbu 236
Chūbu International Airport 96
Chūgoku 32, 236
Chūō Line 120, 182, *210*, 217
Chūō Shinkansen 30, 39, 83–4, 91, 107–8, 118–21, 169, 204, 215; *see also* linear shinkansen
city mergers 52–3, 178, 221, 222, 226
colour 133, 146, 155–8, 199–200, 245, 246, 248
Compulsory Purchase Law 86
COMTRAC 133, 213, 236
Concorde 13, 68, 69, 194
construction costs 73, 87, 94, 100, 102, 180
construction state 71; *see also* public works
cost of the shinkansen 91–6
CTC 133–4, 138, 166, 190, 236
culture: definition of 3, 6, 151–2
cultures at JR companies 150–5, 158, 180, 182, 185

Daiei Fukuoka Hawks 157
Daihatsu 153
Dai-Shimizu Tunnel 33, 78, 166, 213, 214

Dakota 54, 65, 68
Dangan Ressha 20–5, 43–5, 72, 211, 236
Dark Valley period 30, 43, 122, 207, 213–15, 242
dedicated track 208
definition of 'shinkansen' 44–7, 242
deity *see* kami
delays *see* punctuality
derailment 1, 26, 34, 122, 131, 162, 169, 213, 218, 242
Diet 20, 24, 83, 84, 88, 198, 236
documentary 64, 65, 88
Dōrō 149, 150, 177, 236
Dr Yellow 122, *132*, 133, 245
duckbill-platypus 42, 55, 199

E1-series shinkansen 42–3, 58, 156, 169, 242; photographs of *49, 167*; significant dates 216; technical information 95, 176, 234
E2-series shinkansen 42, 58, 63, 156; photographs of *40, 141, 142*; significant dates 216, 217; technical information 80, 176, 193, 235
E3-series shinkansen 42–3, 58, 63; photographs of *40, 98*; significant dates 216, 218; technical information 176, 235
E4-series shinkansen 43, 55, 58, 63, 65, 110, 139, 169, 242; photographs of *111, 141, 179*; significant dates 217, 218; technical information 176, 235
earthquakes 120, 166–71, 176, 177, 204, 214, 237; *see also* Great Hanshin Earthquake; Great Kantō Earthquake; Niigata-Chūetsu Earthquake; UrEDAS
East-i 132, 218, 245
Echigo-Yuzawa 78, 110, 113, 165, 166, *167*, 216, 217, 224, 229
education 4, 6, 9, 12, 13, 90, 149, 184, 195, 240, 248; *see also* training
Eiffel Tower 56, 66
ekiben see bentō
electric multiple unit *see* EMU
Emperor *see* Imperial Family
EMU 23, 41, 236
English language 45, 59, 64, 71, 140, 159, 178, 185, 192, 241, 242, 243, 246, 248
environmental problems *see* earthquakes; Kyōto Protocol; noise pollution; pollution; rain; snow; typhoon; visual pollution; wind

Index 261

Equal Opportunity Law 184, 185
Eurostar 59, 188
EXPO 70 29, 76, 116, 213, 241, 246
EXPO 2005 116
expositions 39; *see also* EXPO 70; EXPO 2005

fare 54, 104, 119, 126, 128, 203, 220
film 59, 61, 64, 69, 88, 136, 243; *see also* documentary; *Shinkansen Daibakuha*; television
finances of JNR 24, 30, 34, 76, 79, 89, 93, 102, 104–6, 126, 128–9
flag 66–7
flights *see* air travel
Flying Scotsman 65, 68
football 66, 127, 178, 194, 244; *see also* 2002 World Cup; Sheffield United; World Cup
Free Gauge Train 100–1, 218
Fukuoka 19, 31, 35, 37, 52, 114, 133, 163, 222, 243, 244; *see also* Fukuoka Airport; Hakata
Fukuoka Airport 37
futsū service trains 15, 17, 45, 125, 238, 245

Gāla-Yuzawa *49*, 113, 116, 151, *167*, 215, 229, 244
gauge 18–23, 28, *36*, 40, 42, 87, 98, 100, 197, 208, 236, 238, 241
gemba kaizen see kaizen
General Education Centre 23, 26, 144, 146
Genesis period 22–7, 43, 211–12
genkan 52, 179
Getty Image Bank 58–9
Gifu 73–4, 236; *see also* Gifu-Hashima
Gifu-Hashima 52, 72–4, 75, 77, 78, 212, 221
Golden Age period 43, 61, 216–18
Gotō Shimpei 20
Great Hanshin Earthquake 147, 167–9, 177, 247
Great Kantō Earthquake 20
Great Wall of China 24,
green carriage 128, 194, 212, 218, 236

Hachinohe 32, 33, 38, 91, 94, 95, 115–16, 165, 200, 217, 218, 223, 242
haiku 152, 180
Hakata 15, 31, 32, 35, 38, 47, 52, 90, 92, 115, 125, 159, 193, 212–18, 222, 246, 247

Hakata-Minami *36*, 215, 229, 244
Hakodate 36, 37, 38, 115, 117; *see also* Shin-Hakodate
Hamamatsu 82, 136, 151, 221
Hamanako *12, 55, 210*
Haneda Airport 37, 96, 172
Hanshin Tigers 152, 157
hara-kiri 45, 182, 242
hardware 130–3, 161
Hayate 48, 49, 50, 127, 156, 192, 218, 223, 242, 245
Hello Kitty 62–3
Higashi-Hiroshima *41, 210*, 215, 222
High Speed Surface Transport *see* HSST
Hikari 47–8, 60, 82, 124, 127, 139, 161, 212–17, 221, 222, 242, 245
Hikari Rail Star 156, 157, 191–4, *191*, 218, 242
Hinomaru *see* flag
hiragana see kana
Hiroshima 2, 38, 74, 122, 125, 172, 195, 214, 222, 236, 244
history of Japanese railways up to 1964 18–22
Hitachi 54, 152, 197, 232–5
HMS Hood 14, 68, 240
Hokkaidō 20, 21, 22, 35–9, 50, 80, 84, 115, 153, 215, 243; *see also* Hokkaidō Shinkansen; JR Hokkaidō
Hokkaidō Shinkansen 32, 35–9, 76, 84, 114–15, 208, 213; *see also* Hokkaidō; JR Hokkaidō
Hokuriku 80, 81
Hokuriku Line 164
Honjō 112, 113, 217, 224, 225
Honjō-Waseda 113, 217, 218, 224, 225
honne 67, 238
Honshū JR Companies 83, 104–5, 216, 236
hot springs *see* onsen
HSST 39

Ikeda Hayato 1, 29
image of the shinkansen 53–65
Imperial Expansion period 20–2, 43, 211
Imperial Family 51, 67, 194
internet 11, 190
Ishida Reisuke 26
Iwakuni 243; *see also* Shin-Iwakuni

JAL 39, 101, 126, 147, 236
Japan 2001 58

Japan Airlines *see* JAL
Japanese Communist Party *see* JCP
Japan Exchange and Teaching Programme *see* JET Programme
Japan Rail Pass 46, 57, 236, 243
Japan Railway Construction Public Corporation *see* JRTT
Japan Railway Technical Service *see* JARTS
JARTS 197, 237
JCP 81, 89, 237
JET Programme 13, 46, 248
jingle 159, 185, 192; *see also* music; song
JNR Settlement Corporation *see* JNRSC
JNRSC 104, 237
JRCC *see* JRTT
JR company backgrounds 14–17
JR Freight 15, 104, 150, 237
JR Hokkaidō 15, 16, 50, 115, 150, 153, 237, 238, 245
JR Rengō 149–50, 237
JR Shikoku 15, 100, 150, 237, 238, 245
JR Sōren 149–50, 237
JRTT 16, 31, 79, 85, 102, 212, 215, 237
Jumbo jet 13, 42, 68, 122, 248

Kagoshima 35, 38, 48, 58, 60, 77, 159, 169, 226, 244
Kagoshima-Chūō 32, 35, *36*, 47, 52, 94, 226
kaizen 151
Kakegawa 127, 215, 221, 245
kami 171, 182
kamikaze 247
Kamonomiya 26, 211
kana 49, 50, 229, 242
Kanazawa 32, 34, 38, 80–1, 125
Kanemaru Shin 83
kanji 44, 50, 59, 192, 237, 242
Kansai 39, 152, 153, 237, 238
Kansai International Airport 37, 96
Kantō 39, 76, 77, 108, 110, 112, 115, 152, 237
Kaohsiung 37, 197, 202
Karuizawa 34, 38, 77, 79–80, 116, 117, 216, 217, 225, 242
katakana *see* kana
Kawanabe Hajime 23, 26
Kawashima Ryōzō 38, 64, 80, 81, 120, 150, 154, 159
Keisei 244
Kimigayo *see* anthem
Kinki (region) *see* Kansai
Kinosaki 118, 166, 177

Kintetsu 73, 121
Kitakyūshū 52, 170; *see also* Kokura
Kita ward, Tōkyō 87–9
KMT 198–204, 238
Kōbe 19, 25, 60, 96, 167, 222, 243, 247; *see also* Shin-Kōbe
Kōbe Earthquake *see* Great Hanshin Earthquake
Kōchi 32, 77, 125
Kodama 47–8, 118, 124, 127, 148, *157*, 212–18, 221, 222, 229, 245
Kokura 52, 122, 125, *132*, 133, 214, 215, 217, 222; *see also* Kitakyūshū
Kokurō 149–50, 177, 238
Komachi 50, *63*, 218, 223, 228
Komoro 104, 116–17
Korea *see* South Korea
Kuomintang *see* KMT
Kuroiso 50–1, 223
Kyōto 19, 29, 85, 116, 125, 147, 179–80, 195, 217, 221, 237, 243
Kyōto Protocol 171
kyūkō service train 17, 45, 47, 78, 238

Lake Hamana *see* Hamanako
Lancaster 67
late train *see* punctuality
LCA 172–3, 238
LDP 75, 81–3, 85, 198, 238
Lee Teng-Hui 201–2, 204
level crossings 23, *33*, 40, 98, *98*, 117, 132–3, 136, 148
Liberal Democratic Party *see* LDP
Life Cycle Assessment *see* LCA
Linear shinkansen 2, 39, 60, *120*, 123, 205, 249; *see also* Chūō Shinkansen
literature 64, 247
local trains *see futsū* service trains
London 7, 37, 59, 67, 95, 128, 162–3, 240, 244, 245
lost decade 43, 204

Machida 20
machizukuri 178–9
Maglev *see* linear shinkansen
Maibara 29, 50, 72, 73, 81, 105, *123*, 212, 221
Maihara *see* Maibara
Mallard 68
Manchuria 20, 21, 48
manga 187, 189
marketing *see* advertisements
Marunouchi 85, 244
Matsudaira Tadashi 23, 26

Index   263

Matsue 32, 125
Matsumoto 120
media 2, 27, 58–9, 88–90, 147, 148, 149, 160, 169, 170, 186, 209; *see also* film; internet; NHK; television
meishi 155, 246
Meitetsu 50, 124
methodology 3–12
METI *see* MITI
Mikawa-Anjō 50, 215, 221, 245
Mikawashima 26
Miki Tadanao 23, 26, 53–4
Minamata 86, 170, 226; *see also* Shin-Minamata
mini-shinkansen 39–40, 42, 95–102, 133, 150, 156, 215, 216, 240, 245
Ministry of Economy, Trade and Industry *see* MITI
Ministry of International Trade and Industry *see* MITI
Ministry of Land, Infrastructure and Transport *see* MLIT
Mishima 148, 212, 216, 221, 241
MITI 75, 238
Miyazaki 39, 214
MLIT 39, 84–5, 103, 119, 151
mobile phone 69, 138, 149, 187, 190
Morel, Edmund 18
Mori Yoshirō 81
Mount Fuji 1, 7, 9, 12–13, 44, 50, 56–9, 57, 66, 69, 118, 170, 243
Mount Ibuki 73
Mount Osutaka 147
Mount Suzuka 73
movie *see* film
music 69, 164, 173, 187, 190, 192

N700-series *28*, 43, 56, 95
Nagano Olympics 34, 77, 80, 116
Nagasaki 32, 38, 241
Nagasaki Shinkansen 32, 35, 38, 84, 115, 213
Nakasendō 19, 38, 119
Nakasone Yasuhiro 4, 12, 75, 78, 107, 149, 188
Nakayama Tunnel 34, 214
Nanjing 207
Nara 38, 84, 121, 237
Narita Airport 32, 86, 87, 89, 96, *175*, 244
Narita Shinkansen 77, 85, 87, 94, 172, 213, 244
Nasuno 49, 50, 51, 216, 223
Nasu-Shiobara 50–1, 194, 216, 223

national economic impact of the shinkansen 118–19
National Railway Museum (York) *see* NRM
New Austrian Tunnelling Method 92
New Hope period 43, 215–16
NHK 27, 88, 177, 212, 244
Nihonzaka 25, 211
Niigata-Chūetsu Earthquake 34, 110, 169
Nikkyōso 149
Nippon Sharyō 54, 152, 232–5
Nishi-Kagoshima see Kagoshima-Chūō
Nixon, Richard 30, 241
Nixon Shocks *see* Nixon, Richard
noise pollution 42, 86, 90, 133, 163, 173–7, 188, 214
Noto Airport 81, 96
Nozomi 48, 51, 82, 108, 116, 124, 127, 138–9, 158, 161, 183, 216, 218, 221, 222, 242, 245
NRM 64–5, 238, 243
NTT 107

Odakyū 53, 54
Ogōri 51–2, 218, 222; *see also* Shin-Yamaguchi
Oil Shock 30, 170
Ojiya 78, 169
Okayama 30, 31, 32, 125, 135, 139, 172, 212, 213, 216, 222, 236
Olympics 67, 207; *see also* Nagano Olympics; Tōkyō Olympics
Ono Banboku 72, 74, 78
onsen 116, 166, 238; *see also* Kinosaki; Komoro
opposition to the shinkansen 85–90
Oū Line 40, 215

Pacific War 2, 21, 23, 48, 182, 198, 207, 242, 247
Palin, Michael 60, 170, 178
pantograph 147, 173, 238
Paris 69, 123
periods of shinkansen history 43; *see also* Boom Years period; Dark Valley period; Genesis period; Golden Age period; Imperial Expansion period; New Hope period; Wilderness Years period
photographs 10–11, 58, 61, 136–8, 149
political interference in JNR days *see seiji eki*
political interference post JNR 79–85

pollution 88, 170–2, 182; see Kyōto Protocol; noise pollution; visual pollution
pork-barrel 2, 5, 71–2, 75, 79, 81
PR see advertisements
precipitation see rain; snow
problems on the shinkansen 146–9
profitability of the shinkansen 104–8
public relations see advertisements
public works 78, 80, 83, 95, 102–3
punctuality 15, 133, 138–43

Rail Star see Hikari Rail Star
railway enthusiast 10–11, 199; see also trainspotter
Railway Technology Research Institute see RTRI
rain 162–4, 166
regional impact of the shinkansen 108–18
religion 6, 181–2; see also Buddhism; *kami*; Shintō
reset generation 146, 149
Robin Hood 14, 240
Rokken 22
role of symbols see symbolism
RTRI 23, 26–7, 39, 95, 100, *123*, 168, 176, 217, 238
Russia 20, 171; see also Russo-Japanese War
Russo-Japanese War 21, 128

safety record of JNR 22, 26, 131
Saigō Takamori 77, 244
Saitama 34, 52, 87–9, 113, 223, 224, 225, 237, 242
Saku 117, 225; see also Sakudaira
Sakudaira 117, 225
Sakuragichō 18, 22
Sakurajima *36*, 58, 169
San'in Line 84, 118, 247
Sanrio 62–3, 156, 166, 243
Sapporo 32, 36–9, 77, 114–15, 163, 212
Sapporo Airport see Chitose Airport
Satō Eisaku 24, 74, 243
schedule see punctuality; timetable
seat pitch see seat size
seat reservation 128, 191
seat size 193, 232, 248
Seibi Keikaku see Seibi Shinkansen
Seibi Shinkansen 30, 32, 39, 76, 82, 100, 101, 119, 122, 238, 242

seiji eki 61, 72–9, 238
Seikan Tunnel 32, 35, 37, 77, 96, 104, 106, 215, 245
Sekigahara 73, 164, 165, *210*
Self Defence Forces 23
service names 47–50
Seto Ōhashi 96, 104, 245
Shanghai 69, 207,
SHC 104–5, 107, 238
Sheffield 240
Sheffield United 178
Shikoku Railway Company see JR Shikoku
Shima Hideo 23, 26, 131, 198, 210
Shima Takashi 197–8
Shima Yasujiro 23, 198
Shimbashi 241
Shimonoseki 21, 32, 211, 222; see also Shin-Shimonoseki
Shinagawa 25, 82, 112, 154, 158–9, *160*, 181, 216, 217, 218, 221, 246
Shinano Railway 104, 116
Shin-Aomori 32, *33*, 36, 115, 217, 242; see also Aomori
Shin-Fuji 50, 215, 221
Shin-Hakodate 32, 115
Shin-Iwakuni 50, 74–5, 222
Shinjō 38, 40, 51, 94, 217, 218, 227, 245
Shinjuku 32, 34, 125, 217
shinkansen see definition of 'shinkansen'; specific line names and specific series names
Shinkansen Citizen Movements 87–90
Shinkansen Daibakuha 64, 145
Shinkansen Holding Corporation see SHC
Shin-Kōbe 50, 222, 247
Shin-Minamata *36*, 178, 226
Shin-Shimonoseki 51, 218, 222
Shin-Shirakawa 52, 114
Shin-Tanna tunnel 24, 25, 72, 211, 241
Shintō 171, 181–2, 184, 242
Shin-Yamaguchi 51, 222
Shin-Yatsushiro 32, 35, *36*, 92, 94, 140, 159, 217, 226, 247
Shin-Yokohama 25, 29, 60, 155, 188, 221, 247
Shiun-Maru 22
Shizuoka 9, 24, 29, 82, 127, 138, 158, 221, 236, 239, 244, 245
Shizuoka Airport 29
sleep 188–9

Index 265

smoking 193, 214, 215, 216, 218, 248
snow 40, 42, 73, 94, 116, 148, 156, 164–6, 212, 247
software (employees) 131, 134, 143, 161
Sogō Shinji 1, 23–4, 26–7, 28, 48, 56, 93, 131, 240
song 61, 88, 158, 192; see also jingle; music
Southern Hyōgo Prefecture Earthquake see Great Hanshin Earthquake
South Korea 20–1, 37, 48, 203, 207–8
speed 21, 22–3, 40, 45–6, 69, 92, 98, 121–5, 127, 160, 163, 168, 232–5; see also ATC
Spitfire 65, 67
sprinkler 165–6, 167
stamps (postage) 60–1, 243
station names 50–3
steam train 21, 23, 51, 62, 67, 68, 83
stereotype 2, 6, 65–6, 145, 153
strike 150–1, 152, 213
suicide 182–3, 247
Super-*Tokkyū* 98–9, 101, 102, 238, 245
symbolism 2, 13–14, 65–70, 89, 198–200

Taipei 37, 197, 199–200, 202
Taiwan 20, 37, 63, 64, 196–206
Taiwan High Speed Railway see Taiwan; THSRC
Taiwan High-Speed Railway Corporation see THSRC
Taiwan Railway Administration see TRA
Takasaki 32, 34, 38, 77–80, 94, 113, 216, 217, 224, 225
Tanaka Kakuei 3, 4, 61, 74, 75–9, 80–1, 119, 125, 244
Tanaka Yasuo 83
Tanigawa 49, 217, 224, 229
tatemae 67, 105, 238
Tazawako Line 40, 216
television 23, 27, 59, 64, 89, 244; see also documentary; film
terrorism 67, 136
tetrapod 247
TGV 58, 122, 123, 188, 208
Third Sector Companies 98, 100, 102–4, 116, 120–1
Thomas the Tank Engine 63
THSRC 56
ticket cost see fare
timetable 51, 127, 139–40, 161, 183, 187, 199
Titanic 68

Toda 87, 90
Tōkaidō Line 19, 22, 24–5, 27, 73, 183, 211
Tōkaimura 147
Toki 48, 49, 214, 217, 218, 224
Tokkyū 17, 33, 45, 47–8, 78, 120, 239, 242, 247, 248
Tokugawa Ieyasu 152
Tokugawa Period 27–8, 118, 244
Tokuyama 53, 74, 222
Tōkyō Airport see Haneda Airport; Narita Airport
Tōkyō Disneyland 157, 158, 195
Tōkyō Olympics 2, 27, 67, 72, 116, 200
tourism 45, 57, 112, 115–18, 162, 181, 183
Toyama 32, 38, 80–1, 91, 125, 236
Tōya-Maru 22, 35
Toyota 123, 136, 153
TRA 197, 199, 202, 239
training 12, 130–46, 151
trainspotter 10–11; see also railway enthusiast
Tsubame 48, 65, 226, 242, 247
Tsubame-Sanjō 52, 113, 224
Tsubasa 50, 216, 223, 227
tsunami 45, 168, 242, 243
Tsurumi 26
Tsushima 21, 242
tunnelling 31, 72, 77, 78, 92, 177
TV see television
typhoon 133, 147, 163–4, 190, 218

unions 88, 107, 149–51, 209
Urasa 74, 77–8, 113, 165, 224, 244
Urawa 87, 113, 223
UrEDAS 168, 169, 215, 239

visual pollution 177–8
volcano 48, 49, 58, 169

Washu-Maru 22
West Coast Mainline 95, 208
Wilderness Years period 43, 211
wind 163–4
window 59, 142, 188, 205
World Bank 24, 202, 211, 214
World Cup 67; see also 2002 World Cup
World Trade Center see 9/11
wrapping 155–7, 178

Yamabiko 48, 50, 214, 215, 223
Yamaguchi 51–2, 236, 244
Yamanashi 39, 83

Yamanote Line 25, 105, 148, 192
Yamato 24, 68
Yanagi Sōetsu 54–5
Yatsushiro *see* Shin-Yatsushiro
Yokohama 18, 19, 25, 60, 96, 221, 239, 243; *see also* Shin-Yokohama
Yokohama Line 20, 23, 25

Yono 87, 89, 113
yume no chōtokkyū see chōtokkyū
Yuzawa 112, 113, 224; *see also* Echigo-Yuzawa; Gāla-Yuazawa

Zero 26, 67
zoku 39, 81, 87, 172